WRECK-CLIFF AND CASCADE — Pictured Rocks.

SAILING ON THE GREAT LAKES

AND

RIVERS OF AMERICA;

EMBRACING A DESCRIPTION OF

LAKES ERIE, HURON, MICHIGAN & SUPERIOR,

AND

RIVERS ST. MARY, ST. CLAIR, DETROIT, NIAGARA & ST. LAWRENCE;

ALSO, THE

𝔆𝔬𝔭𝔭𝔢𝔯, 𝔍𝔯𝔬𝔫 𝔞𝔫𝔡 𝔖𝔦𝔩𝔳𝔢𝔯 𝔕𝔢𝔤𝔦𝔬𝔫 𝔬𝔣 𝔏𝔞𝔨𝔢 𝔖𝔲𝔭𝔢𝔯𝔦𝔬𝔯,

COMMERCE OF THE LAKES, &c.

TOGETHER WITH NOTICES OF THE

RIVERS MISSISSIPPI, MISSOURI AND RED RIVER OF THE NORTH;

CITIES, VILLAGES AND OBJECTS OF INTEREST.

FORMING ALTOGETHER A

COMPLETE GUIDE

𝔗𝔬 𝔱𝔥𝔢 𝔘𝔭𝔭𝔢𝔯 𝔏𝔞𝔨𝔢𝔰, 𝔘𝔭𝔭𝔢𝔯 𝔐𝔦𝔰𝔰𝔦𝔰𝔰𝔦𝔭𝔭𝔦, 𝔘𝔭𝔭𝔢𝔯 𝔐𝔦𝔰𝔰𝔬𝔲𝔯𝔦, &c.

ALSO,

RAILROAD AND STEAMBOAT ROUTES.

WITH MAP AND EMBELLISHMENTS.

COMPILED AND PUBLISHED BY

J. DISTURNELL,

AUTHOR OF THE "INFLUENCE OF CLIMATE," ETC.

PHILADELPHIA:
1874.

𝕰𝖒𝖇𝖊𝖑𝖑𝖎𝖘𝖍𝖒𝖊𝖓𝖙𝖘.

TO MY PATRONS,

TO WHOM THIS WORK IS RESPECTFULLY DEDICATED.

During the Summer of 1843, I visited the Lower St. Lawrence, proceeding to Tadousac, and ascended the far-famed Saguenay River to Chicoutimi,—then this wild and interesting region of country was almost unknown to the American travelling public, although since it has become a most fashionable Summer Resort.

In 1844 the "PICTURESQUE TOURIST" appeared, being a *Guide through the State of New York* and *Canada.* This work was favorably received, and two or three editions were issued.

In 1846 the RAILWAY and STEAMSHIP GUIDE was issued, as a quarterly publication, being the first work of its kind in the United States, and kept up until 1858, when it was discontinued owing to being superseded by monthly publications, which proved a great success,—filling a want required by the Travelling Public.

In the month of September, 1854, I visited Mackinac and the Saut Ste. Marie, and remained until the first of October. Then a horse-railway conveyed passengers and freight from the Lower to the Upper Landing, a distance of one mile; the travel and trade being comparatively small, although the construction of the Ship Canal was in progress, being finished in 1855.

In 1856 I again visited Lake Superior and proceeded to Bayfield, stopping at the intermediate ports, for the purpose of obtaining information for a new work which appeared in 1857, entitled "TRIP THROUGH THE LAKES AND RIVER ST. LAWRENCE." This work was also favorably received, having since passed through four editions.

In 1867 the "INFLUENCE OF CLIMATE IN NORTH AND SOUTH AMERICA" was issued, containing 336 pages, octavo size. *A new Edition is now being prepared for publication, to contain Health Statistics relating to the several States, Territories and principal Cities of the Union, Agricultural Statistics, &c.*

The present Work, entitled, "SAILING ON THE GREAT LAKES AND RIVERS OF AMERICA," with Map and Embellishments, is compiled with much care, hoping it may meet with a hearty approval. The aim of the Compiler has been to faithfully describe the objects of interest and wonders on the INLAND SEAS and the Great Rivers flowing into the Ocean by widely different channels.

J. DISTURNELL.

PHILADELPHIA, *June,* 1874.

TO THE AMERICAN PUBLIC.

The Great, *or* **"NEW NORTH-WEST,"** *embracing the Region of the Great Lakes of America, the Upper Mississippi Valley, the Valley of the Red River of the North, and the Upper Missouri Valley — including in part the Northern portion of the United States and a part of the Dominion of Canada, are included in the above designated territory, lying, for the most part, in the middle of the Continent, between the 45th and 50th degrees of North latitude.*

Within this vast region are the sources of **four** *of the greatest Rivers of America, — the Missouri River rising in the Rocky Mountains, the Mississippi River rising in Northern Minnesota, both flowing into the Gulf of Mexico, — the Red River of the North rising in Dakota and Minnesota, together with the Saskatchewan, flowing into Lake Winnipeg and Hudson Bay, and the main sources of the Great Lakes and the St. Lawrence River, flowing into the Gulf of St. Lawrence, — forming altogether about eighteen thousand miles of Inland Navigation.*

This hitherto neglected portion of the Continent is now attracting the attention of the naturalist and the geographer, owing, in the main part, to its agricultural and mineral products. Already railroad routes have been surveyed and explorations made in various directions, while the NORTHERN PACIFIC RAILROAD, *destined soon to form another Trans-Continental line of travel, has been completed for a distance of four hundred and fifty miles, west of Lake Superior, connecting the heads of all the above mighty bodies of water, flowing South, North and East into the Ocean, at widely different points of the compass. By the above railroad route, following up the Missouri River, the* GOLD FIELDS *of Montana can be reached, as well as the famous* NATIONAL PARK *situated on the headwaters of the Yellowstone River.*

LEWIS *and* CLARK, *two eminent explorers, early in the beginning of the present century, passed over a portion of this wilderness, inhabited only by roving tribes of wild Indians, and faithfully described its wonderful scenery and adaptation for the residence of civilized man. Employees of the American Fur Company,*

iii

and others, followed in after years; but the far-seeing and energetic ASA WHIT-
NEY, *some thirty years since, was the first to project a line of railroad to extend
from the shores of Lake Michigan to the shores of Puget Sound, on the Pacific
coast.*

*After fruitless efforts, running through several years, his plans were abandoned,
although the importance of the subject was brought to the attention of the United
States Government and State Legislatures; also, advocated before the American
Geographical Society as early as the year 1851. After a lapse of twenty years
this great National work was commenced and soon completed across Minnesota and
Dakota Territory to the Missouri River.*

*It is this extended field, made easy of access by means of the Ship Canal at the
Saut Ste. Marie, and Railroads finished to the southern borders of Lake Superior,
and westward to the Upper Missouri River, that is described in the present work,
entitled* **"SAILING ON THE GREAT LAKES."**

*While thousands of American tourists are annually flocking to Europe and the
far ends of the earth, our own country from Maine to Florida, and from the
Atlantic to the Pacific Ocean, including California and Oregon, has been strangely
neglected.*

*In all the above Regions alluded to, in the North-west, health prevails,—the
climate possessing properties that invigorate the whole human frame—giving
vitality and strength to* MAN *that enables him to endure fatigue and labor to a
wonderful extent. Thousands are now living in the Upper Peninsula of Michigan,
Northern Wisconsin and Minnesota that would have long since been in their graves
were it not for the recuperative climatic influence that reigns in the region
bordering on the Great Lakes and Rivers of North America, lying for the most
part in a high latitude, being free from malaria and favored with pure air and
water.*

*The Cities, Villages and Objects of Interest on the Shores of Lakes Erie,
Huron, Michigan and Superior are, also, faithfully described — including the
far-famed Copper, Iron and Silver Region.*   **J. D.**

PHILADELPHIA, May, 1874.

CONTENTS.

PART VIII.

ADVERTISEMENTS.

INLAND SEAS.

THE magnitude and volume of the waters which form the INLAND SEAS, or Great Lakes of America, is so large as to exceed the comprehension of most minds, unless the subject has been closely investigated. Under the above designation is included Lakes Erie, Huron, Michigan and Superior, with Green Bay on the American side and Georgian Bay on the Canadian side of the Boundary Line which separates the United States from the Dominion of Canada. The Straits, or connecting links forming the outlets to this vast body of water, are St. Mary's River, Straits of Mackinac, St. Clair River and Detroit River, — all large and navigable streams, and, by means of the Ship Canal at the Saut Ste. Marie, are capable of floating vessels of 2,000 tons burthen for a distance of 1,200 miles, from Chicago or Duluth to Buffalo, New York. Here navigation for a large class of vessels ceases at the present time; but by the enlargement of the Welland Canal, or the construction of the proposed Niagara Ship Canal, together with the enlargement of the St. Lawrence Canals, ships of a large burthen could sail direct from Chicago or Duluth to Montreal or Quebec, and thence to European ports.

Before many years, no doubt, all the above facilities will be afforded,— thus conferring incalculable benefits to the American and Canadian public, as well as to the commercial world. Then all the Lake ports on both sides of the "Inland Seas," like the Baltic and the Black Seas, or the Mediterranean, will form one continued line of sea-ports, from whence can be shipped, at a low rate, all the agricultural and mineral wealth of a vast region of country teeming with all the products that go to enrich nations, while giving profitable employment to thousands of laborers on land and water.

This great thoroughfare, lying between two friendly nations, should be carefully guarded, so that no impediments could interfere with free and open navigation from the Upper Lakes to the Gulf of St. Lawrence, a distance of upwards of 2,000 miles.

By a late decision of the Supreme Court of the United States, the Upper Lakes, including Lake Erie, with their connecting waters, were declared to be *Seas,* commercially and legally. Congress, under this decision, is empowered to improve the harbors of the Lakes and connecting Straits, precisely as it has power to do the same on the seaboard. This should lead to a vigorous policy in the maintenance of Federal authority, both in improving the harbors and making provision for the safety of commerce, by guarding the channels of communication against any kind of obstruction.

The States washed by the Great Lakes are New York, Pennsylvania, Ohio, Michigan, Indiana, Illinois, Wisconsin, Minnesota, and Ontario, Canada—the boundary line between the United States and the Dominion of Canada running through the centre of Lakes Superior, Huron, St. Clair, Erie, and Ontario, together with the connecting rivers or straits, and down the St. Lawrence River to the forty-fifth parallel of latitude. From thence the St. Lawrence flows in a north-east direction through the Dominion of Canada into the Gulf of St. Lawrence.

The Lakes alone cover an area of upwards of 90,000 square miles, draining a surface of about 400,000 square miles of territory, situated for the most part on the northern confines of the temperate zone, being capable of sustaining many millions of inhabitants.

Already the commerce of the Great Lakes has assumed immense proportions,—during the year 1873, upwards of 37,000 vessels of different classes passed Port Huron, Mich., situated on the river St. Clair, at the foot of Lake Huron.

With all this immense shipping interest involved, it has been proposed by certain moneyed interests to bridge both the Detroit and St. Clair Rivers at several points, for the assumed purpose of affording quick transit from the West to the Atlantic seaboard. This privilege, if granted, would so far injure lake navigation, for all future time, as to place a great impediment on commerce, that, too, to the injury of the producer and consumer, now struggling for cheap transportation. By means of lake, river and canal transportation, millions of money can be saved as compared with railroad charges, — now severely taxing the industries of the country, and causing high prices to be charged for most of the necessaries of life.

All the lake cities and towns, as well as cities on the Mississippi and Atlantic seaboard, should alike take a deep interest in preserving free and uninterrupted navigation on the *Great Lakes,* as well enlarging and constructing Ship Canals,—thus increasing the volume of trade through the legitimate channel which Nature has provided.

CANAL AND RIVER IMPROVEMENTS.

The proposed Canal and River Improvements, in order to connect the waters of the Mississippi River with those of the Great Lakes and the Atlantic Ocean, are of the greatest importance.

In addition to the Ohio River Improvements and the Illinois and Michigan Canal, which are of great advantage to commerce, it is proposed to construct a Canal from near Rock Island, on the Mississippi River, to a junction of the Illinois and Michigan Canal at Hennepin, Ill., a distance of about 80 miles. Another Canal and River Improvement is partially completed extending from Green Bay, via Lake Winnebago and Fox River, to Portage, Wis., situated on the Wisconsin River. The latter stream enters the Mississippi River a few miles below Prairie du Chien. When this work is completed it will furnish the most direct and convenient water route between the Valley of the Upper Mississippi and the basin of the Great Lakes.

By referring to the accompanying *Table*, showing the comparative elevations of cities, etc., on the banks of the Mississippi River and shores of the Great Lakes, it will be seen that there is nearly a level divide between the two great systems of navigation — the Mississippi at Rock Island being elevated 528 feet, and at Chicago, on Lake Michigan, 578 feet.

In addition to the above improvements of water communication, it is proposed to construct the *Huron and Ontario Ship Canal*, to extend from Georgian Bay to Toronto, situated on Lake Ontario, a distance of about 100 miles. The difference of water level on this line is 344 feet, with an intervening summit level, at Lake Simcoe, elevated about 100 feet above Lake Huron. This canal, when completed, will shorten the water communication to the Atlantic seaboard several hundred miles. Then the channel of commerce can flow down the St. Lawrence River to the Ocean, or the trade be diverted across Lake Ontario to Oswego, or other American ports on the above Lake, connecting with canal and railroad lines of transportation running to the seaboard.

TABLE—Showing the Comparative Elevation of Several Places above the Ocean.

	FEET.		FEET.
CAIRO, Illinois	275	CROW WING, Minnesota	1,100
Mouth Ohio River.		BRAINERD, Minnesota	1,140
CINCINNATI, Ohio	500	ITASCA LAKE, Minnesota	1,550
PITTSBURGH, Pennsylvania	700		
ST. LOUIS, Missouri	335	LAKE SUPERIOR	600
ALTON, Illinois	345	*Rapids at Saut Ste. Marie, 18 feet.*	
Near mouth Missouri River.		LAKE MICHIGAN	578
BURLINGTON, Iowa	500	GREEN BAY	578
ROCK ISLAND, Illinois	528	LAKE WINNEBAGO, Wisconsin	748
DUBUQUE, Iowa	570	LAKE HURON	576
PRAIRIE DU CHIEN, Wisconsin	602	GEORGIAN BAY, Canada	576
LA CROSSE, Wisconsin	630	LAKE ST. CLAIR, Michigan	570
PRESCOTT, Wisconsin	670	LAKE ERIE	565
ST. PAUL, Minnesota	685	*Falls of Niagara, 160 feet.*	
ST. ANTHONY, Minnesota	760	LAKE ONTARIO	232

TABLE—SHOWING THE AREA, DEPTH AND ELEVATION OF THE GREAT LAKES OF AMERICA.

GREAT LAKES.	Greatest Length. Miles.	Greatest Breadth. Miles.	Greatest Depth. Feet.	Height above Sea. Feet.	Area. Square Miles.
Superior	450	170	900	600	32,000
Michigan	320	85	700	578	22,000
Huron	250	120	800	576	20,500
Erie	250	65	250	565	9,700
Ontario	180	85	700	232	6,300
Total Area	......	......	......	......	90,500

Sailing on the Great Lakes.

PART I.

NATURAL BEAUTIES AND OBJECTS OF INTEREST.

THE CASTLE-PICTURED ROCKS.

ST. MARY'S STRAIT.

IN modern times the voyager, in sailing from the American shore to Europe or Asia, or in making the circuit of the globe, pursues a trackless path across the broad ocean without being in sight of land for most of the distance. Not so in sailing on the Great Lakes of America — forming altogether by far the greatest expanse of fresh water on the face of the earth. Here, however, we have no "sea-gods" to appease, or leviathans of the deep about which to fabricate marvellous tales, such as the salt water sailors love to narrate. To faithfully describe, however, the magnificent scenery of Lake Superior,— the "*Gitchee Gummee*," or Big Sea water of the Chippewa dialect,— in connection with Indian traditions, would reveal untold wonders.

Our task consists of a plain, unvarnished narration of the interesting objects that surround the Inland Seas — made famous by their extent, pure air and waters, picturesque islands, cultivated banks, and rich deposits of various kinds of mineral — that too in such abundance as to vie with other parts of the world, both as to richness of the ore and the extent and variety of the deposits. Here iron, copper, and silver are being annually produced in marvellous quantities — also other precious minerals — giving profitable employment to many thousand laborers.

13

The field is almost illimitable, covering portions of the Upper Peninsula of Michigan, Northern Wisconsin, and Minnesota, while the Province of Ontario, forming part of the Dominion of Canada, comes in for her share, as yet but partially developed. Add to the above the value of the fisheries and the lumber trade, and you have wealth enough to enrich a nation. Then, again, the health-invigorating climate of the Upper Lake region being added to all the previous advantages in the shape of wealth, and you have a vast virgin country which the ancients would, no doubt, have deified as the abode of gods of a superior order. According to Indian traditions, some of the islands and mountain tops have already been made the supposed abode of their great spirits — partaking, in their version, the character of a Divine Being.

When we consider that the far-famed Niagara cataract has no equal in the world, and that the smallest of the Upper Lakes exceeds in extent that of any other fresh body of water on the globe, (except Lake Baikal, in Russia,) with the imperial *Lake Superior* covering an area of upwards of thirty thousand square miles, having a depth of about one thousand feet near its centre, and standing six hundred feet above the ocean, into which it flows through the River St. Lawrence into the Gulf — its outlet bearing several different names before it enters the main stream among the "Thousand Islands," — then being baptized by different appellations, when applied to its plunging rapids and gently flowing expansions, which often occur in its long course to the ocean — forming the boundary line between two rival nations — these grand features altogether excite our admiration.

The principal rapids on the St. Lawrence, after passing Lake Ontario, are known as the Long Sault, Coteau, Cedar, Cascade, and La Chine, while the expansions are called Lake St. Francis, Lake St. Louis, and Lake St. Peter. It also receives the Ottawa River, the Saguenay River, and several other considerable streams, all of which accumulated waters flow onward in majestic grandeur to the briny ocean — its tide-waters extending above the city of Quebec.

The above rapid sketch of lake and river combined, shows the magnitude and grandeur of the whole system of internal communication which drains an area of about four hundred thousand square miles of territory, affording a most desirable outlet for the products of the North-western States of the Union, as well as for the Dominion of Canada—sea-going vessels being enabled, by means of ship canals, to ascend for fifteen hundred miles above tide-water to the head of Lake Michigan on the south, or Lake Superior on the west, where stand two rival cities of modern date.

The many cities and villages which have sprung into existence, as if by the magician's wand, and adorn the shores of the Great Lakes and river, as well as on the banks of tributary streams, are equally marvellous as the mighty waters which drain several hundred thousand square miles of territory, most of which extent is susceptible of settlement and cultivation, and principally lying within the temperate zone, being favored with four seasons — Spring, Summer, Autumn, and Winter.

There are few countries that have so varied and healthy a climate as the one in which we rejoice; for the United States comprise within their limits, at one and the same time, almost every degree of temperature from zero to summer heat. If one would escape the cold, driving snow-storms of the North, and the gales that pile the ice into the harbors of the Great Lakes, he can, if he will, sit under the blooming orange groves in the South; if he would escape the torrid heats of the summer Atlantic, he can find unvarying comfort on the Upper Lakes or the Pacific shores ; and while thus a wayfarer among all the zones and all the climates, he is always at home in his own country, where health, happiness, and freedom prevail.

While Asia boasts of her Himalaya Mountains, and Europe of her glaciers and mountain peaks, America can justly boast of her lakes and rivers, which are of far greater value to the human race. In addition to the Valley of the St. Lawrence, in an extended view, consider the extent of the rich Valley of the Mississippi, — irrigated by the " Father of Waters," with its many navigable tributaries, forming some fifteen thousand miles of navigable water altogether, flowing gently into the Gulf of Mexico, giving life and vitality to a nation of freemen. This view, when rightly considered, ought to make the American people bless their Creator.

The hand of God is apparent on the bodies of the Great Lakes and the Upper Mississippi Valley, where air, water, and sunshine are blended, so as to invigorate the human frame and make life a pleasure. The very animals are here so formed as to meet the changing seasons with comfort, being warmly clad in winter with their fur-bearing robes, which they shed as the heat increases.

Passing from Summer to Winter, the region of the Great Lakes undergoes a climatic change which it is hard to conceive, and still more difficult to describe. The East is still the land of the " Arabian Nights " with its deserts and heated plains, while in the Western world the scene is changed. We here have to grapple with the secrets of nature, and by observation develop more astonishing truths than the ancients ever conceived of in their flights of fancy and fiction. If Summer here brings its reward in the shape of renewed vegetation and health, Winter does the duty designed by Nature in keeping ever pure and cold the waters which refresh us during warm weather, making the Lakes the great refrigerator for the benefit of the human race. Here the finny tribe love to gambol in the cold waters of these Inland Seas, yielding their share in supplying man with healthy and invigorating food, which goes to sustain the body while improving the intellect.

DE SOTO, while seeking for gold and fountains whose waters were to prolong human life and make man immortal, reached the banks of the turbid Mississippi, near the mouth of the Arkansas River, there to die and be buried in the hot sands that lined its banks, while his followers proceeded southward in hopes of preserving their lives.

FATHER MARQUETTE, at a later period, with the spirit of true discovery, having at heart the good of the aborigines, by whom he was adored, wandered along the shores of the Inland Seas; here hearing of the "Great Mississippi," he extended his travels, reaching De Soto's upper stream, which he descended several hundred miles; but, being informed by the natives that the country was inhabited by hostile Spaniards, returned to die on the shore of Lake Michigan, one of the pure fountains which truly invigorates and gives life.*

Marquette was the first real explorer of the Mississippi, and, after De Soto, the first European who beheld it. It is now proposed to erect a suitable monument to his memory. "The end men propose to themselves is seldom the end they reach. God works through them and plans over them. Marquette meant the Christianization of a handful of savages and the aggrandizement of his king. He opened to the world the gigantic commerce of half a continent. They prepared the soil for the growth of an independent people, greater than any of the past. But though that is our destiny, let us not be deceived. It is not greatness of numbers, but of quality, which alone should occupy this splendid New World. It is not advancement only, but advancement in the spirit of nobleness, which is real progress. We must go, as Marquette went, eager for knowledge, for discovery, even for new and material gain, on the one hand, but, on the other, not less eager for truth, for freedom, for justice, for the helping of every man we meet. And if the proposed monument to this simple priest, who never dreamed of worldly honors, shall keep this thought in the mind of a nation not too prone to translate the doctrine of manifest destiny into spiritual conquest, that monument may well be builded."

Many of the people of Eastern nations, from necessity, utilize the water as well as the land for their places of abode — erecting cabins or floats on their great rivers as their dwellings. Here, in process of time, the same mode may be adopted by striving millions that are sure to congregate on our inland waters, now teeming with life and commerce.

---

* Fathers Marquette and Joliet, on the 17th of May, 1673, started from the mission of St. Ignatius, at Michilimackinac, for the exploration of the Mississippi. On the 19th day of March, 1675, after his return, Father Marquette, the zealous missionary, whose life was devoted to the cause of religion and the welfare of the Indians, died on the east shore of Lake Michigan, at the mouth of the present Marquette River.

When we view the lovely shores and headlands, with the numerous wooded islands, which adorn the Lakes, now mostly uninhabited, we cannot believe the stillness of death will always exist, but that man, from choice or necessity, must ultimately occupy these lovely retreats. The Rhine, with its teeming villages and castellated edifices, will, to a certain extent, be reproduced in this portion of our favored country.

Natural Terraces.

Natural Terraces abound on the borders of the Great Lakes as well as on the banks of the St. Lawrence River, affording delightful sites for cities and villages as well as country residences. Prof. AGASSIZ noticed several in succession on the shores of Lake Superior, while the same feature is to be seen on the beautiful wooded islands of Michipicoten and Grand Island — the latter the "Gitchee Munising" of the Chippewa — one on the Canadian and the other on the American side of Superior. Both of these islands will, no doubt, soon become fashionable resorts, the latter being in the vicinity of the far-famed " Pictured Rocks."

The Island of Mackinac, the most lovely isle on the Upper Lakes, is already the favorite resort of seekers of health and pleasure — rising terrace on terrace. It is elevated from one to three hundred feet above the pure waters of the Straits of Mackinac, and will always attract the attention of the refined and wealthy.

Duluth and *Superior City*, standing at the head of Lake Superior, are both finely situated, the one on a rising ascent and the other on a slightly elevated plateau or terrace.

Bayfield, lying on a large and secure bay, protected by the Apostle Isles, rises by two or three terraces to an elevation overlooking one of the most extensive and grandest scenes on Lake Superior, extending northward through a labyrinth of islands, and southward across Chaquamegon Bay, where the entire fleet of the Lakes might lie in security. *Ashland*, at the head of the above bay, lies on an elevated terrace for the most part, while the water-front is improved by steamboat landings and railroad depots.

Keweenaw Point, from Eagle River to Portage Lake, presents numerous terraces, many of which are already occupied by thriving villages, being occupied by sturdy miners engaged in mining pure copper, for which this section of country is justly famous.

Marquette, the "Iron City" of Superior, is another beautifully situated town, situated on three distinct terraces, rising some two hundred feet above the waters of the Lake, here enlivened with vessels of commerce and the sail-boat of pleasure, while the fisherman's craft may be seen gliding in the far distance, being propelled by sails. At night the scene is enlivened by the ever-watchful mariner's light-house and the fiery blaze of furnaces sending up their lurid flames far above the surrounding country.

Munising, lying on Grand Island Bay, is situated on a low terrace, while immediately behind rises an abrupt hill, forming an elevated plateau, from which descends a silver stream of great beauty. This whole lake front, on the mainland, extending east to Miner's Castle and River, rises by terraces of steep ascent until the Pictured Rocks are reached; then an abrupt precipice, with beetling crags and caves, is to be seen, which is safe to approach in calm weather, but dreaded by the mariner during severe storms.

The River St. Mary, or Strait, connecting Lakes Superior and Huron, with its numerous islands, is terraced near the Rapids, where stands the ancient and romantic settlement of the Saut Ste. Marie. St. Joseph Island, attached to Canada, is an elevated piece of land of great extent, rising gradually above the water's edge.

The terraces formed by the St. Clair and Detroit Rivers are most beautiful, when compared with the low banks of St. Clair Lake, where extensive marshes abound. For the most part, the heavy growth of forest trees are cut down along the banks of the above broad streams, forming the outlet to Lakes Superior, Michigan, and Huron, while in the back-ground rises the majestic oak and other trees of the forest. *Detroit* occupies a fine terrace, rising gradually from the shores of Detroit River. Here steamers and sail-vessels are seen continually passing.

Lake Erie, on its south-eastern shore, presents many fine elevations, where flourishes the grape and other kinds of fruit, as well as on the lovely islands which adorn its western terminus. The terrace on which *Cleveland* stands, elevated about one hundred feet above the waters of the Lake, is one of the most beautiful sites for a city on all the waters of the Upper Lakes, while other localities are nearly equally favored. The city of *Buffalo* stands on a gently rising terrace, unrivalled for healthy situation, convenience and beauty.

From the head of Lake Superior to the city of Quebec, the Great Lakes and River St. Lawrence present the most uniform, clean, and lovely banks imaginable, being continually laved with pure, clear waters.

Pleasures of Travel.

There are thousands of people in the Eastern, Northern, and Southern States that desire, during warm weather, to flee from a hot to a cool climate to enjoy health and pleasure, but unfortunately do not know whither to direct their steps. Some visit the sea-shore, others the fashionable watering-places and the more mountainous portions of the country, in hopes of obtaining a desirable location to enjoy themselves — often without any satisfaction other than a relaxation from business pursuits.

The aim of all tourists should be to obtain enjoyment, health, and knowledge. This can be best obtained by selecting a cool retreat, such as mountains afford, or the islands or shores of the Great Lakes of America — more particularly Lake Superior.

Comparing all sections of Europe or America, Nature has done more to render the Great Lake region cool and healthy than any other portion of the globe, including the shores of Lakes Huron, Michigan, and Superior. All that is at present wanted is good Hotel accommodations. Knowing where to go, however, is but half the knowledge wanted. You next want to know *how to travel and how to live.* The best mode is to prepare to go direct to some given point, by the most comfortable conveyance, not forgetting to observe all the objects of interest on the intended route.

If bound for Lake Superior, proceed to some of the ports on Lake Erie, or start from Detroit, and embark on a favorite steamer for Mackinac or Saut Ste. Marie. The latter interesting place — the gateway to the Lake Superior region — can also be reached from Chicago, passing Mackinac, or from Collingwood, Canada, the latter route passing lovely lake and river scenery, where may be seen numerous islands and islets.

A Word to the Wise, etc.

We quote from a late letter-writer: — " This is the season (during the warm summer months) for universal recreation for men of sedentary occupations, such as ministers, merchants, lawyers, teachers, students, etc., but very poor taste and decrepit judgment are often displayed in making choice of a location for this popular object. The purpose is frequently completely frustrated through this defect, and the weary pilgrim in search of necessary recuperation, physical and mental, returns to his place of business, after weeks of unprofitable sojourn at some crowded resort, disgusted, cross, and more debilitated than if he had remained at his desk in a musty, sultry office, or behind the counter in the stifling atmosphere of his counting-room. Why do they not betake themselves to some retired retreat, like the Island of Mackinac, or Bayfield, or its vicinity, on the shores of Lake Superior, where Nature presents her grandest beauties; bury themselves in the sombre depths of the primeval forest, or hasten to some picturesque lake-side hamlet, where they can effectually dismiss the shadows of care, bid adieu to the harassing turmoils of the commercial world, and throw the whole spirit of their being into the most delicious yet harmless abandon. Boating, trout-fishing, and other sports can here be enjoyed without the corrupting influences that usually pervade the sea-bathing resorts."

SELECTED POETRY.

THE RIVER ST. LAWRENCE.

A ride down the St. Lawrence during the summer season is regarded as a *sine qua non* by all pleasure-seekers in this region. The noble river abounds in variegated and fascinating scenery, not the least of which are the islands and the rapids. "The Thousand Islands" are entered upon soon after leaving Kingston, and continue for a number of miles down the river. They are interesting only as suggestive of those convulsions of nature that must have produced them. The other islands of the river, of which there are many, are more attractive in their quiet beauty. Of the rapids, the most noted are those of the "Long Sault" and the "Lachine." The passage of these rapids forms the most exciting part of the trip. Most of these rapids are circumnavigated by canals, as, owing to the shoalness of the water, laden vessels cannot pass them, and they are only passed downward by light draught steamers, all upward bound vessels passing through the canals. Besides the Welland Canal, there are eight of these canals on the St. Lawrence. A writer, catching the inspiration of the above scene, improvised the following lines referring to the leading features of the trip from Kingston to Montreal:

Down the St. Lawrence River
  We took our morning ride —
Sweet fragrance from the summer flowers
  Exhale on every side;
And Nature wore her fairest dress,
  As we went out the bay,
And in the wildest grandeur
  The "Thousand Islands" lay.
The din of dusty cities now,
  With all their toil and strife,
Was all forgotten for a while —
  We knew a higher life.
We drank the inspiration
  From rocks, and sky, and sea,
And felt an exultation
  Like prisoners set free.
We leaped amid the cascades,
  All thrilled with wildest glee,
And rocked among the rapids
  As in a stormy sea.

Fair maidens clung to lover's arms,
   And brave men held their breath —
It seemed like rushing recklessly
   Into the arms of death.
Anon the " Kingston " dropped her prow,
   Then plowed her even way,
Just as the stormy petrel
   Sits down amid the spray.
We scream with perfect ecstasy —
   We cannot tell it now —
The lustre flashed from every eye,
   Shone on each manly brow.
Life gives few hours so brimfull,
   'Twill dwell in memory long —
Green spots in retrospection,
   The richest theme for song.
Oh! ye who feel the load of life,
   Crazed by its blight and wiles,
Go, gather health and bloom again
   Among the *Thousand Isles;*
You 'll feel your heart grow young again,
   And flow forth full and free,
As you see these blending rivers
   Roll onward to the sea.

SAIL ROCK (PICTURED ROCKS), LAKE SUPERIOR.

From the far Saut of Sainte Marie he wanders,
   On, ever on, the white foam in his track,
By night, by day, sails fleet before the wind,
   Until he sees the head of Fond du Lac;
Yet finds not there the rest he seeks with yearning;
   Frown all the cliffs — and he must wander forth
Over the waves again, by south winds driven,
   Past the dark Palisades into the north.

There stands the haunted arch of Spirit River;
   There, in the storm, is seen the misty shape
Of Manitou, who guards the great Superior,
   Rising above the heights of Thunder Cape;
And seeing him, the guilty one, approaching,
   The voices of the surf rise in a roar
Below the porphyry cliffs, sounding a summons,
   To call the spirits to the lonely shore.

Down, down, they troop through the ravines of iron,
  Over the rocks where virgin silver shines;
Up, up, they roll the surf, a seething barrier,
  And marshal on the beach their shadow-lines.
He cries, he weeps, he prays with arms extended:
  "Have mercy upon me, a soul unblest—
I come not for your stores of shining treasure,
  I only beg — I only pray for rest.

"Aged am I, and worn with countless journeys,
  Over the lake forever must I stray;
In the whole south I cannot find a landing,
  Keweenaw's copper arm thrusts me away;
I sail, and sail, yet never find a harbor, —
  Stern is the east, and sterner is the west, —
Oh, grant me but one foothold on the north shore,
  So can I die at last, and be at rest!"

But no! They drive him off with jeers and shouting,
  Before their ghostly glee the cursed one quails;
Forth from the silver rocks of haunted northland,
  Not daring to look back, away he sails;
And sails, and sails, yet never finds a landing,
  Though fairest coasts and isles he passes by;
And hopes, and hopes, yet never finds a foothold
  On any shore where he can kneel and die.

Weary and worn, through many a red man's lifetime,
  Over the lake he wanders on and on,
Till up through Huron, with red banners flying,
  Come white men from the rising of the sun;
The Saut they name from Sainte Marie with blessing,
  The lake lies hushed before their holy bell,
As, landing on the shore of Rocky Pictures,
  They raise the white cross in *le grand Chapelle.*

As the first white man's hymn on great Superior
  Sounds from the rocky church not made with hands,
A phantom-boat sails in from the still offing,
  And at its bow an aged figure stands;
The worn cords strain so full the sails are swelling,
  The old mast bends and quivers like a bow,
Yet calm the windless sky shines blue above them,
  And calm the windless waves shine blue below.

The boat glides in, still faster, faster sailing,
  Like lightning darting o'er the shrinking miles,
And, as he hears the chanting in the chapel,
  For the first time in years the lone one smiles;
At last, at last, his feet are on the dear shore,
  The curse is gone, his eyes to Heaven rise;
At last, at last, his mother earth receives him, —
  At last, at last, with thankful heart he dies.

The poor worn body, old with many lifetimes,
  They find there lying on the golden sands,
But lifting it with wonder and with reverence,
  It crumbles into dust beneath their hands;
The poor worn boat, grown old with endless voyages,
  Floats up the coast, unguided and alone,
And, stranding 'neath the cliffs, its mission over,
  By the Great Spirit's hand is *turned to stone.*

You see it there among the Rocky Pictures,
  The mainsail and the jib just as they were;
We never passed it with a song or laughter
  In the gay days when we were voyagers;
The best among us doffed our caps in silence,
  The gayest of us never dared to mock
At the strange tale that came down from our fathers,
  The pictured legend of the old *Sail-Rock.*
                              Constance Fenimore Woolson.

TO THE MISSISSIPPI.

Tell me whither, in such haste, thou goest,
  Ever whirling, boiling, turbid river!
Art thou destined as thou proudly flowest,
  Always thus to freely flow forever?
Strange Mississippi! thou art at thy source
  Clear and pure, controllable in motion;
What kindred streams have urged thee on thy course
  T' lose thyself so angrily in ocean?

Ere man in his primeval habit stood
  And marked the boundaries where thou hast strayed
Through tangled forests, cane and cottonwood;
  Through prairie, mountain pass and everglade;

In undetermined pathways of thy own
  Thy course has ever been, as it shall be,
For ages past, for years to come unknown,
  As wild, as irresistible, as free.

Instructive memories of other date
  Along thy banks, a thousand miles, are cast,
Which to the curious traveller relate
  Historic records of the hidden past;
Which tell of changes that were slowly wrought
  By thy destructive, devastating tide;
Which mark the character of human thought,
  A nation's progress and a nation's pride.

Where now the opulence of man is spread,
  And art and industry their gifts bestow,
The painted savage, numbered with thy dead,
  Was the sole monarch a few years ago.
Where, undisturbed, the sea-fowl flapped its wings,
  The mariner his canvas has unfurled,
And commerce to a thrifty people brings
  Th' accumulated riches of a world.

From the rough hills of the inclement North,
  As undeterred by distance as by time,
Thy swelling mass of waters issue forth
  To bear earth's bounty to this sunny clime.
How often on thy willing bosom borne,
  Full-freighted vessels have I loved to scan,
Each on its peaceful mission, steering on,
  To cheer the intercourse of man with man!

As now along thy southern banks I range,
  And note the changes that I find in thee,
I am reminded of the greatest change,
  That surely hast and must come over me.
Perhaps, for ages, thou wilt onward move,
  In all thy strength, magnificence and pride,
When, separated from the friends I love,
  I shall be sleeping coldly at thy side.

Thy edd'ing stream that whirls in ceaseless strife;
  The wrecks that on thy shifting sands are seen,
Are but a history of human life;
  Of what my joys, my hopes, my fears have been.
But, unlike thee, oh, may my cares subside
  Ere the dull grave invites me as its guest!
And may my soul in peaceful humor glide,
  Into a haven of eternal rest.

                                  MAUD.

PART II.

EARLY FRENCH DISCOVERIES IN THE REGION OF THE GREAT LAKES.

THE discovery of the 'Great West,' or the Valleys of the Mississippi and the Lakes," says a late writer, "is a portion of our history hitherto very obscure. Those magnificent regions were revealed to the world through a series of daring enterprises, of which the motives, and even the incidents, have been but partially and superficially known.

"In 1641, Isaac Jogues and Charles Raymbault, Jesuit missionaries, preached the Faith to a concourse of Indians at the outlet of Lake Superior. Then came the havoc and desolation of the Iroquois War, and, for years, further exploration was arrested. At length, in 1658, two daring traders penetrated to Lake Superior, wintered there, and brought back the tales they had heard of the ferocious Sioux, and of a great western river on which they dwelt. Two years later, the aged Jesuit Ménard attempted to plant a mission on the southern shore of the Lake; but perished in the forest, by famine or the tomahawk. *Allouez* succeeded him, explored a part of Lake Superior, and heard, in his turn, of the Sioux and their great river, the '*Messipi*.' More and more the thoughts of the Jesuits, and not of the Jesuits alone, dwelt on this mysterious stream. Through what regions did it flow; and whither would it lead them; to the South Sea, or the 'Sea of Virginia;' to Mexico, Japan or China? The problem was soon to be solved, and the mystery revealed."

Father James Marquette, in the spring of 1668, arrived at the Saut Ste. Marie; Father Dablon in 1669, when the first permanent settlement was made on the soil of Michigan. During the same year Father Marquette visited La Pointe, where he found several Indian villages composed of the Huron tribe. *

To Fathers Marquette, Dablon, Jolliet, and La Salle, Jesuit missionaries, are due the credit of the full discovery of the Great Lakes and the Upper Mississippi River, between the years 1668 and 1678.

Marquette visited Michilimackinac in 1670, and spent a winter there before the establishment of his mission. Point St. Ignace, on the north side of the Straits,

---

* The Jesuits and fur-traders, on their way to the Upper Lakes, had followed the route of the Ottawa, through Canada, or, more recently, that of the Georgian Bay. Iroquois hostility had long closed the Niagara portage and Lake Erie against them.

was selected as the most suitable spot for the proposed mission, and there, in 1671, a rude and unshapely chapel was raised, as "the first Sylvan Shrine of Catholicity" at Mackinac. This primitive temple was as simple as the faith taught by the devoted Missionary, and had nothing to impress the senses, nothing to win by a dazzling exterior the wayward children of the forest. The new mission was called St. Ignatius, in honor of the founder of the Jesuit order, and to this day the name is perpetuated in the point upon which the mission stood.

The French take Possession of the Country.

"During the year of 1671 an event occurred of no common interest and importance in the annals of French history in America, but which, after all, was not destined to exert any lasting influence. Nicholas Perrot had been commissioned as the agent of the French Government to call a general Congress of the Lake tribes at the Falls of St. Mary. The invitations of this enthusiastic agent of the Bourbon dynasty reached the tribes of Lake Superior, and were carried even to the wandering hordes of the remote North and West. Nor were the nations of the South neglected. Obtaining an escort of Pottawatomies at Green Bay, Perrot, the first of Europeans to visit that place, repaired to the Miamis at Chicago on the same mission of friendship and diplomacy.

"In May, the day appointed for the unwonted spectacle of the Congress of Nations arrived. St. Lusson was the French official, and Allouez his interpreter. From the head-waters of the St. Lawrence, from the Great Lakes, from the Mississippi, and even from the Red River of the North, envoys of the wild republicans of the wilderness were present; and brilliantly clad officers from the veteran armies of France, with here and there a Jesuit missionary, completed the vast assembly. A cross was set up, a cedar post marked with the French lilies, and the representatives of the wilderness tribes were informed that they were under the protection of the French king. Thus, in the presence of the ancient races of America, were the authority and the faith of France uplifted in the very heart of our Continent. But the Congress proved only an echo, soon to die away, and left no abiding monument to mark its glory."

The aborigines inhabiting the Lower and Upper Peninsulas of Michigan, in early times, of which we have any knowledge, were the Chippewa or Ojibway tribe of Indians, a branch of the numerous Algonquin family. In the more southern part of the territory, however, were found scattered tribes of Hurons or Wyandots, Miamis, Ottawas, Winnebagoes, Pottawatomies, and other tribes, living in peaceful contiguity. The Chippewas still retain and occupy their former hunting-grounds, extending from Georgian Bay or Lake Huron, both shores of Lake Superior, and westward to the head-waters of the Mississippi.

French Account of the Ancient Mines and Miners of Lake Superior.

In the Relacion for 1666–'67, Chap. ii., page 32 *et seq.*, entitled " *Relacion de la Mission du St. Esprit, aux Outaovecs dans le Lac Tracy dite aupravant le Lac Superieur.* Journal du voyage du Pere Claude Allouez dans le Pais de Outaouacs," we find these passages:

"The savages respect this lake as a divinity, and make sacrifices to it, on account perhaps of its magnitude, for it is two hundred leagues long and eighty wide ; on account of its goodness in furnishing them with fishes, which nourish all those people where there is but little game. There are often found beneath the water pieces of copper all formed, and of the weight of ten and twenty pounds. I have seen them many times in the hands of the savages ; and as they are superstitious, they keep them as so many divinities, or as presents from the gods beneath the water, who have given them as pledges of good fortune. On that account they keep the pieces of copper enveloped among their most precious furniture. There are some who have preserved them for more than fifty years, and others who have had them in their families from time immemorial, and cherish them as household gods."

Some time since a large mass of copper, like a rock, was seen with the point projecting out of the water. This afforded passers-by an opportunity of cutting off pieces. Nevertheless, when I went there it was not to be seen. I believe the storms, which are here very violent, and like those on the sea, had covered the copper rock with sand. Our savages wished to persuade us that it was a divinity, and had disappeared, for some reason which they did not mention.

" *De la Mission du Sainte Esprit a la Pointe de Chagaouamigong dans le Lac Tracy ou Superieur* — chap. xi., des proprietez et Raritez."

From the above work I have translated the following interesting description of the form of Lake Superior, and of the copper found there :

"The Lake has nearly the form of a bended bow, of more than eighty leagues in length. The southern side represents the string; and a long tongue of land which springs from the centre of the southern shore, and projects upwards of twenty-five leagues into the lake near to its middle, is the arrow. [Keeweenaw Point.]

"The northern coast is bordered with frightful crags, which are the termination of the prodigious chain of mountains which take their rise at Cape Tourment, above Quebec, and extend to this place, traversing more than six hundred leagues in extent, and losing themselves in the farther extremity of the Lake. There are very few islands in the Lake, and they occur mostly on the northern shore. This great expanse of the waters gives room for the winds, which agitate the lake with as much violence as they do the ocean."

On page 26 is a chapter headed "Mines of Copper which are found on Lake Superior."

Up to the present time it was believed that these mines were found on only one or two of the islands; but since we have made a more careful inquiry, we have learned from the savages some secrets which they were unwilling to reveal. It was necessary to use much address in order to draw out of them this knowledge, and to discriminate between truth and falsehood. We will not warrant, however, all we learned from their simple statement, since we shall be able to speak with more certainty when we have visited the places themselves, which we count on doing this summer, when we shall go to find the "wandering sheep" in all quarters of this great Lake. The first place where copper occurs in abundance after going above the Saut is on an island about forty or fifty leagues therefrom, near the north shore, opposite a place called "Missipiconatong."

The savages say it is a floating island, which is sometimes far off and sometimes near, according as the winds move it, driving it sometimes one way and sometimes another. They add that, a long time ago, four Indians accidentally went there, being lost in a fog, with which this island is almost always surrounded. It was long before they had any trade with the French, and they had no kettles or hatchets. Wishing to cook some food, they made use of their usual method, taking stones which they picked up on the shore, heating them in a fire, and throwing them into a bark trough full of water, in order to make it boil, and by this operation cook their meat. As they took up the stones, they found they were nearly all of them pure copper. After having partaken of their meal, they thought of embarking, fearing to remain lest the lynxes (loups cerviers) and the rabbits (lievres), which are in the place as large as dogs, (!) would come and eat up their provisions, and even their canoe. Before leaving they collected a quantity of these stones, both large and small ones, and even some sheets of copper; but they had not gone far from the shore before a loud voice was heard, saying in anger, "Who are these robbers who have stolen the cradles and playthings of my children?" The sheets of copper were the cradles, for the Indians make them of one or two pieces of wood (a flat piece of bark with a hoop over one end), the child being swathed and bound upon the flat piece. The little pieces of copper which they took were the playthings, such pebbles being used by Indian children for a like purpose. This voice greatly alarmed them, not knowing what it could be. One said to the others, it is thunder, because there are frequent storms there; others said it is a certain genii whom they call "Missibizi," who is reputed among these people to be the god of the waters, as Neptune was among the Pagans; others said that it came from Memogovissiousis — that is to say, seamen, similar to the fabulous Tritons, or to the Sirens, which live always in the water, with their long hair reaching to their waists. One of our savages said he had seen one in the water; nevertheless, he must have merely imagined he did. However, this voice so terrified them that one of these four *voyageurs* died before they reached land. Shortly after, a second one of them expired; then a third, so that only one of them remained, who, returning home, told all that had taken place, and died shortly afterwards. The timid and superstitious savages have never since dared to go there for fear of losing their lives, believing that there are certain genii who kill those who land there; and within the memory of man no one has been known who has set foot on that shore, or even coasted along its shores, although the island is within sight, and even the trees are visible upon another island called Achemikonan.

There is both truth and error in this story, and this is most probably the explanation: These four savages were poisoned by the water which they boiled with red-hot copper, which, by the intensity of the heat, gave off a poison, etc. It is not a poison which acts immediately, and on one as soon as it will on another, as happened in this case. It may be that when they were taken ill, they more readily imagined they heard a voice; perhaps they heard an echo, such as are very common among the rocks which border this island; or perhaps they made this fable since, not knowing to what to attribute the death of these Indians. When they said it was a floating island, it is probable they may have been misled by the vapors which surround it,—they being rarefied or condensed by the variable action of the sun's rays, made the island appear sometimes near and sometimes far off. It is certain, however, that it is a common belief among the Indians that there is an abundance of copper on this island; but they dare not go there. We hope to begin our discoveries upon it this summer.

Advancing to a place called the Grand Anse [Great Bay], we meet with an island three leagues from land, which is celebrated for the metal which is found there, and for the thunder which takes place, because they say it always thunders there. [Thunder Cape.] But further toward the west, on the same north shore, is the island most famous for copper, called the "Minong," (the good place.) [Isle Royale.] This island is twenty-five leagues in length; it is seven leagues from the main land and sixty from the head of the Lake. Nearly all around the island, on the water's edge, pieces of copper are found mixed with pebbles, but especially on the side which is opposite the south, and principally in a certain bay which is near the north-east exposure to the great Lake. There are shores "tous escarpez de terre glaize," and there are seen several layers or beds of copper, one over the other, separated or divided by other beds of earth or rocks. "In the water is seen copper sand, and one can take up in spoons grains of the metal big as an acorn, and others fine as sand." [This description probably refers to Rock Harbor.] This island is almost surrounded with islets, which are said to be composed of copper, and they are met with even to the main land on the north. One of them is two gun-shots from "Minong"—it is near the middle of the island, and the end which looks towards the north-east. Farther out on this side there is another island, called "Mauitouminis," on account of the copper which abounds on it; and it is said that those who were there on one occasion, on throwing stones, made it resound like an explosion.

Advancing to the Lake and returning one day's journey by the south coast, there is seen on the edge of the water a rock of copper which weighs seven hundred or eight hundred pounds, and is so hard that steel can hardly cut it; but when it is heated, it cuts as easily as lead. Near Point Chagaouamigong [Chaquamegon], where a mission was established, rocks of copper and plates of metal were found on the shores of the islands.

Last spring we bought of the savages a sheet of pure copper, two feet square, which weighed more than one hundred pounds. We do not believe, however, that the mines are found on these islands, but that the copper was probably brought from "Minong," [Isle Royale,] or from other islands, by floating ice, or over the bottom of the Lake by the impetuous winds, which are very violent, particularly when they come from the north-east.

Returning still towards the mouth of the Lake, following the coast on the south, at twenty leagues from the place last mentioned, we enter the river called "Nautounagan," [Ontonagon,] on which is seen an eminence where stones and copper fall into the water or upon the earth; they are readily found. Three years since, we received a piece which was brought from this place, which weighed a hundred pounds, and we sent it to Quebec, to M. Talon. It is uncertain exactly where this was taken from; some think it was taken from the forks of the river, others that it was from near the lake, and dug up from the soil.

Proceeding still farther, we come to the long point of land which we have compared to the arrow of the bow (Keweenaw). At the extremity of this there is a small island, which is said to be only six feet square, and all copper!

We are assured that copper is found in various places along the southern shore of the Lake. All the information we obtained from others it is not necessary for us to detail; but it seems necessary that more exact researches should be made, and this is what we shall endeavor to effect. If God prospers us in our enterprise, we shall speak next year with more certainty and knowledge.

The Relacion of 1670–71 contains the remarks of Pere Ablon. In the second part he gave an account of the copper mines in page 61: "We would remark, by the way, that copper is found in all parts of this Lake, although we have not as yet sufficiently exact knowledge, for want of thorough explorations; nevertheless, the plates and masses of this metal which we have seen, weigh each a hundred or two hundred pounds, and much more. The great rock of copper of seven hundred pounds, and which all the travellers saw near the head of the Lake, besides a quantity of pieces which are found near the shores in various places, seem not to permit us to doubt that there are somewhere the parent mines, which have not been discovered."

Father Marquette and his Discovery of the Mississippi River.

Our hero, JAMES MARQUETTE, was born of the noble family of Marquette, in the northern part of France, in the year 1637. In the city of Laon, on the meandering, vine-clad, olive-girt little River Oise, where his ancestors had raised and exerted a wide influence for generations before him, he also spent his youthful days till the age of seventeen; he then attached himself to the Society of Jesus, or Jesuits, and with them pursued a course of study preparatory to the priesthood. When invested with orders, having a strong preference for the missionary work, he determined to make the wilds of America the fields for his future labor. So, in the summer of 1666, he sailed for Canada, on the 20th of September, and landed at Quebec, then but a trading-post. He was soon recognized, by his courage and zeal, as a man specially fitted to advance the efforts that had been made among the Algonquins on Lake Superior, but which had been nearly extinguished by the incursions of the warlike Iroquois. Accordingly, Father James Marquette, on the 10th of October, of the same year that he landed, embarked again, but now in a bark canoe, conducted by a couple of the Mission Indians, for the "Three Rivers," there to commence the study of the language, and soon after he proceeded to the Ottawa Mission on Lake Superior. At first he was stationed at the Saut Ste. Marie, but in a few months it was deemed expedient to remove him farther on to the mission at La Pointe.

The bands that lived about the south-western shores of the Lake were annually visited by the Illinois, who came from their broad prairie-homes on the "Great Mississippi," for the purpose of obtaining by barter the trinkets and chattels distributed by the French. The glowing description given by these of the "Great Water," the Mississippi, that flowed south through many nations, and emptied itself, after many moons' journeyings, into the salt sea, awakened in Father Marquette an ardent desire to explore this mysterious river, find whether or not the sea it emptied into was the Western Ocean, and carry to the Southern nations the Gospel of peace.

The Hurons at La Pointe were about abandoning their homes, from fear of the Sioux warriors; and our missionary, though loth to remove farther from his cherished desire of going south-west to the Illinois and other Mississippi nations, yet accompanied them back in their canoes to Mackinac, and there rebuilt the old church that had been for a time deserted.

In the summer of 1672, the Comte de Frontenac, Governor of Canada, clearly perceiving the importance of an exploration of the Mississippi River and its outlet, appointed Sieur M. Jolliet, accompanied by Father Marquette, to make the discovery; but on their return, Jolliet's journals and reports were lost by the upsetting of their canoe below the Sault St. Louis, near Montreal, and thus the only remaining account of the expedition was that prepared by Marquette.

In two bark canoes, their entire outfit a stock of corn and dried meat, with five Indian attendants, Jolliet and Marquette, on the 17th of May, 1673, started from the Mission of St. Ignatius, at Michilimackinac, for the exploration of the Mississippi. Coasting along the shore, with every precaution to avoid surprise, they entered and went up Green Bay, and ascended the Fox River for a distance of 160 miles to its source, in a level prairie flat, but a little distance from the springs of the Wisconsin, which flowed into the Great Waters they were in search of. Having carried their canoes over the narrow portage, they continued their voyage down the shallow river, often quite hid from sight by the growth of wild oats, through which they had to open a way for their canoe, as one would through the thicket. As they descended it grew broader, and dashed about among reeds and sandy shoals. About 30 leagues below its source they found what they took to be an iron mine; and somewhat farther on, about 120 miles below the portage, on the 17th of June, "with," says Marquette, "a joy that I cannot express," they entered the Mississippi River. Down its gentle current they glided, by the unique though varied scenes, with countless herds of buffalo and deer on its shores, and innumerable fish in its waters, until, in some ten days, for the first time since they left the Lakes, they perceived some indications of humanity. From the river-side a winding footpath led off through the prairie; following this, Jolliet and Marquette soon came to an Indian village, in which they were cordially received, and which proved to be of the Illinois, the very people among whom Father Marquette had so long desired to plant a mission. They strongly urged our adventurers not to proceed farther, for danger would encompass them on every side; but, nothing daunted, again they embarked, and after a journey southward of some 60 miles, they came to the river Onabonbigan, or Ohio, a little after which they discovered what they supposed to be a very rich iron mine.

On they still went, through several nations of hostile Indians, encountering dangers of every kind, until they came among the Akamsea or Arkansas Indians, nearly where De Soto had breathed his last, 130 years before. From this tribe they learned they were only ten days' journey from the sea, where were stationed traders who appeared much like themselves, and came and went in great ships. Judging these correctly to be Spaniards, our travellers were in doubt whether it would be best for them to push on to the mouth of the river or not. They had already ascertained for a certainty that the Mississippi emptied, not, as was supposed, on the eastern coast of Virginia, or through California into the Western Ocean, but into the Gulf of Mexico, from which they certainly could not be far; that mouth they knew to be held by the Spaniards, with whom they were on no friendly terms. Should they happen to fall into the hands of these Spaniards, they could anticipate nothing less than to be held as prisoners, since not only were their respective countries at war, but that the results of their explorations might not be carried back to the French, and thus induce encroachments on the territory held in the name of Spain. Thinking it then more prudent to return, that the fruits they had already gathered might not be at once lost by an effort to grasp too much, on the 17th of July they left the village Akamsea, and commenced pulling back their canoes up the Mississippi current. They took, on returning, however, a different course. Having ascended the Illinois River, they crossed over the portage to the Chicago River, and thence down to what was then called Lake Illinois, but which has since changed its name to Lake Michigan. Coasting along the shore, they returned to Green Bay, and there, at the Mission of St. Francis Xavier, Father Marquette, on account of the enfeebled and shattered state of his health, spent the ensuing winter and summer of 1674.

This was in reality the first exploration of the Mississippi River. Ferdinand De Soto, it is true, generally has the credit of having first discovered it as early as 1541; but, in the first place, whatever expeditions he made were for the purpose of gain and plunder, and so a great deal that would have demanded the notice of one with more liberal and unselfish aims, was quite passed over by him; and then the accounts and reports of his travels that still remain are of such an unreliable character, that but little dependence can be placed in them.

In accordance with his promise to the nations on the Mississippi River, Father Marquette embarked, in the month of November, 1674, to take among them another journey, more exclusively than the first, of a religious character. Though detained on the way by illness, he reached the Illinois nation, on the Mississippi, and commenced a mission in their midst, as he had long desired; but he was obliged, the following spring, on account of his declining health, to commence his return, that he might, if possible, die where some Christian brother could give him an appropriate burial; but in this he was disappointed. His health and strength continued failing rapidly, until, on the 15th of May, 1675, on the shore of Lake Michigan, just within the mouth of a little river that bears his name, he was lifted out of his canoe and placed under a shed of bark and twigs, but to be borne thence to his grave on an eminence overlooking both lake and river. Subsequently, the Kishabon Indians, once of the mission at La Pointe, dug up and unrolled the remains, and dissecting and washing the bones, according to their custom, put them neatly into a box of birch bark, and bore them, with a convoy of thirty canoes, to the house of St. Ignatius, at Michilimackinac, where they were interred with all due ceremonies, not to be disturbed again, most likely, till the last day.

Discovery and Settlement of Mackinac.

OLD MICHILIMACKINAC.

First visited by the *Courriers du Bois* and Jesuit missionaries in 1620.

Permanent settlement in 1671, by Father MARQUETTE, an eminent Jesuit missionary, who, four years previous, in 1667, visited the Saut Ste. Marie, and extended his journey to La Pointe, on one of the Apostle Islands, Lake Superior, where he located an Indian mission. In 1669 he came to Point St. Ignace, in the Straits of Mackinac, and established another Indian mission. Two years thereafter, he located a mission and trading-station at Old Michilimackinac, or " Pequotenonge" of the Chippewa dialect.

"This place and its vicinity is the most noted in these regions for the abundance of its fisheries; for, according to the Indian saying, 'this is the home of the fishes.' Elsewhere, although they exist in large numbers, it is not properly their 'home,' which is in the neighborhood of Michilimackinac."

Old Michilimackinac was for many years the metropolis of the Chippewa and Ottawa tribes of Indians, the country being claimed by the French, who traded with the Indians, it being the principal rendezvous of all the tribes in this part of the country.

The Indians remained on friendly terms with the French until 1760, when the English took possession of the country after the capture of Quebec and capitulation of the French forces in Canada.

In 1761, the English built a palisade fort at Old Michilimackinac, and traded with the Indians, many of whom were very hostile. In May, 1763, the garrison was surprised, and most of them massacred by the Indians. Out of twelve English posts above Montreal, nine were similarly surprised and captured by the combined Indian forces under the celebrated Indian chief PONTIAC. Niagara, Detroit, and Du Quesne, or Pittsburgh, alone narrowly escaped a similar fate. After the above massacre, Old Michilimackinac was abandoned by the English, and the Island of Mackinac selected as a permanent settlement in 1764.

English Accounts of Lake Superior.

The earliest English traveller who visited the shores of Lake Superior, of whom we have any account, is ALEXANDER HENRY. His work is entitled "Travels and Adventures in Canada and the Indian Territories, between the years 1760 and 1776, in two parts, by Alexander Henry, Esq." New York, 1809. 8vo.

Henry was a trader, who, soon after the conquest of Canada by the English, set out on a trading voyage to Fort Mackinac. He arrived there while the fort was in possession of the English troops, and was present when the dreadful massacre of the whole garrison was effected by the Indians, who took the fort by a most ingenious stratagem. Henry was saved by being suddenly, and most unexpectedly, adopted as a brother by one of the conquering Indians, and who most carefully guarded him from harm during the subsequent carousals of the Indians, concealing him in a cave on the opposite island — Michilimackinac — the place now called Mackinaw. Referring to his most interesting and evidently truthful narrative for an account of his voyages and perils, and for his general description of the country, I shall limit myself in this review to a few extracts showing the amount of knowledge he possessed of the existence of copper, and other metals or ores, on the shores of Lake Superior. He says:

"On the 19th of August, 1765, we reached the mouth of the Ontonagon River, one of the largest on the south side of the Lake. At the mouth was an Indian village, and three leagues above, a fall, at the foot of which, sturgeon, at this season, were obtained so abundant that a month's subsistence for a regiment could be taken in a few hours. But I found this river chiefly remarkable for the abundance of virgin copper which is on its banks and in its neighborhood, and of which the reputation is at present more generally spread than it was at the time of my first visit.

"The attempts which were shortly after made to work the mines of Lake Superior to advantage will very soon claim a place among the facts which I am about to describe.

"The copper presented itself to the eye in masses of various weight. The Indians showed me one of twenty pounds. They were used to manufacture this metal into spoons and bracelets for themselves. In the perfect state in which they found it, it required nothing but to beat it into shape. The 'Piwatie,' or Iron River, enters the Lake to the westward of the Ontonagon, and hence, as is pretended, silver was found while the country was in the possession of the French.

. . . . "On my way back to Michilimackinac I encamped a second time at the mouth of the Ontonagon River, and now took the opportunity of going ten miles up the river with Indian guides. The object for which I more expressly went, and to which I had the satisfaction of being led, was a mass of copper of the weight, according to my estimate, of no less than *five tons*." [This is the copper rock now on the ground near the War Department, in Washington.] "Such was its pure and malleable state, that with an axe I was able to cut off a portion weighing a *hundred pounds.* On viewing the surrounding surface, I conjectured that the mass at some period or other had rolled from the side of a lofty hill which rises at its back."

. . . . "The same year [1767] I chose my wintering grounds at Michipicoten, on the north shore of Lake Superior.

"At Point Mamainse the beach appeared to abound in mineral substances, and I met with a vein of lead ore where the metal abounded in cubic crystals.

"Still coasting along the Lake, on the north shore, I found several veins of copper ore of that kind which the miners call 'gray ore.'"

Near Michipicoten Bay he says he found on the beach several pieces of virgin copper, of which many were remarkable for their form, some resembling leaves of vegetables, dendritic copper, and others, animals.

Niagara Frontier.

An interesting and instructive address upon the "Niagara Frontier," was recently read before the Buffalo Historical Society by O. H. Marshall, Esq., which not only embraced sketches of the early history of the frontier, but other historical facts of great interest. To M. DE LA SALLE, the intrepid voyageur and discoverer of the mouth of the Mississippi River, he concedes the honor of being the first European that visited the Cataract of Niagara.

On the 6th day of December, 1678, La Salle, in his brigantine of ten tons, doubled the point where Fort Niagara now stands, and anchored in the sheltered waters of the river. The prosecution of his bold enterprise at that inclement season, involving the exploration of a vast and unknown country in vessels built along the way, indicates the indomitable energy and self-reliance of the discoverer. His crew consisted of sixteen persons, under the immediate command of Sieur de la Motte, and as they entered the noble river the grateful Franciscans chanted the *"Te Deum laudamus."* "The strains of that ancient hymn of the church," says Mr. Marshall, "as they rose from the deck of the adventurous bark and echoed from shore and forest, must have startled the watchful Senecas with the unusual sound as they gazed upon their strange visitors. Never before had white men ascended the river. On its borders the wild Indian still contended for supremacy with the scarcely wilder beasts of the forest. Dense woods overhung the shore, except at the site of the present fort or near the portage above, where a few temporary cabins sheltered some fishing parties of the Senecas. All was yet primitive and unexplored."

La Salle, however, was shortly after compelled, through the jealousy of the savages, to retreat, and no regular defensive work was undertaken until 1686, when, on the representations of De Nouville, a fort was built and garrisoned by one hundred men. But the next season the fort was dismantled, and no efforts were taken for its reconstruction until 1725, when, by consent of the Iroquois, Fort Niagara was commenced in stone, and finished the following year.

Such is in brief the history of the commencement of the settlement of a section of country whose interest is enhanced not only by the events of the war of 1812, but by the possession of one of the greatest natural curiosities in the world. The history of the origin of the name is even more curious. The stream, in which La Salle anchored, he called by its Indian name, Niagara, which is perhaps the oldest of all the local geographical names that have come down to us from the Aboriginals. The name itself, however, seems not at first to have been thus written by the English, since it passed through almost every possible alphabetical variation before its present orthography was established. The original name, as pronounced by the neutral nation, was *On-gui-aah-ra;* by the Mohawks *Nyah-ga-rah*, and by the Senecas *Nyah-gaah.* In 1657 the name appeared on Samson's map of Canada spelled *Ongiara;* and in 1688 it made its first appearance as *Niagara* on Coronelli's map, published in Paris. But this final spelling was not reached until the word, as Dr. O'Callaghan informs us, had been spelled in thirty-nine different ways. The word itself is probably derived from the Mohawks, through whom the French had their first intercourse with the Iroquois. The Mohawks say the word means " neck," in allusion to its connecting the two lakes. And this leads to the noting of an exceedingly curious fact, viz., that the word Niagara has no reference to the Cataract which is now supposed to bear the original Indian name. The term had, in the minds of the Indians, reference only to the river — and it is sometimes applied by the early historians not only to the river, but to a defensive work and groups of Indian cabins which stood near the present village of Lewiston. Nor, indeed, does it appear that those tribes dwelling around the Falls at the time of the discovery knew them by any distinctive name. The first historical notice of Niagara Falls appears in the journal of Jacques Cartier, who states that the Indians whom he met in the Gulf of St. Lawrence in 1535, alluded, in their description of the interior, merely to a " cataract and portage at the extremity of Lake Ontario." Seventy-eight years afterwards, Champlain published an account of his voyage in Canada, illustrated by a map of the country, on which the several lakes, as far west as Lake Huron, are laid down in the edition of 1632, though in a very erroneous outline. It distinctly shows the River Niagara, interrupted by a waterfall, and intersected by an elevation of land, answering to the mountain ridge at Lewiston. It contains no specific name for the cataract, but calls it *Saut d'eau,* or waterfall. And this circumstance is the more remarkable, since the early French explorers were very particular to give the Indian names of all places that they visited. After the discovery, however, the Senecas appear to have given it the name of " *Det-gah-shoh-ses,*" signifying "the place of the High Fall." "They never," says Mr. Marshall, " call it Niagara, nor by any similar term, neither does that word signify in their language ' thunders of waters,' as affirmed by Schoolcraft."

Niagara, is a word of Iroquois origin, and in the Mohawk dialect is pronounced Nyagarah, — the orthography, accentuation and meaning of which are variously given by different authors. It is highly probable that this diversity might be accounted for and explained by tracing the appellation through the dialects of the several tribes of aborigines who formerly inhabited the neighboring country.

Niagara River.

" Majestic stream ! what river rivals thee,
    Thou child of many lakes, and sire of one —
    Lakes that claim kindred with the Inland Seas, —
Large at thy birth as when thy race is run ! "

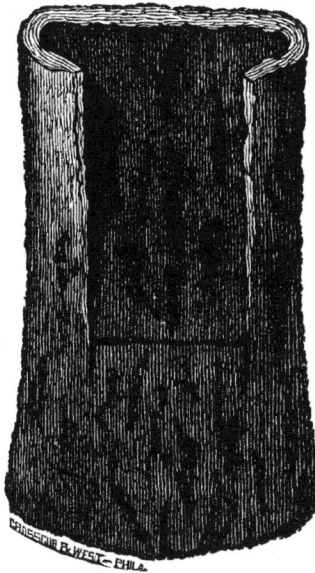

INDIAN RELICS.

The drawings represent two Ancient Relics found on the shore of Lake Superior. They consisted of three finely formed copper instruments for war purposes. They were found in 1871, near the mouth of the Lake Superior Ship Canal, eight miles above Houghton, being taken from an Indian grave. The first is a copper axe, with an eye for a handle, which was decayed; it is ingeniously prepared, with a curvature to admit the wood so as to be firm and secure. The second and most interesting is a large copper spear, two feet in length and weighing five pounds, with an eye for the purpose of affixing a handle. It shows some slight inequalities from corrosion and wear. The third is a copper knife, fifteen and a half inches long, having an elongation for a handle of wood, weighing one pound.

The above interesting relics are in the possession of ISAAC OTIS, Esq., Supt. of the construction of the Portage Lake Ship Canal.

Ancient relics and pit-holes have also been discovered in other parts of Keweenaw Point, showing that the Indians, or native inhabitants, were early acquainted with the value of copper, being used for tools and warlike purposes. They, in fact, became early acquainted with the existence of silver, all of which discoveries they kept secret from the white man — fearing they would be sacrificed by Evil Spirits, if the place of deposit was divulged. Many strange tales or legends are in existence in regard to these Copper and Silver Mines.

These facts, in connection with the early discoveries made by the French Jesuits and Traders, go to show that for several hundred years the Copper Region of Lake Superior has been known to both natives and explorers.

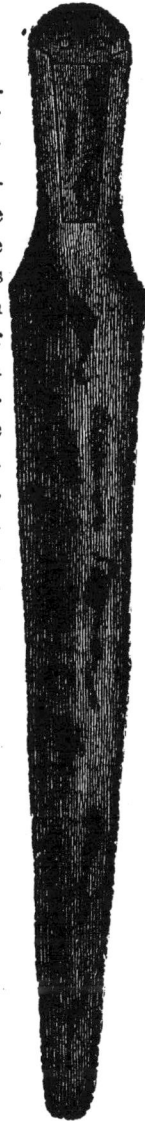

Historical Events in Chronological Order.

1641. *Fathers Raymbault* and *Jogues* visit Saut Ste. Marie, and establish a mission among the Chippewas.

1644. *Iroquois War* commenced against the Hurons who were in alliance with the French.

1658. *Lake Superior* visited by two French fur-traders.

1660. *Father Ménard* visited Lake Superior and attempted to plant a mission on the southern shore.

1666–67. *Pere Cloude Allouez* visits Lake Superior and explores its shores, discovering large deposits of copper.

1668. *Father James Marquette* visits Saut Ste. Marie, where a mission was established.

1669. *Father Dablon,* superior of the mission, erects a church at the Saut. This was the first permanent settlement made on the soil of Michigan.

During the same year *Father Marquette* repaired to La Pointe, situated on one of the Twelve Apostle Islands, where he found several Indian villages.

1670–71. *Pere Dablon* visits the copper mines of Lake Superior.

1670. *Father Marquette* visits the Island of Michilimackinac, inhabited by the Huron tribe of Indians.

1671. *Point Iroquois* selected by Marquette as a suitable place for a mission.

During the same year a *Congress of Nations* assembled at the Saut Ste. Marie, attended by numerous Indian tribes, and St. Lusson, Perrot, Allouez, and others on the part of France. A cross was erected, with imposing ceremonies, and the Indians were informed that they were under the protection of the French king.

1673. *Father Marquette* and *Sieur M. Jolliet* started from the mission of St. Ignatius, at Michilimackinac, on the 17th of May, for the exploration of the Mississippi, and entered Green Bay on their way to the Wisconsin River, which flowed into the "Great Waters." On the 17th of June they entered the Mississippi River, and glided down its gentle current to the mouth of the Arkansas River, returning to Green Bay at the close of the same year.

1674. During this year *Marquette* again proceeded to the Mississippi and visited the Illinois Nation, intending to establish a mission in their midst, but on account of his declining health returned to Lake Michigan.

1675. May 15th, *Father Marquette* died on the east shore of Lake Michigan, near the mouth of a river which bears his name.

1679. *Cavalier de La Salle* and *Hennepin*, the journalist of the expedition, on the 7th day of August, set sail on the waters of Lake Erie on board a vessel named the Griffin, bound on a voyage of discovery to the Mississippi. They arrived at Mackinac in August, and in September sailed for Green Bay.

1680. *Hennepin* dispatched to discover the sources of the Mississippi.

1682. *La Salle* constructed a vessel of a size suitable for the purpose of descending the Mississippi to the Gulf.

1687. *La Salle* was assassinated by one of his own men.

1688. *Baron La Houtan* visited Michilimackinac.

1695. *M. de La Motte Cadillac* commanded at this post.

1699. *Cadillac* was authorized to establish a fort at Detroit, Mich. This he accomplished in 1701.

1750. The French established a military post at the Saut Ste. Marie, for the purpose of excluding the English, as far as possible, from obtaining a foothold on Lake Superior.

1721. *Father Charlevoix*, the historian of New France, visited Mackinac. From the date of his visit down to 1760, when it passed forever out of the hands of the French, the records of this post are meagre and comparatively devoid of interest, although advantageously situated for trade with the Indians.

1756. War between England and France.

1759. *Quebec* captured by the English, and Canada surrendered.

1760. *Old Mackinac* and all the French forts surrendered to the English.

1761. *Alexander Henry*, an English fur-trader, visited Mackinac.

1763. The Indians under *Pontiac* rise against the English, and capture the fort at Old Mackinac, murdering the garrison in cold blood; killing about 70 men.

1776. The *Revolutionary War* commences with England.

1780. The English occupy the *Island of Mackinac*, and erect a government house and fort.

1783. *Peace* made with England, and independence acknowledged.

1795. The British give up *Fort Mackinac* to the Americans, who retained possession until the war of 1812, when it was taken by the English, and again relinquished in 1815, as well as Saut Ste. Marie, Mich.

PART III.

EARLY NAVIGATION, STEAMBOAT TRAVELLING, ETC.

THE sailing vessel is the child of antiquity. It plays an important part in the early history of all nations, and was especially at home in the old valley of the Nile and on the broad waters of the Mediterranean Sea, — " it formed the primitive cradle of the human race." Steamers belong exclusively to modern times — they represent progress, speculation and haste. " Man is the god of the steamboat; it depends upon him for its every movement. The sailing vessel, on the other hand, relies upon its good genius, that mysterious and invisible force which comes from on high" — " the wind cometh and goeth where it listeth, but no one knows the course thereof."

Formerly the only mode of conveyance on the Lakes and Rivers of America was by means of the Indian bark canoe, or the bateau and Mackinac boat as constructed by the early French voyageurs. Then came sail-craft, wafted by the winds, and barges, towed or poled along the shores and through the streams by the force of human strength. The fur-traders and the Jesuits, in search of discoveries, were the first pale-faces to adopt the above modes of navigation.

Now all is changed, although the sail vessel for a time came into common use; but with the introduction of steam-power on the Hudson River, and its use on Lake Erie in 1818, then the steamer *Walk-in-the-Water* was seen ploughing the waters of the Inland Seas, extending her trips through Mackinac Straits toward Green Bay and other early lake ports. This period culminated in the construction of magnificent floating palaces, which were usually crowded with emigrants, both native and foreign, rushing toward the *Far West*.

The side-wheel steamers, from the period of their introduction till about 1860, running from Buffalo through Lakes Erie, Huron and Michigan to Milwaukee and Chicago were, many of them, first-class vessels, carrying great numbers of passengers and an immense amount of freight. Then the Lake travel was in its glory. The steamers running from Buffalo to Cleveland, Toledo and Detroit, connecting with railroads, were truly magnificent. The popular commanders were then almost idolized and looked upon as fresh-water admirals —

40

each vieing for the reputation of running the fastest steamer. This strife often ended in disasters which consigned many a passenger to a watery grave.

Steamboat travelling to Lake Superior is of a recent date. In 1855, the St. Mary's Ship Canal was finished, and immediately the travel assumed large proportions, although several steamers had been carried over the portage at Sant Ste. Marie, and were running on the Lake previous to the canal being finished. These steamers were the Julia Palmer, Independence,* Sam Ward, Manhattan,* Monticello,* and Peninsula.*

Next in order of time the fleet *Iron Horse* was seen dragging its ponderous train along the shores of the Great Lakes, giving a death-blow to the comparatively slow steamers of those days — the former running at twice the speed of the latter conveyance. Still the great natural channels afforded by means of the Lakes and River St. Lawrence were not totally neglected, and gradually the steam propeller was successfully introduced — now carrying passengers and freight, at a cheap rate, from the head of Lake Michigan and Lake Superior to the foot of Lake Erie, and by means of the Welland and St. Lawrence Canals to ports on the Lower St. Lawrence River. Sail vessels, of a large class, also are used profitably to transport freight of different kinds on many of the American rivers.

The agricultural and mineral products of the United States have assumed such large proportions, that large fleets of propellers, barges and sail-vessels will always be required to aid commerce in the trans-shipment of produce and manufactured articles to domestic and foreign ports.

What is now wanted to encourage water travel is swift and commodious Steamers, built in a safe and staunch manner, so as to avoid shipwreck and loss of life, carrying passengers and freight at a low rate. Then will they successfully compete with railroads during the season of navigation, which usually lasts for seven or eight months. Much depends upon the owners and commanders of these steamers in order to make them a success. Let floating-palaces again be built, and, as far as emigrant and pleasure travel is concerned, adopt the European plan to furnish berths and meals aside from the passage money, leaving the traveller to economize or live in luxury while on the trip around the Lakes, or from point to point.

The inducements to select this healthy and delightful mode of conveyance will last as long as water runs, — for while the land and cities become defiled, it is impossible to destroy or mar the grandeur of the Great Lakes with their flowing fountains of pure water. Old Ocean and Inland Ocean are alike above man's control, as was exemplified by Xerxes in ancient times, who vainly commanded the ocean to recede.

* Wrecked on Lake Superior.

Navigation of Lake Superior.

"Between the years 1800 and 1810, large schooners were on Lake Superior, engaged in the service of the Hudson's Bay and American Fur Companies. A schooner called the *Recovery*, belonging to the British North-western Company, was one of those so employed. On the breaking out of the war, great fears were entertained for the safety and preservation of this vessel, and it is said this was accomplished by stratagem. In one of the deep, narrow bays on the north-east end of Isle Royale, which was then within the jurisdiction of the British, this vessel was secreted, after having her spars taken out. Here, entirely covered over with boughs of trees and brushwood, she is reported to have lain until the termination of hostilities between the two nations, and was then brought out from her hiding place and again put in commission. Subsequently she was run down the St. Marie rapids, and placed in the lumber trade on Lake Erie, under the command of Capt. Fellows. Her owner, I believe, was Mr. Merwin, of Cleveland. The fragments of the *Recovery* were for a long time visible near Fort Erie, opposite Buffalo. •

"Another schooner, named the *Mink*, was also one of those on Lake Superior previous to the war. This craft was also brought down over the rapids, and was employed in the general trade of Lake Erie for several years, under the command of Tom Hammond, an officer who served in Perry's fleet. This vessel, after being used here some time, was finally sunk in Riviere Rouge, which empties into Detroit River, a few miles below Detroit, where her fragments remain. A third vessel, which had been in the same service with the *Recovery* and *Mink*, also undertook to pass down the rapids, but in so doing struck a rock and went to pieces. Her name is not given. From that time until 1822, Lake Superior was navigated only by a solitary sail, a small craft, which also passed down the rapids, and soon became extinct among the young fleet then springing into existence on the Lower Lakes. In 1834 the fur business revived, and orders were issued by the American Fur Company for the commission of a large vessel for Lake Superior. To Messrs. Ramsay Crooks and O. Newberry, of Detroit, were given the management of the matter, and the command, when ready for service, entrusted to Capt. Charles C. Stanard. The *Astor* was the first American vessel that was launched upon Lake Superior.

"The Astor was a schooner of 112 tons, and was built by the American Fur Company, in the summer of 1835. Her builder's name was G. W. Jones. Her frame timbers and planks were got out at Charleston, Ohio, in the fall of 1834, and were shipped on board the schooner *Bridget* from that place, in April, 1835, and arrived at Saut Ste. Marie on the 1st of May. The timbers were then carried to the head of the rapids, where the Astor was built. Her keel was laid on the 17th of May, and the vessel was ready to launch about the 1st of August, and she sailed on her first voyage on the 15th of August, on her upward bound trip to La Pointe. On the 26th of August, Capt. Stanard discovered the celebrated rock, which has since excited so much curiosity, and has been so great a source of annoyance to the navigators of Lake Superior. Capt. Stanard did not go to it at that time, as it was near night, and the weather thick and the Lake rough. But in the fall of that season he went on it. When first discovered, it appeared to be a bateau capsized, and the sea breaking over it, with a rough Lake, and the weather so thick that he was unable to make out what it was until within half a mile.

"Capt. Charles C. Stanard sailed the Astor until the close of the season of 1842; after which time his brother, Capt. Benjamin A. Stanard, sailed her until she went ashore and was wrecked, at Copper Harbor, on the 21st of September, 1844. No lives were lost; cargo mostly saved. At the time of the gale, the Astor lay at anchor in Copper Harbor. When it came to blow very hard, her cable broke, and she went ashore. Her hull is still to be seen on a low conglomerate cliff in Copper Harbor, immediately south of the entrance.

"The American Fur Company had two small vessels built, of about 20 tons each, in the year 1837, one of which was so poorly constructed that it was never launched. The other, named the *Madaline*, was sailed by Capt. Angus, and was employed principally in the fishing trade — built by a Frenchman. In 1838, the same Company built the schooner *William Brewster*, of 73 tons. She was launched some time in August, and sailed in September, Capt. John Wood, master. In September, 1842, the American Fur Company, supposing that she would rot before she could pay for herself on Lake Superior, the Brewster was run down the rapids, and subsequently put in service on Lake Erie. The timbers of the Brewster were got out at Euclid, Ohio, and carried up above the Saut, where the vessel was built. The fleet on Lake Superior, previous to the opening of the ship canal, was composed of the following: — Steamboat Julia Palmer, 280 tons; Propeller Independence, 280 tons; Schooner Napoleon, 180 tons; Algonquin, Swallow, and Merchant, about 70 tons each; Uncle Tom, Chippewa, Fur Trader, Siskowit, 40 tons; and White Fish, 50 tons."

The favorite steamers that followed on the opening of the Ship Canal, and now withdrawn from the trade, or lost by shipwreck, etc., were the Arctic No. 1, Lady Elgin, Illinois, North Star, Iron City, Planet, Lac la Belle, Pewabic, North-west, Northern Light, Coburn, and Meteor.

The popular steamers and propellers at present running (1873) from Lake Erie to Lake Superior, are Keweenaw,* St. Paul, Arctic No. 2, Atlantic, Pacific and Winslow.

The Peerless, and two other steamers running from Chicago to Duluth, are also favorite vessels.

The Canadian steamers running from Collingwood, through Georgian Bay and North Channel, to Lake Superior, are the Chicora, Cumberland, Algoma and Frank Smith. The Acadia, Manitoba, and City of Montreal run from Sarnia, Can., to Fort William, and other ports on the Lake.

Most of the above vessels are propellers, built in a substantial manner, and finished off for the accommodation of travellers seeking health and pleasure.

* The Steamer KEWEENAW alone bears the name of a prominent object on Lake Superior, while many other prominent points might be selected, instead of giving the names of empires, cities, or individuals.

The increase of steamboat tonnage on Lake Superior was at first rather slow, but soon after the opening of the *Lake Superior and Mississippi Railroad*, August, 1870, the tonnage rapidly increased; and now, 1873, there are running ten steamers from Buffalo, Erie, Cleveland, and Detroit; five from Chicago and Milwaukee; four from Collingwood, Can., and three from Sarnia, Can. The American steamers (after passing through the Ship Canal) run along the South shore of Lake Superior to Duluth; while the Canadian steamers usually run along the North shore to Silver Islet and Fort William, or Prince Arthur's Landing, — here connecting with a line of travel to Fort Garry, Manitoba.

The American steamers connect with the Peninsula Railroad of Michigan, and the Marquette, Houghton and Ontonagon Railroad at Marquette; with the Wisconsin Central at Ashland; and with the Lake Superior and Mississippi Railroad, and Northern Pacific Railroad at Duluth, Minnesota.

With an increase of the production, and the utilization of coal and iron, the civilized world will command a power that Hercules never conceived of in ancient times. It is in fact to be the lever that will give an ascendancy to national wealth and power. Already is commerce enriched by the shipment of stores of these valuable minerals — one flowing westward and the other eastward, to meet the demands of trade in the different sections of our extended country. A late writer says, when speaking of

"OUR COAL AND IRON WEALTH."

"Mineral ores abound in almost every section of our extended country. Every State seems to have its representative of either coal or iron. The most important and generally distributed, however, is iron, and to estimate the amount of business involved in its utilization for manufactures, as well as its production and manipulation, is impossible. The amount of capital to-day employed in the iron interest cannot be computed, and after the labor of forty years has been devoted to its development in this country, it still appears in its infancy, and vast areas of that immense field of iron ore extending from New York to Alabama are comparatively undisturbed. Countless millions of dollars will have yet to be invested to bring from the bowels of the earth its hidden treasures, and generations to come will be the investors. The close proximity of our coal fields to our ore deposits, each facilitating and increasing the development of the other, promises the richest returns to those investing in either.

"As week after week we chronicle the continued development of our Western coal and iron fields, we become more and more deeply impressed with the reality of these truths. Every day brings forth some new enterprise calling in the aid of capital and mechanical skill to keep us a busy, striving and wealthy nation. It does not appear as though there ever could be a scarcity of iron in this country, or even in the world, as is threatened. For though our coal and iron belts are so lightly developed, and increased facilities for mining appear from day to day, a contemporary calculates that the 'number of persons employed in the primary production of iron in the United States is 140,000 — 58,000 of whom work in rolling-mills; 42,000 in preparing ore and fuel; 25,000 in preparing fuel for rolling-mills; 12,500 in blast-furnaces, and 2,500 at forges and bloomeries. Add the 800,000 engaged in manufacturing articles of iron, and we have a total of 940,000.

" 'The approximate value of the pig iron manufactured last year was $75,000,000. Adding to this the product of the rolling-mills and forges, the amount is $138,000,000. Adding again the value of articles manufactured of iron, and the value of the iron manufacture of the country for the year is $900,000,000. Of rails we produced in 1853 but 87,000 tons, and in 1869 the amount had risen to 580,000 tons. Of steel rails we laid in the latter year 50,000 tons, 15,000 tons of which were of domestic manufacture, and it is further estimated that the quantity of steel rails laid this year will reach 150,000 tons.'

" The increase in consumption of iron in this country alone, during the past few years, has been immense. As wood becomes scarce, iron is employed in the manufacture of everything, from the insignificant iron lifter for stove covers to the beautiful iron front for buildings. The increase in production has not kept pace with this increased consumption, because the necessity for iron was not felt until the surplus stock of this and other markets had been exhausted." — *Coal and Iron Record.*

The Lake Superior Iron region is now producing above 1,000,000 tons of this useful metal, which must rapidly increase as railroad and shipping facilities are afforded.

The product of Iron alone will vastly benefit the commerce of the Great Lakes, and add wealth and strength to the nation.

Commerce of the Lakes.

REPORT of Vessels of all classes passing Port Huron, Michigan, and through St. Clair and Detroit Rivers, during the season of navigation for 1873.

RECAPITULATION.

Date.	Barges.	Barks.	Brigs.	Schoon's.	Scows.	Steamers.	Total.
April	22	.....	...	7	2	78	109
May	760	322	48	1,409	276	1,959	4,774
June	944	371	57	1,526	521	2,170	5,589
July	1,090	351	53	1,737	486	2,402	6,119
August	1,028	373	57	1,968	313	2,454	6,393
September	891	358	52	1,907	488	2,377	6,073
October	806	274	48	1,643	432	2,322	5,225
November	243	93	16	760	136	1,311	2,559
December	3	.....	...	3	4	37	47
Total	5,787	2,142	331	10,960	2,858	15,110	37,188

REMARKS.

184 days, average one vessel every ten minutes; and during the most busy part of the season, one in every five minutes.

NOTE. — The above Lake Vessels, it is estimated, carried during the season 10,000,000 tons of freight, and many thousand passengers to and from the ports on the Great Lakes.

Flow of Water in the Rivers forming the Outlet to the Great Lakes.

During the past few years observations have been made, under the direction of the Superintendent of the Lake Survey, upon the flow of water in the several rivers which connect the Great Lakes of America. The following are the results:

RIVERS.	Length, Miles.	Depth, Feet.	Maximum velocity, miles per hour.	Discharge, cubic feet, per second.
St. Mary............................	65	10 to 100	1.30	100,783
St. Clair............................	40	20 to 60	3.09	233,726
Detroit.............................	27	12 to 60	2.71	236,000
Niagara	35	....	2.32	242,494
St. Lawrence.....................	760	....	1.00	319,943
River navigation................	927			
River and Lake navigation...	2,800			

REMARKS.

The Enlargement of the St. Mary's Ship Canal and the improvements of the St. Clair Flats, affording, as proposed, 16 feet of water, should prevent all projects of ever bridging the above great outlet to the Ocean, from Lake Huron to Lake Erie, through which now flows the united commerce of the United States and Canada.

List of Steamers in the Lake Superior Trade,

BEFORE AND SINCE THE OPENING OF THE ST. MARY'S SHIP CANAL IN 1855.

American Steamers.

BUILT.	NAME.	TONS.
1836	Julia Palmer	280
1843	Independence	280
1847	Sam Ward	450
	Manhattan	
	Monticello	
	Napoleon	180
	Albany	
	Peninsula	
1845	Superior	567
1847	Baltimore	500
1848	Queen City	1,000
1851	Arctic No. 1	867
	Lady Elgin	1,037
	Northerner	800
1852	Cleveland	574
1852	Traveller	603
1852	Michigan	
	Gen. Taylor	
1853	E. K. Collins	950
1853	Garden City	
1854	North Star	1,100
	Illinois	926
	Concord	
1855	Iron City	700
	Planet	1,154
1856	Dubuque	
	Comet	
	City of Toledo	
	Favorite	
	Gazelle	
	Queen of the Lake	
	Sunbeam	
1857	City of Cleveland	800
1858	Northern Light	800
1862	Meteor	957
1863	Pewabic	960

American Steamers.

BUILT.	NAME.	TONS.
1863	Arctic No. 2.*	
	Atlantic.*	
	Pacific.*	
1864	Iron Sides	
	Lac la Belle	
1866	North-west	1,200
	Sea Bird	600
	City of Madison †	487
	Cuyahoga †	727
	Ontonagon †	682
	Norman †	545
	Mineral Rock †	719
1866	Keweenaw *	800
1869	St. Paul *	1,000
1870	R. G. Coburn	900
1871	China	1,240
1871	India	1,240
1871	Japan	1,240
	St. Louis	
	Winslow *	1,137
1872	Peerless †	1,275

Canadian Steamers.

BUILT.	NAME.	TONS.
1856	Gore	
1860	Plough Boy	400
1862	Rescue	
	Collingwood	
	Algoma ‡	416
1864	Waubuno ‡	200
	Manitoba ‡	450
	Acadia ‡	
1869	Chicora ⸿	550
1871	Cumberland ‡	418
1872	Frank Smith ‡	460
1873	City of Montreal ‡	300

* Sailing from Buffalo, Cleveland and Detroit to Lake Superior.
† Sailing from Chicago and Milwaukee to Lake Superior.
‡ Sailing from Canadian ports to Lake Superior.
⸿ The *Chicora* was built in Liverpool, England, in 1864, for a blockade-runner, sailing to Charleston, S. C.; purchased in 1867 by Milloy & Co. of Toronto, Can.; taken to Quebec and cut in two pieces and passed through the St. Lawrence Canals to Buffalo, N. Y. Here she was rebuilt and put on the Collingwood Line, in 1869. Her engines are marine, oscillating, of superior workmanship, affording great speed.

Sailing on the Lakes — Summer Weather.

During the months of June, July and August the weather usually is calm and delightful on the Great Lakes, affording pleasurable sensations while sailing on the upward trip toward Lake Superior, the " *ultima thule* " of the seekers of health and pleasure.

Starting from Buffalo, or Cleveland, the tourist passes over the waters of Lake Erie, entering the Detroit River at Amherstburg, Can., where is a flourishing town. Here the *Chicago and Canada Southern Railroad* crosses the river to the Michigan shore by means of a steam ferry. Ascending this beautiful stream several islands are passed, mostly lying on the West or American shore, where stands Wyandotte, a flourishing manufacturing village. Here are located the most extensive Iron Works in Michigan ; also, Copper and Silver Smelting Works.

Before arriving at Detroit by steamer, Sandwich and Windsor are passed, the latter place being connected with Detroit by a steam ferry. Here terminates the *Great Western Railway* of Canada, connecting with the *Michigan Central Railroad,* and other railroads running through Michigan.

Tuesday, July 1st, 1873.—After suffering from the heat for two weeks in Chicago and Detroit, the thermometer ranging from 80° to 90° Fahr., we left Detroit on board a steamer for Saut Ste. Marie, 365 miles distant, passing through St. Clair River, Lake Huron, and St. Mary's River. Thermometer standing at 75° Fahr.

July 2d was a cloudy day, with rain on the St. Clair River and at Port Huron, situated near the foot of Lake Huron, where stands Fort Gratiot.

July 3d was a delightful clear day on Lake Huron, with a moderate southerly wind, the steamer spreading her canvas to the breeze. The Michigan shore constantly in sight, while to the north was seen nothing but the waters of the Lake. Thermometer 65° Fahr. during the middle of the day, growing colder in the evening, when the wind veered round toward the North. When off Saginaw Bay, out of sight of land, the Lake presented a calm surface, while steamers and sail-vessels were constantly in sight, the latter being wafted on their course by a light wind.

Friday, July 4th, was a comfortable day, with sunshine and rain, the thermometer varying from 65° to 70° Fahr. while running on St. Mary's River toward the Saut. Here is to be seen lovely lake and river scenery, with beautiful wooded islands lying on the American and Canadian shores. The healthy and bracing atmosphere of the Lake region is here felt with all its pleasant sensations of delight during the summer months. Arrived at Saut Ste. Marie at 10 A. M. Here stands Fort Brady, garrisoned by United States troops, while the national flag waves over the ground, and a park of artillery may be seen near the water's edge. This being the anniversary of our nation's independence, a salute of thirty-seven guns was given at noon, — sending an echo to the opposite side of the river, where may be seen the emblem of England's glory waving on the Canadian shore. A more delightful day could not be desired than the 4th at the Saut, while to the south rain and thunder-storms prevailed.

July 5th. — Clear cold day at Saut Ste. Marie, Michigan, situated on northern boundary of the United States in 46° 30′ north latitude; thermometer 56° Fahr., making a fire desirable in sitting rooms.

July 6th. — Clear pleasant day on land; thermometer 60° Fahr., falling to 55° on Lake Superior as evening approached. The Lake was unruffled while presenting a dark-blue color, the waters sparkling with purity as the steamer ploughed over its surface. In the far distance on the north was seen Caribou Island, some thirty miles off, while the shores of the Upper Peninsula of Michigan were visible from White Fish Point to Point au Sable. The sun went down in grandeur, presenting a magnificent view from the deck of the steamer. Pass the Pictured Rocks during the evening.

July 7th. — Cool rainy day in Marquette, the thermometer standing at 60° Fahr., which is considered cold for the season in this latitude, vegetation being unusually backward. Steamers usually arrive and depart from this port several times daily.

July 8th. — Cool pleasant day at Marquette; thermometer 62° Fahr.

July 9th. — Cloudy and rainy day; thermometer 63° Fahr. At this low temperature it is comfortable for invalids, while the health-inspiring air tends to assist digestion, and causes strangers to sleep as soundly as infants.

July 10th. — Cool bracing air on Lake Superior; thermometer 60° Fahr. This low temperature has prevailed for the past five days, rendering a fire and overcoat almost necessary for those unacclimated to cold weather. In cool weather the Northern Lights often assume a brilliant aspect common to Lake Superior.

July 11th. — The sun rose on Keweenaw Bay with an unclouded brilliancy, which proved to be a delightful day; thermometer 68° Fahr. at noon. This was one of the charming days which usually prevail during July in this latitude.

July 12th. — Another fine day for the Lake Superior region, with a cloudless morning and a refreshing shower in the evening; thermometer 70° Fahr.

July 13th. — Warm morning, with wind from the South, and a hazy atmosphere, with sunshine and cloud; thermometer 80° Fahr. At 4 P. M., a severe rain-storm, with thunder and lightning, visited Marquette, the thermometer falling to 70° Fahr.

July 14th. — Sunshine, cloud and rain at Marquette; the weather assuming the coquettish character common to Lake Superior during the summer months, when flying clouds usually scud low, giving out a chilling influence. Ther. 60° to 65° F.

July 15th. — Delightful day experienced during a trip to Escanaba, Delta county, Michigan; thermometer 75° Fahr. At this place, 70 miles south of Marquette, the weather is perceptibly warmer; at the former place the mean summer temperature is 65½° Fahr., and at the latter place 62° Fahr.

July 16th. — The day was ushered in with a terrific thunder-storm, the rain pouring down in torrents, causing great damage to the railroad track and the streets in Marquette; thermometer 65° Fahr.

July 17th. — Cloudy morning, with clear, warm weather at noon, the thermometer ranging from 70° to 80° Fahr., it being one of the hottest days of the season. *July 18th*, thermometer 56°; *July 19th*, thermometer 55°; *July 20th*, 62° Fahr.

The above is a correct statement of the weather usually experienced on Lake Superior during the months of July and August.

4

Duluth—Lake Superior.

TABLE showing Daily and Monthly Mean of Barometer and Thermometer, and amount of Rainfall, with the prevailing direction of wind, for the month of July, 1873, at DULUTH, Minn., N. lat. 46° 48′; W. lon. 92° 06′.

DATE.	Mean Daily Barometer.	Wind.	Mean Daily Thermometer	Rainfall, Inches.	Remarks.
July 1......	29·722	N.W.	64·5	......	Clear.
" 2......	29·832	N.E.	55·7	......	Clear.
" 3......	29·657	S.W.	71·5	·29	Cloudy.
" 4......	29·737	S.W.	70·2	·30	Clear.
" 5......	30·115	N.N.E.	59·5	......	Clear.
" 6......	30·032	N.N.E.	55·7	......	Clear.
" 7......	29·995	N.E.	58·7	·29	Cloudy.
" 8......	30·020	N.E.	63·0	......	Cloudy.
" 9......	29·967	N.W.	63·7	·21	Cloudy.
" 10......	30·062	N.E.	60·5	......	Clear.
" 11......	29·957	N.E.	57·5	......	Clear.
" 12......	29·720	S.	64·2	......	Cloudy.
" 13......	29·717	N.	69·7	1·22	Cloudy.
" 14......	29·957	W.	69·7	·81	Fair.
" 15......	29·972	N.E.	60·5	·06	Fair.
" 16......	29·662	N.E.	61·0	1·57	Cloudy.
" 17......	29·595	W.	64·7	·92	Cloudy.
" 18......	29·975	N.	62·2	·08	Cloudy.
" 19......	30·017	N.W.	63·2	·21	Cloudy.
" 20......	30·010	N.E.	66·5	......	Clear.
" 21......	29·847	S.W.	73·5	......	Fair.
" 22......	29·808	S.W.	76·2	......	Cloudy.
" 23......	30·010	N.	76·7	......	Clear.
" 24......	29·770	N.E.	63·5	·11	Cloudy.
" 25......	29·815	N.W.	71·2	......	Clear.
" 26......	29·952	N.E.	66·5	......	Clear.
" 27......	29·960	N.E.	63·0	......	Fair.
" 28......	29.790	N.E.	63·0	·80	Cloudy.
" 29......	29·885	N.W.	68·5	......	Clear.
" 30......	29.920	N.E.	60·0	·53	Cloudy.
" 31......	29·880	N.E.	57·0	1·10	Cloudy.
Monthly mean for July .	......	......	64.5Fr.		
Total Rainfall	......	......	......	8·50	

Summer Months.	Mean Temperature.	Total Rain.
June, 1872,	59.6 Fahrenheit.	4.46 inches.
July, "	66.8 "	5.83 "
August, "	68.1 "	2.84 "
Mean Temperature......64.8	"	13.13 "

PART IV.

MINERAL REGION OF LAKE SUPERIOR.

THE extent and value of the Mineral Region lying on the South and North shores of this Inland Sea are truly astonishing, when we take into view the different rich deposits of iron ore, copper, and silver, as found to exist in different localities.

The *Iron Region* is the most extensive, commencing at the Iron Mountain, 12 miles west of Marquette, and extending south to the Menominee River and north to near Keweenaw Bay, while in a western direction it stretches past Michigamme Lake to the Penoka iron range in Wisconsin, lying south of the new town of Ashland — a total distance of about 150 miles. Other iron deposits exist in Northern Minnesota and in the Province of Ontario, Canada.

As you mount up from Escanaba or Marquette, the principal outports for iron ore, to the table lands surrounding the Iron Mountain, 800 feet above the level of the Lake, you are surrounded by the richest deposit of iron ore in the world.

The *Silver Region*, as yet known, is more limited. Virgin silver as well as mass copper abounds on the South Shore in the Northern Peninsula of Michigan, also in Minnesota bordering on Lake Superior; but by far the richest mines have recently been discovered and worked on the North Shore, or Canada side.

Silver Islet is said to be one of the most productive mines in the world, yielding annually about one million dollars of the precious metal. The ore is obtained by blasting and being raised 100 or 200 feet from the mine, extending under the bed of the Lake. Other mines around Thunder Bay, and in the vicinity of Pigeon River, are attracting much attention. *Lead* also abounds on the North shore of Lake Superior near Black Bay and Thunder Bay.

The newly discovered *Gold Region* in Canada, lying 50 or 60 miles west of Prince Arthur's Landing, is supposed to be of great value, as well as the region in the vicinity of Vermillion Lake, in the State of Minnesota. Time alone is wanted to further develop this Eldorado.

The *Marquette, Houghton and Ontonagon Railroad*, finished to Lake Michigamme, 32 miles, thence extends to L'Anse, 32 miles farther, will branch off and follow the Iron Range into the State of Wisconsin, there connecting with other railroads now in progress of construction. Then Ashland will be added to the shipping ports of Lake Superior.

The *Copper Region* is still more extensive and scattered, including Keweenaw Point, running from near Copper Harbor, in a south-west direction, past Hancock and Houghton to Rockland, situated back of Ontonagon, thence westward past Lake Gogebec to the Porcupine Mountains, a distance of about one hundred miles. The Calumet and Hecla copper mines at the present time are the most productive.

There are also deposits of copper in Northern Wisconsin, Minnesota and on Isle Royale, attached to Michigan. The Copper Region of the North Shore is equally extended, running from Mamainse Point, opposite White Fish Point, past St. Ignace and Thunder Cape to the mouth of Pigeon River, a distance of 300 or 400 miles.

First Discovery of Iron Ore in Marquette County.

The United States Geologist, Professor Charles T. Jackson, in his report to the Secretary of the Interior made in 1849, says that during his first visit to Lake Superior, in the summer of 1844, he obtained from a trader at Saut Ste. Marie a fine specimen of specular iron ore, which he had received from an Indian chief. He also learned, at the same time, that this chief knew of a mountain mass of ore, somewhere between the head of Keweenaw Bay and the head-waters of the Menominee River.

It is not more than twenty-eight years since the first iron ore was taken from what is now known as the Jackson Mountain, and less than half a ton of it packed upon the backs of half breeds, and carried to the mouth of the Carp River, and from thence transported in canoes to the St. Mary's River. Yet in that short interval the development of the iron mines in this locality has been so rapid, that they now contribute the ores from which is made about one-fourth of all the iron manufactured in the United States. About a quarter of a century ago, the district which now supplies the ore for upwards of one hundred and fifty furnaces, and which boasts of a population of not less than sixteen thousand people, was an unexplored wilderness, never penetrated save by the wild Indian and devoted missionary. Over a million tons of this ore have been shipped annually for the past few years, and each succeeding year is destined to increase the amount in a rapid ratio; and yet the Lake Superior Iron District is but in its infancy, and only needs to be fully developed to become the great iron centre of the North-west, and of the Union.

The Jackson Company was organized in June, 1845, with a view to operations in the Copper District, and P. M. Everett, Esq., one of the original incorporators, came to Lake Superior the same summer, and located what is now the Jackson Iron Mine, under a permit from the Secretary of War. To Mr. Everett is due the credit of being the pioneer in the discovery and development of the Lake Superior Iron Mines. Others may have visited the Jackson Mountain about the same time, but there is no evidence that any of them discovered the existence of its hidden treasures.

The first opening in the Iron District was made by the Jackson Company in the fall of 1846, in the summer of which year they commenced the erection of a forge on Carp River, about three miles east of Negaunee. The forge was put in operation in the spring of 1847, and the first ore mined at the Jackson was there manufactured into blooms. The first blooms were sold to E. B. Ward, Esq., and from them was made the walking beam of the steamboat "Ocean." For further information, see second edition "History of the Lake Superior Iron District."

Classification of the Iron Ores.

The most valuable, so far as developed, is the *specular hematite*, which is a very pure anhydrous sesquioxide, giving a red powder, and yielding in the blast-furnace from 60 to 70 per cent. of metallic iron. The ore occurs both slaty and granular, or massive.

The next in order of importance is the *soft* or *brown hematite*, which much resembles the hematite of Pennsylvania and Connecticut. This ore is generally found associated with the harder ores, from which many suppose it is formed by partial decomposition or disintegration.

The *magnetic ore* of the district has thus far only been found to the west of the other ores — at the Washington, Edwards and Champion Mines — at which none of the other varieties have been found, except the specular, into which the magnetic sometimes passes, the powder being from black to purple, then red.

The *flag ore* is a slaty or schistose silicious hematite, containing rather less metallic iron, and of more difficult reduction than either of the varieties above named.

Michigamme Iron District.

Work has already actively commenced at the Michigamme mine, and considerable ore shipped from it and the Spurr Mountain the present season (1873). The following description of this range possesses interest at this time:

"In the fall of 1868 attention was directed to this range, and what is now known as the *Spurr Mountain* was discovered, or rather, I should say, rediscovered, on the north half of the south-west quarter of section twenty-four, town forty-eight, north of range thirty-one west. There is a large outcrop of pure magnetic ore (the largest I ever saw of this kind of ore) occurring in an east and west ridge one hundred and eighteen feet above the surface of Michigamme Lake. The direction of the bed is due east and west, dipping to the south at a high angle. It presents a thickness of thirty feet of first quality of merchantable ore, and facilities for commencing to mine which I have never seen surpassed. The exposure along the range is short, owing to the covering of earth, but the magnetic attractions, which are very strong, continue east and west for a long distance, determining the position of the range with great precision. This ore is of the same character as the magnetic ore of the Champion and Washington mines, differing only in being softer from the effects of the weather. A specimen collected for analysis in October,

1868, by breaking indiscriminately numerous fragments from all parts of the outcrop, and from the loose masses, with the view of obtaining a safe average, afforded Dr. C. F. Chandler, of the School of Mines, New York, the following constituents:

	Per cent.
Oxide of iron	89·21
Pure metallic iron	64·60
Oxygen with the iron	24·61
Oxide of manganese, a trace.	
Alumina	2·67
Lime	0·67
Magnesia	0·19
Silica	6·28
Phosphoric acid, a trace.	
Sulphur	·35
	99·37

"The percentage of iron given, I am confident, is not above what will be obtained by working the ore in the furnace, if mined with ordinary care. In richness it is unquestionably first class.

"On the north side of Spurr Mountain deposit is the well known 'mixed ore' of the Marquette region, which is invariably found associated with the pure ores. It is a banded rock of pure ore and quartz (white or red), and is usually most abundant at the base of the formation. This material is seen in great abundance at all the old mines, the best exposure being at the Cleveland Knob, where there is a small mountain of 'mixed ore' with pure ore flanking it on either side. Still farther north, and therefore geologically below the Spurr Mountain deposit (for the whole series dips south), is a thick bed of greenstone or diorite, being the same rock which underlies all the workable deposits of hard ore in the Marquette region, and which is conspicuous at the Lake Superior, Jackson and Champion mines. At the Spurr Mountain this greenstone rises into a bold ridge, which can be traced for miles east and west, always on the north side of and parallel with the iron belt. Passing to the south, hence hanging wall side of the deposit, we find the same quartzite overlying the ore which is found at all the old mines, and which is so conspicuous at the New York and Cleveland, where it dips to the south just as at the Spurr Mountain. The perpendicular distance from the underlying greenstone to the overlying quartzite, which would represent the thickness of the entire iron formation, is two hundred feet, as near as can be made out; which is made up of pure ore, 'mixed ore,' and, usually, in part of a magnesian schist, such as occurs in abundance at the west end of the Champion mine, as well as at the older mines."

Penoka Iron Range in Ashland County, Wisconsin.

A late writer says: — "I have had examined a large number of iron deposits in Michigan, Wisconsin, Mississippi, Kansas and Iowa. We found large beds of specular, hematite and other valuable ores. But in all our investigations we found no beds of ore so good as the Penoka Range.

"I sent one good geologist, one chemist and three axe-men, who spent two weeks in tracing and examining that range. The result of the expedition was entirely satisfactory. There is an unlimited supply of fine magnetic ore, the surface specimens of which closely resemble the first Swedish ores, from which the best and finest steel is made. The analysis gives from fifty to seventy-five per cent of iron.

" It is my opinion that the ore from this range will be highly valuable for mixing with the specular ores of Marquette.

" I am very anxious to have these ores brought to market, as I can use with advantage several thousand tons annually.

" I hope you may induce capitalists to take hold of this most valuable deposit and develop it for the good of this country. The Range is not less than twelve miles long, and will yield ore enough to span the world with railroads. I have had analysis made of a fine specimen of Swedish ore, and the ore of your Range is richer than the Swedish by about 10 per cent. There is no doubt but your ore will make as good cutting steel as any made in the world.

"Respectfully yours,

E. B. W."

Extent of the Lake Superior Iron Mines.

Since the opening of the Jackson Mine in 1846, developments have gradually extended westward toward Lake Michigamme, about 20 miles west of Negaunee. On all sides of Lake Michigamme — north, south and west — and on the Menominee River, southward, immense deposits have been discovered, compared to which the famous " Iron Mountain " of Missouri sinks into insignificance.

The Future Iron Centre.

" The manufacture of iron, or the proper place for its manufacture, in order to enable this country to hold its place among nations, in view of the threatened reduction of the tariff, leads the *Mining Journal* to consider what most vitally concerns our own home interests. Our main interests are in IRON, — in the hills, in the mined ores, and in the products of iron. Without it, we are of small account; with it, and with the care taken of it that the situation demands and makes imperative, we are peer of all. We of the Lake Superior Iron District have the iron ore, and the fuel in juxtaposition to convert it into merchantable forms, and the *manufacturing* as well as the mining must hereafter be done in this part of the country. In the matter of the fuel we have only to say that our forests of hard wood have hardly been touched, and peat beds, in embryo, promise everything asked of them for the future ; and when it becomes necessary, in the course of the next decade, to make the balance of lake freights heavier from the East than to the West, we can load the lake crafts with coal to smelt our ores at home.

" At present, by far the greater portion of our iron ore is shipped to the furnaces of Ohio and Pennsylvania. This ore is loaded with a percentage of from 35 to 40 per cent. of waste material, on which the same freight must be paid as on the metallic iron. From these furnaces and rolling-mills the manufactured iron is shipped to its various markets, some of it being returned to the place from which it started in the raw state, and the only reason given for not manufacturing our own iron, is the want of fuel and our inability to procure it. We will reverse the proposition. Suppose that Ohio and Pennsylvania furnaces were obliged to depend upon their own home supplies for ore ? That they have enough of it is true ; that the quality is not good enough to make merchantable iron is equally true. Depending upon the Lake Superior district for their supply of ores to mix with their own in order to bring their iron to a grade which would command a sale in an open market, they are obliged to pay double freight on the raw material. It would not matter so much with the furnaces in Ohio and Pennsylvania at the present time, if they were only shutting down temporarily. The outlook for them is not a good one, i. e. for the future exclusive control of the iron trade of the country. Our railroad communications now opening will give us, when completed, an equal chance for the great Western markets. We can ship to the seaboard during the season of navigation at nearly as low rates as the Pittsburgh manufacturers — and that Marquette county will ultimately become the grand iron manufacturing centre of the United States, is a question that admits of scarcely a doubt."

Iron Mining Companies,

SITUATED IN THE COUNTIES OF MARQUETTE, HOUGHTON AND MENOMINEE.

NAME.	KIND OF ORE.	T. N.	R.W	LOCATION.
Ada....................	Hematite........................	47	26	Negaunee.
Albion *..............	....	47	27	Stoneville.
Allen †	Hematite	47	26	Negaunee.
American ‡..........	....	...	...	Michigamme.
Bagaley †	Hard Hematite................	47	26	Negaunee.
Barnum † ··········	....	47	27	Ishpeming.
Berea ‡	....	...	...	Michigamme.
Breitung.........	....	39	29	
Breen and Ingalls †	....	39	28	Menominee co.
Buckeye	....	48	30	Champion.
Burt	Hematite........................	...	...	Negaunee.
Calhoun and Spurr	Hematite	47	26	"
Canon...................	....	47	30	
Carr † ·············	....	47	26	"
Cascade †	Granular and Hematite	47	30	
Champion *	Magnetic and Specular......	47	29	Champion.
Chippewa.............	Magnetic and Specular......	47	30	Humboldt.
Cleveland * †	Red Specular and Granular	47	27	Ishpeming.
Edwards *	Magnetic and Specular.....	47	28	Humboldt.
Emma † ....….....	Hard Hematite................	47	26	Negaunee.
Empire	....	48	30	
Everett * ....,.....	....	...	...	Humboldt.
Excelsior †...........	....	...	...	Ishpeming.
Fall River ‡	....	...	...	L'Anse.
Foster†	Hematite........................	...	...	Negaunee.
Franklin *	....	...	...	"
Goodrich †	....	47	27	Stoneville.
Grand Central † ...	Hard and Soft Hematite ...	47	26	Negaunee.
Green Bay †.........	Hematite........................	47	26	"
Gribben †	Red Specular...................	47	26	"
Harlow *	....	...	...	Marquette.
Harney ‡	....	...	...	Michigamme.

* Shipped from Marquette. 　　　　　† Shipped from Escanaba, Delta co.
‡ Shipped from L'Anse, Houghton co.

Iron Mining Companies. — (*Continued.*)

NAME.	KIND OF ORE.	T.N.	R.W	LOCATION.
Himrod *	Hematite	47	26	Negaunee.
Horne	....	47	26	
Howell Hoppock †	....	47	27	Ishpeming.
Hungerford	....	47	29	
Hussey	Hard Hematite	...	...	Negaunee.
Iron Cliffs †	Hematite	...	...	"
Iron King ‡	....	...	...	Michigamme.
Iron Mountain......	....	47	27	Ishpeming.
Jackson †	Granular and Hematite	47	27	Negaunee.
Jefferson............	....	...	...	"
Keystone *...........	....	47	29	Champion.
Kloman *...... *....	....	46	29	Humboldt.
Lake Angeline * †..	Hematite and Specular......	47	26	Ishpeming.
Lake Superior * ...	Soft Hematite and Specular	47	27	"
McComber * †	Soft Hematite.................	47	26	Negaunee.
Magnetic.............	Black Magnetic	47	30	Ishpeming.
Marquette *	Red Specular..................	47	27	Negaunee.
Mather..............	....	...	...	"
Michigamme ‡......	Magnetic	48	30	Michigamme.
Michigan	....	47	28	Clarksburgh.
Negaunee	Hard Hematite	47	26	Negaunee.
New England	Hematite and Specular......	47	27	Ishpeming.
New York †	Hard Specular.................	47	27	"
Parsons........	....	...	...	
Pendill	....	...	...	"
Pioneer..............	....	47	27	Negaunee.
Pittsburgh and L. Superior †	....	47	26	"
Quartz..............	....	...	...	
Quinesaik............	....	40	30	Menominee co.
Red Chalk †	....	...	...	Negaunee.
Republic *	Magnetic and Specular......	46	29	Humboldt.
Riverside	....	47	30	
Rolling Mill †	Hematite	47	26	Negaunee.
Rowland †	....	47	26	"
Saginaw †	Red Specular	47	27	Stoneville.
Sheldon	....	48	31	
Shenango *...........	....	47	27	Ishpeming.
Smith †	Specular and Hematite......	45	25	Forsyth.
Spurr Mountain ‡..	Magnetic	48	31	Michigamme.
Stewart ‡	....	48	31	L'Anse, Houghton co.
Taylor ‡	....	...	...	" "
Teal Lake	....	48	26	
Washington *.......	Magnetic and Specular......	47	29	Humboldt.
Winthrop * †	Hard Ore	47	27	Ishpeming.

Iron Product for the Year 1873.

MINES.	Gross Tons.	Value.	MINES.	Gross Tons.	Value.
Jackson	113,892	$797,246	Rowland	1,405	$9,832
Cleveland	132,082	924,576	Himrod	2,074	14,519
Lake Superior	170,988	1,196,918	Marquette	2,148	15,036
New York	70,882	496,176	Shenango	8,658	60,608
Republic	105,453	738,170	Albion	1,189	8,320
Champion	72,783	509,478	Carr	1,656	11,592
Washington	38,015	266,103	Bagaley	1,276	8,935
Kloman	21,065	147,457	Howell Hoppock.	1,240	8,678
Cascade	20,507	143,551	Emma	7,138	49,964
Barnum	48,077	336,536	Smith	9,329	65,302
Foster	27,372	191,605	Grand Central	6,630	46,407
Lake Angeline	43,934	307,535	Gribben	4,518	31,623
Pittsburgh	21,498	150,490	Goodrich	3,258	22,808
Edwards	31,730	222,111	Home	1,090	7,636
Spurr	31,934	223,536	Green Bay	950	6,649
Michigamme	28,966	202,765	New England	181	1,268
Keystone	10,426	72,984	Allen	510	3,570
McComber	38,970	272,788	Iron Mountain	113	789
Winthrop	33,547	234,826	Hungerford	145	1,016
Saginaw	37,139	259,978	Magnetic	79	551
Rolling Mill	11,319	79,235			
Michigan	3,212	22,484	Total	1,167,379	$8,171,652

Furnaces.

The following TABLE shows the aggregate production of the several Furnaces in the Marquette District for 1873, with the value of the Iron ($45) at furnace:

FURNACES.	Gross Tons.	Value.	FURNACES.	Gross Tons.	Value.
Pioneer	7,098	$319,410	Munising, Schoolcraft Co	2,237	$100,665
Collins	2,000	90,000	Grace	7,800	351,000
Michigan	4,467	201,015	Beecher	710	31,950
Greenwood	4,416	198,710	Beecher (Muck bar)	428	25,680
Bancroft	4,100	184,500	Lake Superior (peat furnace)	500	21,500
Morgan	6,324	184,580	Escanaba, Delta Co	2,175	97,875
Champion	3,949	177,705	Menominee	2,500	108,000
Deer Lake	3,447	155,115			
Fayette, Delta Co	10,696	481,320	Total	71,507	$3,224,235
Bay	8,760	394,200			

Total Amount and Value of Iron Produced.

	Tons, Iron Ore.	Tons, Pig Iron.	Value.		Tons, Iron Ore.	Tons, Pig Iron.	Value.
1856......	7,000		$28,000	1866......	296,972	18,437	$2,405,960
1857......	21,000		60,000	1867......	466,076	30,911	3,475,820
1858......	31,035	1,629	249,202	1868......	507,813	38,246	3,992,413
1859......	65,679	7,258	575,529	1869......	633,238	39,003	4,968,435
1860......	116,908	5,660	736,496	1870......	856,471	49,298	6,300,170
1861......	45,430	7,970	419,501	1871......	813,379	51,225	6,115,895
1862......	115,721	8,590	984,977	1872......	952,055	57,595	9,188,055
1863......	185,257	9,813	1,416,935	1873......	1,167,379	71,507	11,395,887
1864......	235,123	13,832	1,867,215				
1865......	196,256	12,283	1,590,430	Total...	6,712,792	423,257	$55,770,920

Shipment of Iron Ore, 1873.

The following is the shipment of Iron Ore and Pig Iron from the ports of Escanaba, Marquette, and L'Anse:

MARQUETTE.

Iron ore......................... 526,264 tons.
Pig Iron .... 25,997 "

Total Iron ore and Pig Iron. 552,261 "

ESCANABA.

Iron ore......................... 479,714 "
Pig Iron....................... 9,248 "

Total Iron ore and Pig Iron. 488,960 "

L'ANSE.

Total Iron ore.................. 60,899 tons.

Total shipments from the three ports for 1873:

Iron ore..................... 1,066,877 tons.
Pig Iron....................... 35,245 "

Total......................1,102,122 "

Copper Mines of Lake Superior.

During the past thirty years the Copper Region of Lake Superior has attracted the attention of capitalists and scientific minds — year by year increasing in interest and value, while building up mining villages and causing improvements of a durable character. Keweenaw Point, in an extended view, runs southward from Lake Superior, embracing the counties of Keweenaw, Houghton and Ontonagon, — these counties comprise the Copper Districts. The product of pure copper during the year 1873 exceeded any previous year since the mines were opened.

Upwards of one hundred mining companies have been formed during the above period, but only about thirty are now worked to profit, while upwards of seventy openings are for the most part lying idle. The Cliff, Quincy and Minnesota mines formerly proved very rich in copper, but of late the Calumet and Hecla have far exceeded all other mines in this region.

Statement of Copper (Mineral) Product from 1845 to 1873.

	Tons.
1845 to 1854.............................	7,642
1854 to 1858.............................	11,312
1858 to 1859.............................	4,100
1859..	4,200
1860..	6,000
1861..	7,500
1862..	9,962
1863..	8,548
1864..	8,472
1865..	10,791
1866..	10,376
1867..	11,735
1868..	13,049
1869..	15,288
1870..	16,183
1871..	16,071
1872..	15,166
1873..	18,638
Total.......................	195,033

Approximate Statement of Ingot Copper Produced and its Value.

	Tons.	Value.
1845 to 1858.....	13,955	$9,000,500
1858...............	3,500	1,886,000
1859...............	3,500	1,890,000
1860...............	4,800	2,610,000
1861...............	6,000	3,337,500
1862...............	8,000	3,402,000
1863...............	6,500	4,420,000
1864...............	6,500	6,110,000
1865...............	7,000	5,145,000
1866...............	7,000	4,760,000
1867...............	8,200	4,140,000
1868...............	9,985	4,592,000
1869...............	12,200	5,368,000
1870...............	12,946	5,696,240
1871...............	12,857	6,171,360
1872...............	12,132	7,774,720
1873...............	14,910	8,200,500
Total.......	149,985	$84,503,820

Copper Product for the Year 1873.

PORTAGE LAKE DISTRICT.

	Tons.	Lbs.
Calumet and Hecla...	11,551	1,938
Quincy.....................	1,600	180
Franklin and Pewabic	671	1,673
Atlantic	464	701
Houghton	285	......
Schoolcraft	270	1,520
Isle Royale..............	143	1,417
Sumner	77	......
Concord	84	300
Albany and Boston ...	50	......
Quincy Tribute, Can...	25	635
Mesnard	5	1,587
Total	15,229	1,951

KEWEENAW DISTRICT.

	Tons.	Lbs.
Central	1,031	1,983
Copper Falls	834	927
Phoenix	350	......
Cliff	326	137
Delaware	209	500
St. Clair	55	742
American	24	380
Amygdaloid............	19	303
Petherick	10	34
Total	2,860	1,006

ONTONAGON DISTRICT.

	Tons.	Lbs.
Ridge........................	150	113
National...................	131	318
Minnesota...............	103	1,700
Bohemian	50	500
Flint Steel...............	45	1,356
Knowlton	39	1,864
Rockland.................	16	460
Mass.......................	6	868
Adventure...............	3	1,238
Fremont	.....	700
Total	547	1,117

RECAPITULATION, 1873.

	Tons.	Lbs.
Portage Lake District	15,229	1,951
Keweenaw Point "	2,860	1,006
Ontonagon District...	547	1,117
Grand Total.......	18,636	4,074

Silver Mining Companies,

IN ONTONAGON COUNTY, MICHIGAN.

Argentine, Township, 51 N., 41 W.	Ontonagon and Lake Superior.
Hancock.	Porcupine.
Iron River.	Scranton.
Mammoth.	South Shore.
Ontonagon, Township, 51 N., 41 W.	Superior, Township, 51 N., 42 W., Sec. 24.

The above new Silver Mines are located near Iron River, which empties into Lake Superior east of the Porcupine Mountain. They are supposed to be rich in silver and very extensive. Some of the ore assayed has yielded over $1,500 coin value per ton of 2,000 pounds.

Silver Mining on Lake Superior — North Shore.

BY PETER M'KELLAR.

"The Thunder Bay District occupies the portion of Canada bordering on Lake Superior, north of the United States boundary. It is mountainous in character, with bold cliffs rising from 300 to 1,400 feet above the lake and valleys. The valleys are numerous and generally fertile, some being of considerable extent, such as that of the Kaministiquia River, so that it is capable of supporting a much larger agricultural population than the people in general have any idea of. The many low and lofty islands, strewn along the coast and in the bays, render the scenery picturesque and beautiful.

"The geological formations of this section are: — The Laurentian, Huronian, and the Upper Copper Rocks, named and described by Sir Wm. Logan (see his Geology of Canada), afterwards described by others, and especially by Professors R. Bell and E. J. Chapman. The Laurentian and Huronian are the crystalline or azoic rocks, and the Upper Copper Rocks are supposed to be the equivalent of a part of the lower Silurian, and are divided into two divisions — the Upper and the Lower beds. The Laurentian occupy the Height of Land principally, touching the lake shore in but few places within this district. It consists of granite, gneiss, syenite, and micaceous schists, almost entirely. Its veins of quartz and spar carry copper and iron pyrites, also galena and zinc blende occasionally, but from my experience I am not favorably impressed in regard to its metalliferous qualifications.

"The Huronian series lay generally between the Silurian and Laurentian, striking occasionally in a north-easterly direction, in a broad belt or trough, back toward the Height of Land into the Laurentian. The principal area occupied by these belts stretches westward from Thunder Bay through Shebandowan Lake, thence on to the American boundary. It consists of greenish and greenish-gray strata, with a dip nearly vertical. The principal portions have a slaty structure, consisting of chloritic, argillaceous, talcose, silicious, dioritic, and fine grain micaceous slates, with interstratified beds of massive diorite. It is from these slates we are expecting great results in mining. It is only three years since the first silver mine (3 A), or any other mine (except the little Pic Iron), was discovered in the

Huronian, in this section. The following summer, the Jackfish Lake Gold and Silver Mine was discovered in the same series, lying nearly 100 miles to the west of 3 A.

"Next in ascending order is the lower beds of the Upper Copper Rocks. They occupy the coast and islands, with the exception of a little piece in Thunder Bay, where the older rocks come in from the east side of this bay westward beyond the American boundary, showing also at a few points farther east, underlying the Upper Beds. They consist of layers of chert, dolomite, and iron ore, the latter being near the base, with thick beds of clay, slate, and gray argillaceous sandstone shales, interstratified with beds of columnar trap. The intersecting veins carry silver, galena, zinc blende, and copper and iron pyrites, and other metals in small quantities. Until the discovery of the silver and gold lodes above referred to in the Huronian series, these slates were considered the silver bearing formation of the country. The Silver Islet, Thunder Bay, Shuniah (now the Duncan), Silver Harbor, Prince's Bay, Spar Island, Jarvis Island, McKellar's Island, Lambert's Island, Thompson Island, McKenzie, Trowbridge, 3 B, and McKellar's Point veins are all silver bearing, and intersect these slates, most of them being undeveloped.

"The upper beds of the copper rocks occupy the principal part of the coast, and almost all the islands from Thunder Bay to the east end of Nepigon Bay. They consist of sandstones, conglomerates, indurated marls, and some interstratified soapstone, crowned by an immense thickness of trappean beds, most of which are amygdaloidal in character. The quartz and spar veins which traverse the sedimentary or lower portion, hold galena, copper ores, and zinc blende in very considerable quantities, also gold and silver, as shown by Professor Chapman in his Report of the Black Bay Lode, now the North Shore Mine. The Silver Lake, Cariboo, and the above, are the principal lodes known in these strata. The above trappean beds are the famous native copper-bearing rocks of the South Shore and Isle Royale. At the former place the workable lodes which conform with the strata dip at a high angle, and are wonderfully rich. On our side these rocks dip at low angles, and, as far as I can understand, all the explorations and work for native copper were made on the intersecting veins instead of the bedded or conformable ones. It may be that the difference of dip may have something to do with their richness, it being generally considered that strata with a high dip are more favorable for mineral veins than those with a low dip. However, since the discovery of silver in Thunder Bay, little or no attention has been paid to these copper rocks by the explorer, so that there is no knowing what they may turn out yet.

"Native copper with associated nuggets of silver is the principal metal found in these rocks, but occasionally the sulphurets and other ores are met with in small quantities. The veins, in passing through them into the sandstone, seem to drop the native copper, it being replaced by the sulphurets. Many years ago the Montreal Company and others spent a considerable sum of money mining in these rocks without success; but that is of little importance, as there was ten times more spent in the South Shore before the mines proved productive.

"In the winter of 1867 and 1868, the Government placed an annual tax of two cents per acre on all the patented land on Lake Superior, which worked much good, as parties holding much land, and having to pay such heavy tax, set to work to explore and find out its value. The result was the discovery of silver in the Jarvis Island and in the far-famed Silver Islet, which in all probability would still be unknown, and for years to come in the hands of the Montreal

Mining Company, as it had been for many years previous. In the spring of 1868, this Company secured the valuable services of the well known Mr. Thomas Macfarlane, with a party of men, to examine and report on their lands on Lake Superior. His first explorations were made on the Jarvis location, situated about 22 miles to the south of Fort William, on which he discovered silver in a large lode of heavy and calcareous spars and quartz, on the Island of the same name, which lies one and a half miles off the shore. The Island is about twenty chains wide where the vein crosses, a considerable portion being deeply covered with earth. The vein is well defined, eight to ten feet wide, with a dip of about fifty degrees to the east-north-east. The silver shows in leaves generally, but also in strings and small nuggets, through the spar, and the black carbonaceous zinc-blende matter. Mr. Macfarlane sunk a shaft about twelve feet deep in the summer of 1869, taking out some fine silver ore. Again in 1870 he followed it down some twenty feet more, taking out a considerable quantity of ore.

"This location passed into the hands of the Ontario Mineral Land Company about the 1st of September, 1870, along with the Silver Islet and all the other lands owned by the Montreal Mining Company on Lake Superior. This Company, in June and July of 1871, employed a small party mining on this lode for a month or two, when they sold it for $150,000 to Messrs. R. F. McEwen of London, and Simon Mandlebaum of Detroit. Since then the mine has been worked with considerable energy. The first shaft has been sunk about 100 feet, and two others about 70 feet each, the first 30 to 40 feet being an alluvial deposit. The ten fathom level was being driven to connect Nos. 1 and 2 shafts when I was there last winter, and I presume it is through before this. They drove a winze down below this level 35 to 40 feet, in which they found a rib of ore resting against the hanging wall. It increased from an inch or so to over a foot in thickness in the 35 feet. The ore consists of spar charged with argentiferous zinc blende. I heard it assayed from $100 to over $200 per ton. I have no doubt, when followed, it will be found very rich in silver in places. The vein is found to be very large, well defined, and regular. Very little silver was found in driving the drift save the indication — it being seen as a sulphuret coating the blende. They commenced in the fall, sloping away the ground next No. 1 shaft, and according to the reports they are taking out rich ore. I saw a specimen from it before leaving Fort William in the fall, which was charged with both the glance and native silver. Captain Plummer informed me that he had broken fine specimens of silver out of the lode under the water, on the west side of the island. The true fissure character of the lode is quite evident by its appearance. Again, it intersects the immense diorite dyke of Silver Island, so that there will be no danger of its giving out, or losing its character by sinking. A number of good, substantial houses were erected in the fall of 1872, and everything is in good working order, with a force of 20 to 30 men at work. On the whole, it may be considered a mine of much promise, although it has turned out but little silver for the work done. It seems to belong to the same class of veins as that of the Silver Islet, which has been proven to a depth of 300 feet already, and shown to carry the precious metal in such large quantities. They bear in the same direction, carry the same minerals and metals, and intersect the same formation, with the comparison of size being in favor of the former. Even at Silver Islet a good deal of work has been done where it shows very little silver.

"A month or so after the discovery of the Jarvis, Mr. Macfarlane turned his attention to the Wood's Location, which lies a few miles to the east of Thunder

Cape. During his examination, he discovered the Silver Islet Lode, the silver being first noticed by one of the party named Morgan. The Islet was about 75 feet long, rising 6 to 8 feet above the Lake, and lying about half a mile from the main shore. The vein crosses the Islet in two branches, about 20 feet apart, each 4 to 6 feet wide, bearing about N.N.W. and S.S.E., with a dip nearly vertical. The veinstone consists of calcareous or bitter spar, of a reddish-white color, with some quartz, etc. The rich streak of ore consists of spar, fine-grain galena, and occasionally zinc blende. Through the whole, especially the galena, the native silver is more or less thickly disseminated in fine strings, etc., the sulphuret of silver being occasionally present, also small quantities of nickel and cobalt. This streak is 2 or 3 inches wide generally, but in places it spreads out to a foot or more. It becomes poor and disappears in places, and again comes in; and in sinking, in some of the layers between two floors it is found to be rich, and in others poor. These floors, if I remember rightly, dip at a low angle to the north, and are generally 2 or 3 feet apart. The west branch, or vein that showed the most silver at surface, is the one upon which the work is being carried on. It crossed at the west end of the island, being almost entirely covered by water, though shallow, for 50 feet or more along the lode, so that the men had to work in the water in taking out the ore. The Islet lies on the above-mentioned dyke of diorite (Macfarlane's band), which is rather coarse in texture away from the walls. A portion running along the middle shows a reddish or brown appearance, owing to the color of the feldspar, the rest being greenish-gray in color, with white or greenish-white feldspar, and dark-green hornblende, etc. It differs in appearance from any of the numerous trap dykes I have seen in the country, and is very wide. Here it intersects the slightly inclined bluish or greenish-gray shales, which seem to occupy a position near the summit of the lower beds of the Upper Copper Rocks.

"They succeeded in taking out several hundred weights of the ore, which Mr. Macfarlane brought with him to Montreal in the fall, and had it thoroughly tested, various grades yielding from $1,000 to over $3,000 per ton. They built a house or two on the main land, and one on the Islet, and left a party of seven or eight men to mine and take out timber for crib work, during the winter. They commenced to sink a shaft on the dry part of the Islet, in the country rock, with the intention of intersecting the lode at the depth of 30 feet, where the water would not trouble them. By the time they had got down about 18 feet, a heavy storm came and filled it up, piling the ice high above it, knocking the floor of the boarding-house through, and nearly carrying the whole thing away, leaving the men in great danger. That put a stop to that work; but some time in February or March the shallow water had frozen solid to the rock, and the miners cut through it, and succeeded in taking out, according to their own estimates, about $25,000 worth of ore in about a week, when a storm broke it up again. Mr. Macfarlane was in Montreal during the winter, and made an estimate of the amount necessary to place the mine on a safe footing for carrying on the works, and, as I understood, asked the Company for $50,000 to do it; but they refused, although the evidences of its richness were so strong. No doubt they had the opinion of some pretended mining men who had been in the country, which was to the effect that all these veins would 'play out' in sinking 30 feet. People will not be deceived any more by such a theory, as time has shown its simplicity. The men continued doing what little they could on the Islet on calm days, and taking out timber, etc., until the mine was sold in September following.

"In the winter and spring of 1870, Mr. Macfarlane entered into negotiations with American capitalists for the sale of Silver Islet, with the consent of the Company; and Captain Wm. B. Frue, of Portage Lake, south shore of Lake Superior, a person of long experience in mining, went with Mr. Macfarlane, on the opening of navigation, to see it. He was so favorably impressed with its appearance that he and Mr. A. H. Sibley, the above-mentioned capitalists, became very sanguine, but were in danger, for some time, of being thrown to one side altogether, as the Montreal Company had entered into an agreement with an English Company for all their lands on Lakes Huron and Superior, including Silver Islet. However, the English and American Companies came to an understanding of being equally interested in carrying out the agreement; but when the first day of payment arrived, the English Company backed out, and the Americans stepped in and took the whole, paying $50,000 in gold on the 1st of September, 1870, the rest in three instalments, making in all $125,000, the amount of the purchase."

Included in said contract is an oblong tract, about five miles in length coastwise, and two miles in width, with the adjacent islands (including Silver Islet), containing sixty-four hundred acres, designated and known as the "Wood's Location." It is situated five miles easterly from Thunder Cape, and twenty miles northerly from the eastern end of Isle Royale, in the Province of Ontario, Can. A corporation was formed, styled the "Silver Islet Mining Company of Lake Superior," with a capital stock of $2,000,000, divided into 20,000 shares, of $100 each.

"Immediately on closing the bargain, Mr. Sibley telegraphed from Montreal to Captain Frue, at Houghton, Michigan, who chartered a steamer to transport some thirty men, with the necessary supplies, and heavy timber, for a breakwater, to Silver Islet. Arriving safely, he immediately commenced building the coffer-dam, etc., and by the beginning of October they were enabled to commence mining. Again the works were interrupted for a week or so, in the first part of November, by a storm, which carried away part of the coffer-dam. After all the difficulties they had to contend with, they had shipped over $90,000 by the close of navigation. They continued mining, with few interruptions, until a severe storm in the first part of March had carried away nearly half the crib-work, filling the mine with water and ice. The works were again restored, and mining re-commenced some time in May (1871) following.

"The mine had to be worked entirely from one surface opening or shaft, which was a great disadvantage, as but few miners could be employed for a long time. The shaft was built of timber, and made water-tight from the solid rock to the height of 15 to 20 feet above the water, so that the water could not get into the mine, even if it should get over the crib-work. From this opening the mine was continually enlarging by sinking and by driving along each way on the lode, taking care to leave a strong backing to prevent the Lake from breaking in. By the time they had got down some 80 or 100 feet, they had gained distance enough on each side to sink winzes, which would be connected at certain distances, or at each level, by galleries to the main shaft, which was then closed in all the way up. This would purify the air by causing a circulation, and allow them to employ more miners to advantage than they could do when it was all one open cut. On this principle the works have been expanding lengthways on the lode and in depth, and now they are down over 300 feet below the surface of Lake Superior. When I passed there last fall they had cut a strong vein of water, which at first threatened to fill up the mine; but when they got the new engine to work, as I was told, it was

5

able to keep the water down; but a larger pump is required, which they sent for, but it was frozen in on the south shore, and they will have to wait until navigation opens. However, they are able to carry on the mining in the other parts of the mine, and I see by the last letters from there that it is looking as well as ever. In the winter of 1871 they drove a cross-cut some 30 feet to cut the east vein, which they said was large, and looked very well, but it made too much water, so they had to close it up. At surface, improvements have been continually going on. They have enlarged the area of the Islet from less than a sixth of an acre to more than two acres, and built 10 or 12 good buildings on it. On the main shore they have built extensive docks, a large store, church, school-house, and about 40 dwelling-houses. I suppose the whole population is over 300, there being on an average about 130 laborers employed on the location.

"I have no doubt the value of the product of this mine is greater, for the amount of ground opened, than that of any other mine, although the annual product of some mines is much greater, owing to its peculiar situation. In 1870 the product was 172,825 pounds — value of same, $105,328, which gave an average of about 61 cents per pound, or $1,218 per ton of 2,000 pounds. In 1871 the product was 969,454 pounds — value of same, $645,397, at a rate of 66½ cents per pound, or $1,330 per ton; showing an average yield of a ton and a half per day.

"By excluding all expenditure not properly belonging to the Silver Islet Location, such as the price paid for the seventeen locations, the taxes, and the expenses of the explorations on the other location, and the law fees relating to the title, etc., we have a total expenditure of about $430,000 for 1870, 1871, and 1872, while the product of the mine equals $797,448 for 1870 and 1871. Why the product for 1872, which we know was very large, is excluded from the report, is more than I can tell, unless it was for the purpose of keeping down the reputation of the mine, so that those parties who were trying to break their title would not be so sanguine. Since the date of the above report we know that the mine has been continually turning out rich ore, and paying dividends to the shareholders, besides expending large sums in permanent improvements.

"Had this mine been on the main land, and worked under favorable or ordinary circumstances, what a rich mine it would be. Even where it is, with all its disadvantages, it is producing largely, and has been a very profitable mine to the stockholders, and no doubt further developments will show many such mines in this extensive metalliferous section."

The product from Silver Islet Mine for 1870–71–72 and 73, no doubt exceeds $2,000,000, although the exact amount is unknown to the public.

Silver Mining Companies, North Shore, Lake Superior.

LOCATED ON THUNDER ISLAND BAY AND ITS VICINITY.

NAME.	NAME.	
Algoma,	Spar Island,	The extent and richness of the Thunder Bay Silver District is as yet but partially known, although several silver mines are being worked. It is supposed that the Mineral Region extends westward to Lake Shebandowen, a distance of about one hundred miles.
Cornish,	Silver Harbor, or	
Howland,	Beek,	
International,	Silver Islet,	
Jarvis Island,	Thompson's Island,	
McKellar's Island,	Thunder Bay,	
Ontario,	Trowbridge,	
Prince's Bay,	3 A Mine,	
Shuniah,	3 B Mine.	

GRAND PLEASURE EXCURSION
FROM NEW YORK TO LAKE SUPERIOR & ST. PAUL,
via NIAGARA FALLS.

TABLE ROCK.

STOPPING PLACES.		MILES.
NEW YORK to ALBANY, *(Railroad or Steamer,)*		**145**
ALBANY TO ROCHESTER, *(Railroad,)*	229—	**374**
ROCHESTER TO BUFFALO	69—	**443**
BUFFALO TO CLEVELAND, *via Lake Shore Route*	183—	**626**
ROCHESTER TO NIAGARA FALLS, *(Susp'n Bridge,)*	75—	**448**
SUSPENSION BRIDGE TO HAMILTON, Canada	43—	**491**
HAMILTON TO DETROIT, Mich.	187—	**678**
DETROIT TO CHICAGO, *via Mich. Central R. R.*	284—	**962**
Detroit to Port Huron, *(Steamboat Route,)*	73—	**751**
PORT HURON TO DE TOUR, *(Lake Huron,)*	225—	**976**
DE TOUR, *(Mouth St. Mary's River,)* to SAUT STE. MARIE	50—	**1,026**
SAUT STE. MARIE TO MARQUETTE, *(Lake Superior,)*	170—	**1,196**
MARQUETTE TO PORTAGE ENTRY	75—	**1,271**
PORTAGE ENTRY TO COPPER HARBOR	63—	**1,334**
COPPER HARBOR TO ONTONAGON	92—	**1,426**
ONTONAGON TO BAYFIELD, Wis.	88—	**1,514**
BAYFIELD TO DULUTH	88—	**1,602**
DULUTH TO ST. PAUL, Minn., *(L. S. & Miss. R. R.)*	154—	**1,756**
ST. PAUL TO CHICAGO, *(Direct Railroad Route,)*	—	**410**
CHICAGO TO NEW YORK, *via Detroit*	962—	**1,404**

☞ THIS RAILROAD and STEAMBOAT ROUTE from the **City of New York** to **St. Paul,** Minn., via Niagara Falls, Lakes Huron and Superior,—passing the Island of Mackinac, the Saut Ste. Marie, and the Pictured Rocks,—a total Distance of 1,756 Miles, affords the invalid, and seeker of pleasure, during the Summer months, one of most healthy, interesting, and **Grand Excursions** on the Continent of America.

GRAND EXCURSION.

STOPPING PLACES and OBJECTS OF INTEREST IN THE ROUND TRIP FROM
BUFFALO TO DULUTH, ST. PAUL, &c.

DISTANCES

PORTS, ETC.	MILES.	PORTS, ETC.	MILES.
BUFFALO, N. Y.	0	*Ontonagon* 60	1,066
DUNKIRK	42	LA POINTE, Wis 80	1,146
ERIE, Pa 48	90	*Bayfield* 3	1,149
ASHTABULA, Ohio 41	131	SUPERIOR CITY 80	1,229
CLEVELAND, Ohio 54	185	*DULUTH*, Minn 6	1,235
MALDEN, Can 100	285	Lake Superior & Mississippi Railroad.	
DETROIT, Mich 20	305	FOND DU LAC 16	1,251
Lake St. Clair 7	312	(Dalles of the St. Louis River.)	
Port Huron 68	380	*Thomson* 8	1,259
Point au Barque and Light 70	450	Junc. Northern Pacific R. R 1	1,260
Thunder Bay and Light 75	525	*Hinckley* 53	1,313
De Tour, Mich 85	610	White Bear Lake 65	1,378
CHURCH'S LANDING 41	651	*ST. PAUL* 12	1,390
Saut Ste. Marie 14	665	Lake Pepin 66	1,456
White Fish Point and Light 40	705	LAKE CITY 25	1,481
Pictured Rocks 80	785	*Winona* 75	1,556
MARQUETTE 50	835	*La Crosse*, Wis 40	1,596
Portage Entry 80	915	*Prairie du Chien* 84	1,670
(HOUGHTON, 14 Miles.)		*DUBUQUE* opp. *Dunleith* 70	1,740
Keweenaw Point 50	965	*Freeport*, Ill 68	1,808
COPPER HARBOR 15	980	*CHICAGO* 121	1,929
EAGLE HARBOR 16	996	MICHIGAN CITY, Ind 55	1,984
EAGLE RIVER 10	1,006	*DETROIT* 229	2,213

STEAMBOAT AND RAILROAD ROUTES CONNECTING THE UPPER LAKES WITH THE MISSISSIPPI RIVER,

This GRAND EXCURSION embraces 1,585 Miles of Lake and River Navigation and 628 Miles Railroad Travel.

RETURNING VIA THE MISSISSIPPI RIVER TO DUBUQUE.

RAILROAD CONNECTIONS, &c.

From MARQUETTE the *Peninsula Division* of the *Chicago & Northwestern Railroad* convey Passengers, via Green Bay, to CHICAGO, ST. LOUIS, &c.

From DULUTH the *Northern Pacific Railroad* conveys Passengers to the Mississippi and Missouri Rivers, 450 miles, there connecting with Steamers on the Upper Missouri,—making another GRAND EXCURSION of great interest,—ascending the Missouri to *Fort Benton*, Mont., by Steamers.

This road also forms a Line of Travel to *Fort Garry*, Manitoba.

DISTANCES FROM EASTERN CITIES TO PORTS ON LAKE ERIE.

CITIES, ETC.	MILES.
Baltimore to Cleveland, Ohio, via Baltimore & Ohio R. R.	478
Philadelphia to Cleveland, Ohio, via Pennsylvania Central Railroad	505
Philadelphia to Erie, Pa., via Philadelphia & Erie Railroad	451
New York to Erie, Pa., via Catawissa Route	486
New York to Buffalo, via Erie Railway	423
New York to Buffalo, via New York Central R. R.	443
Boston to Buffalo, via Boston & Albany R. R. and New York Central R. R.	498

ST. LAWRENCE RIVER AND LAKE NAVIGATION

The St. Lawrence navigation extends from the Straits of Belle Isle, near the coast of Labrador (N. Lat. 51° 30′, W. Long. 55° 30′), to Duluth, Minn., at the head of Lake Superior, a distance of about 2,500 statute miles, by direct course.

The *Canadian Canals* on this route are the Lachine Canal, 8½ miles; Beauharnois, 11½ miles; Cornwall, 11½ miles; Farran's Point, 1 mile; Rapid Plat, 4 miles; Galops, 7½ miles, (on the St. Lawrence River), and the Welland Canal, (between Lakes Ontario and Erie), 27 miles. Their united length is 71 miles, and total lockage is 537 feet, passing through 54 locks.

The *St. Mary's Ship Canal* 1 mile in length and 18 feet lockage (two locks), avoiding the Rapids and uniting the waters of Lake Huron and Superior, was constructed by a company with the aid of the United States Government. Lake Ontario is elevated 234 feet above the highest tidal flow of the St. Lawrence, at Three Rivers: Lake Erie, 564 feet; Lake Huron, 574 feet; Lake Michigan, 576 feet, and Lake Superior, 600 feet above the ocean.

TABLE OF DISTANCES.

MILES.	PORTS, &C.		MILES.
2,540	STRAITS of BELLE ISLE,		00
1,870	Mouth Saguenay River,		670
1,730	QUEBEC,	140	810
1,645	Three Rivers,	85	895
1,560	MONTREAL,	85	980
1,551	Lachine,	9	989
1,536	Beauharnois Canal,	15	1,004
1,492	Cornwall,	44	1,048
1,440	OGDENSBURG, N. Y.,	52	1,100
1,402	KINGSTON, Can.,	38	1,138
	LAKE ONTARIO.		
1,242	TORONTO,	160	1,298
1,202	Mouth Welland Canal,	40	1,338
1,175	*Welland Canal,*	27	1,365
	LAKE ERIE.		
895	Malden, Can.,	280	1,645
875	DETROIT, Mich.,	20	1,665
800	Port Huron, Mich.,	75	1,740
	LAKE HURON.		
570	Point de Tour,	230	1,970
520	SAUT STE. MARIE,	50	2,020
480	White Fish Point,	40	2,060
	LAKE SUPERIOR.		
350	MARQUETTE, Mich.,	130	2,190
270	Keweenaw Point,	80	2,270
170	Ontonagan,	100	1,370
86	Bayfield,	84	1,454
00	DULUTH, Minn.,	86	2,540

NOTE.—This distance could be shortened 450 miles by the completion of the proposed *Huron and Ontario Ship Canal.*

Distance from CHICAGO to the Straits of Belle Isle, 2,400; from TOLEDO, Ohio, 1,700 miles, from BUFFALO, 1,365 miles.

Lower St. Lawrence and Saguenay Rivers.

The Trip down the noble St. Lawrence River, passing through the Rapids to the City of Montreal, 160 miles, is one of the most exciting character. The excursion from Montreal to Quebec, 170 miles, is also deeply interesting, passing through Lake St. Peter. The river thus far being from a half to one mile in width. The Trip from Quebec to the far-famed Saguenay River, 140 miles, is another deeply interesting excursion, passing Murray Bay, 80 miles; Kamouraska, 100 miles; Riviere du Loup, 120 miles, arriving at Tadousac, 140 miles, where the river is 30 miles wide. At the Watering Places on the Lower St. Lawrence, are several well kept Hotels, being much frequented during the Summer Months.

BUFFALO AND NIAGARA FALLS TO MONTREAL AND QUEBEC,

PASSING through LAKE ONTARIO, the THOUSAND ISLANDS, and down the RAPIDS of the ST. LAWRENCE RIVER.

American Side.	Objects of Interest, etc.	Canada Side.
BUFFALO ....Miles. *New York Central R.R.*. BLACK ROCK........... 3 *Buffalo & Niagara Falls R. R.* TONAWANDA........... 8–11 Schlosser's Landing....10–21 **NiagaraFallsVillage** 1–22 NIAGARA CITY......... 2–24 *New York Central R. R.* **Lewiston**........... 4–28 Youngstown.......... 7–35 Fort Niagara........... 1–36 Oak Orchard Creek.... CHARLOTTE, Outport for **Rochester**..........80–116 Pultneyville .. ..20–136 Sodus Bay............. **OSWEGO**..........40–176 Mouth of Oswego River. Stoney Point & Island..34–210 SACKET'S HARBOR.....12–222 **Cape Vincent**......20–242 *Watertown & Rome R. R.* **Clayton** ....13–233 **Alexandria**.........16–249 Morristown..........20–269 **OGDENSBURGH** 11–280 *Northern Railroad*, 118 miles to Rouse's Point.	**Foot of Lake Erie.** ERIE CANAL. **Niagara River.** Grand Island, A. Navy Island, C. Goat Island, A. **Falls of Niagara.*** **Suspension Bridge.** Rapids and Whirlpool. *Head of Navigation.* **Lake Ontario,** 180 miles long AMERICAN and CANADIAN STEAMERS leave Lewiston daily, during the season of navigation, for Toronto, Charlotte, Oswego, Kings- ton, and other Ports on Lake Ontario, passing down the St. Lawrence River to Ogdensburgh and Prescott, from thence to Montreal, passing through all the Rapids, having a total de- scent of over 200 feet, af- fording the most interesting excursion on the Continent. THOUSAND ISLANDS. Wolf, or Grand Island, Can. Howe Island, Gore Isl'd. " Well's Island, New York. Admiralty Islands. Navy Islands. Old Friends Group, and other groups.†	**Fort Erie**...........Miles. *Buffalo & Lake Huron R. R.* WATERLOO............. 3 *Steam Ferry*.......... CHIPPEWA..........17–20 Table Rock. .... 2–22 CLIFTON .... 4–24 *Great Western Railway.* Brock's Monument...... QUEENSTON............. 4–28 **Niagara**............. 8–36 Fort Massasauga..... Port Dalhousie.........12–48 **Hamilton** .. ..32–80 **TORONTO** (direct)...42–78 *Grand Trunk Railway.* Bowmanville..........43–121 PORT HOPE ....19–140 COBOURG............. 8–148 Long Point...........52–200 Amherst Island........30–230 **KINGSTON** ......12–242 Fort Henry.......... **Gananoque**.........18–238 Mallorytown..........18–256 **Brockville**.........12–268 *Brockville ,and Ottawa R. R.* **Prescott**............12–280 *Ottawa and Prescott R. R.*, 53 miles to Ottawa City.

* Situate in North latitude 43° 6′, and West longitude 2° 6′ from Washington, being 594 miles above Quebec.

† There are two channels through these numerous and romantic Islands, known as the *American*, and *Canadian Channels.* The former passes near Cape Vincent, Clayton, Alexandria, etc., being for the most part in American waters, attached to the State of N. York.

American Side.	Objects of Interest, etc.	Canada Side.
Chimney Island........5–285		Windmill Point.........2–282
		Isle aux Moutons.......
		Drummond's Island.....
Tibbet's Island.........		Duck Island...........
Isle aux Galops........2–287	GALOPS RAPIDS, 14¾ feet descent.	Canal, 1 mile.
LISBON................7–294		Point Cardinal.........6–288
		MATILDA..............6–294
Ogden's Island........4–298	RAPID PLAT, 11½ feet descent.	Point Iriquois..........5–299
		Canal, 2 miles.
WADDINGTON..........2–300		WILLIAMSBURG.........2–301
	RAPID DEPLAU.	Canal, 4 miles.
Goose Neck Island.....		Chrysler's Farm...,.....4–305
		Cat Island.......... ..
Chrysler's Island.......		Faren's Point..........6–311
LOUISVILLE LANDING..12–312		Canal, 3 miles.
Long Sault Island......4–316		DICKINSON'S LANDING ..5–316
	LONG SAULT RAPIDS, 48 feet descent.	
Barnhart's Island......		Canal, 11½ miles.
		Sheek's Island.........
SOUTH SIDE...........		NORTH SIDE...........
45th degree North lat..	Boundary Line between the	Cornwall..........10–326
ST. REGIS.............12–328	United States and Canada.	
Squaw Island........18–336		St. Regis Island.......
	LAKE ST. FRANCIS, 25 miles in	LANCASTER...........14–340
Beauharnois Canal, 11	length.	
miles..............24–360		Coteau du Lac.....17–357
MacIntyre Island......		Giroux Island.........
Maple Island..........		French Island.........
Thorn Island..........	COTEAN RAPIDS.	Fish Island...........
Pig Island............		Isle aux Vaches.... ..
Broad Island..........5–365		
La Pierre Island.......		
Isle l'Ail.............	CEDAR RAPIDS	CEDAR VILLAGE...... 10–367
ST. TIMOTHY..........		Isle aux Quacks......
Isle aux Nois..........	SPLIT ROCK RAPIDS.	Isle de la Grand Chûte..
		Point aux Moulin......
Beauharnois........8–373	CASCADE RAPIDS.*	Isle aux Cascade.......5–372
	Mouth of the Ottawa River.	Isle Perrot...........
	LAKE ST. LOUIS.	Mouth Ottawa River....

The North, or Canadian Channel, extends from Kingston, passing near Ganan oque. Several light-houses, or beacons, have been erected by the Canadian authorities to mark this intricate channel, which is studded with beautiful groups of islands—the *Fiddler's Elbow*, the *Sisters*, and the *Scotch Bonnet*, being passes, or groups of islands, of the most romantic character.

* The above four rapids are ascended by means of the *Beauharnois Canal*, 11½ miles in length, with locks, overcoming a descent of 84 feet.

American Side.	Objects of Interest, etc.	Canadian Side.
CAUGHNAWAGA.......17–390		Lachine............18–390
Montreal & Plattsburgh R.R.		Canal, 8½ miles.
	LACHINE RAPIDS, 44¾ feet de-	Isle aux Heron.........6–396
Isle aux Diable........	scent.	Nun's Island..........
LA PRAIRIE............8–398	VICTORIA BRIDGE.	*Grand Trunk Railway.*
Champlain & St.Lawrence R.R.	*Head of Navigation.*	MONTREAL........8–404
LONGUEIL............. 8–406	St Helen's Island.	North latitude 45° 30′.
Varennes............13–417	St. Theresa Island.	L'Assumption.........
Verchere		St. Sulpice............
		La Vitre.............
Sorel...............20–447	Group of Islands.	BERTHIER..,........ ...43–447
River St. Francis......	LAKE ST. PETER, 25 miles in	
DOUCETTES...........40–489	length.	Fond du Lac ....
Branch Grand Trunk R. R.	*Head of Tide Water,* 90 miles	Three Rivers.......42–489
	above Quebec.	
Becancour		Magdalen.............
Gentilly...............		
St. Piere.............		BATISCAN............15–504
		St. Marie.............
Dechellons............		St. Anne............10–514
Lothinier.............	*Richelieu Rapids,* 45 miles	Point aux Trembles....
St. Croix.............	above Quebec.	St. Augustine.........
St. Antoine...........		Cape Sante 30–544
Chaudiere River........		Cape Rouge..........20–514
Grand Trunk Railway.		Wolfe's Cove 8–572
Point Levi......... 574		**QUEBEC**..........2–574
Steam Ferry.	ISLAND of ORLEANS.	North latitude 46° 49′.

Trip to the Lower St. Lawrence and Saguenay Rivers.

The noble ST. LAWRENCE RIVER, which is about one mile wide opposite Quebec, extends a distance of about 400 miles when it empties into the Gulf, widening to 100 miles and upwards before reaching the Island of Anticosta.

The far-famed SAGUENAY RIVER, its largest tributary, enters from the West about 140 miles below Quebec, the St. Lawrence here being about 30 miles wide.

At *Murray Bay,* 80 miles; *Kamouraska,* 100 miles; *Riviere Du Loup,* 120 miles: Ca-couna, 126 miles; and the *Tadousac,* 140 miles; and at other resorts, or Watering Places, along the Lower St. Lawrence, are well kept Hotels, where sea-bathing and fishing can be enjoyed by visitors seeking health and pleasure during the summer months.

During warm weather, Steamers run every few days from MONTREAL and QUEBEC for the Lower St. Lawrence and Saguenay Rivers, affording one of the most romantic and healthy excursions on the Continent of America.

W. ORR & CO.

CEDAR RAPIDS.—St. Lawrence River.

MAGNITUDE OF THE LAKES, OR "INLAND SEAS."

NOTHING but a voyage over all of the great bodies of water forming the "INLAND SEAS," can furnish the tourist, or scientific explorer, a just idea of the extent, depth, and clearness of the waters of the Great Lakes of America, together with the healthy influence, fertility, and romantic beauty of the numerous islands, and surrounding shores, forming a circuit of about 4,000 miles, with an area of 90,000 square miles, or about twice the extent of the State of New York—extending through eight degrees of latitude, and sixteen degrees of longitude—this region embracing the entire north half of the temperate zone, where the purity of the atmosphere vies with the purity of these extensive waters, or "Inland Seas," being connected by navigable rivers or straits.

The States, washed by the Great Lakes, are New York, Pennsylvania, Ohio, Michigan, Indiana, Illinois, Wisconsin, Minnesota, and Upper Canada—the boundary line between the United States and the British Possessions running through the centre of Lakes Superior, Huron, St. Clair, Erie, and Ontario, together with the connecting rivers or straits, and down the St. Lawrence River to the 45th parallel of latitude. From thence the St. Lawrence flows in a northeast direction through Canada into the Gulf of St. Law-rence. The romantic beauty of the rapids of this noble stream, and its majestic flow through a healthy and rich section of country, is unsurpassed for grand lake and river scenery.

Lake Superior, the largest of the Inland Seas, lying between 46° 30' and 49° north latitude, and between 84° 30' and 92° 30' west longitude from Greenwich, is situated at a height of 600 feet above the Gulf of St. Lawrence, from which it is distant about 1,800 miles by the course of its outlet and the St. Lawrence river. It is 460 miles long from east to west, and 170 miles broad in its widest part, with an average breadth of 85 miles; the entire circuit being about 1,200 miles. It is 900 feet in greatest depth, extending 300 feet below the level of the ocean. Estimated area, 31,500 square miles, being by far the largest body of fresh water on the face of the globe—celebrated alike for its sparkling purity, romantic scenery, and healthy influence of its surrounding climate. About one hundred rivers and creeks are said to flow into the lake, the greatest part being small streams, and but few navigable except for canoes, owing to numerous falls and rapids. It discharges its waters eastward, by the strait, or river *St. Mary*, 60 miles long, into Lake Huron, which lies 26 feet below, there being about 20 feet descent at the Saut Ste Marie, which is overcome by means of two locks and a ship canal. Its outlet, is a most lovely and romantic stream, embosoming a number of large and fertile islands, covered with a rich foliage.

Lake Michigan, lying 576 ft. above the sea, is 320 miles long, 85 miles broad, and 700 feet deep; area, 22,000 square miles. This lake lies wholly within the confines of the United States. It presents a large expanse of water, with but few islands, except near its entrance into the Straits of Mackinac, through which it discharges its surplus waters. The strait is 30 or 40 miles in length, and discharges its accumulated waters into Lake Huron, on nearly a level with Lake Michigan. At the north end of the lake, and in the Straits, are several large and romantic islands, affording delightful resorts.

Green Bay, a most beautiful expanse of water, containing several small islands, lies at about the same elevation as Lake Michigan; it is 100 miles long, 20 miles broad, and 60 feet deep; area, 2,000 square miles. This is a remarkably pure body of water, presenting lovely shores, surrounded by a fruitful and healthy section of country.

Lake Huron, lying at a height of 574 feet above the sea, is 250 miles long, 100 miles broad, and 750 feet greatest depth; area, 21,000 square miles. This lake is almost entirely free of islands, presenting a large expanse of pure water. Its most remarkable feature is Saginaw Bay, lying on its western border. The waters of this lake are now whitened by the sails of commerce, it being the great thoroughfare to and from Lakes Michigan and Superior.

Georgian Bay, lying northeast of Lake Huron, and of the same altitude, being separated by islands and headlands, ies wholly within the confines of Canada. It is 140 miles long, 55 miles broad, and 500 feet in depth; area, 5,000 square miles. In the *North Channel*, which communicates with St. Mary's River, and in Georgian Bay, are innumerable islands and islets, forming an interesting and romantic feature to this pure body of water. All the above bodies of water, into which are discharged a great number of streams, find an outlet by the River *St. Clair*, commencing at the foot of Lake Huron, where it has only a width of 1,000 feet, and a depth of from 20 to 60 feet, flowing with a rapid current downward, 38 miles, into

Lake St. Clair, which is 25 miles long and about as many broad, with a small depth of water; the most difficult navigation being encountered in passing over " *St. Clair Flats,*" where now about 14 feet of water is afforded. *Detroit River,* 27 miles in length, is the recipient of all the above waters, flowing southward through a fine section of country into

Lake Erie, the *fourth* great lake of this immense chain. This latter lake again, at an elevation above the sea of 564 feet, 250 miles long, 60 miles broad, and 204 feet at its greatest depth, but, on an average, considerably less than 100 feet deep, discharges its surplus waters by the Niagara River and Falls, into Lake Ontario, 330 feet below; 51 feet of this descent being in the rapids immediately above the Falls, 160 feet at the Falls themselves, and the rest chiefly in the rapids between the Falls and the mouth of the river, 35 miles below Lake Erie. This is comparatively a shallow body of water; and the relative depths of the great series of lakes may be illustrated by saying, that the surplus waters poured from the vast *basins* of Superior, Michigan, and Huron, flow across the *plate* of Erie into the deep *bowl* of Ontario. Lake Erie is reputed to be the only one of the series in which any current is perceptible. The fact, if it is one, is usually ascribed to its shallowness; but the vast volume of its outlet—the Niagara River—with its strong current, is a much more favorable cause than the small depth of its water, which may be far more appropriately adduced as the reason why the navigation is obstructed by *ice* much more than either of the other great lakes.

The ascertained temperature in the middle of Lake Erie, August, 1845, was temperature of air 76° Fahrenheit, at noon— water at surface 73°—at bottom 53°.

Lake Ontario, the *fifth* and last of the Great Lakes of America, is elevated 234 feet above tide-water at Three Rivers on the St. Lawrence; it is 180 miles long, 60 miles broad, 600 feet deep.

Thus *basin* succeeds *basin*, like the locks of a great canal, the whole length of waters from Lake Superior to the Gulf of St. Lawrence being rendered navigable for vessels of a large class by means of the Welland and St. Lawrence canals— thus enabling a loaded vessel to ascend or descend 600 feet above the level of the ocean, or tide-water. Of these five great lakes, Lake Superior has by far the largest area, and Lake Ontario has the least, having a surface only about one-fifth of that of Lake Superior, and being somewhat less in area than Lake Erie, although not much less, if any, in the circuit of its shores. Lake Ontario is the safest body of water for navigation, and Lake Erie the most dangerous. The lakes of greatest interest to the tourist or scientific traveler are Ontario, Huron, together with Georgian Bay and North Channel, and Lake Superior. The many picturesque islands and headlands, together with the pure dark green waters of the Upper Lakes, form a most lovely contrast during the summer and autumn months.

The altitude of the land which forms the water-shed of the *Upper Lakes* does not exceed from 600 to 2,500 feet above the level of the ocean, while the altitude of the land which forms the water-shed of Lake Champlain and the lower tributaries of the St. Lawrence River rises from 4,000 to 5,000 above the level of the sea or tide-water, in the States of Vermont and New York.

The divide which separates the waters of the Gulf of Mexico, from those flowing northeast into the St. Lawrence, do not in some places exceed ten or twenty feet above the level of Lakes Michigan and Superior; in fact, it is said that Lake Michigan, when under the influence of high water and a strong northerly wind, discharges some of its surplus waters into the Illinois River, and thence into the Mississippi and Gulf of Mexico—so low is the divide at its southern terminus.

When we consider the magnitude of these Great Lakes, the largest body of fresh water on the globe, being connected by navigable Straits, or canals, we may quote with emphasis the words of an English writer: "How little are they aware, in Europe, of the extent of commerce upon these 'Inland Seas,' whose coasts are now lined with flourishing towns and cities; whose waters are plowed with magnificent steamers, and hundreds of vessels crowded with merchandise! Even the Americans themselves are not fully aware of the rising importance of these great lakes, as connected with the Far West.

TRIBUTARIES OF THE GREAT LAKES AND ST. LAWRENCE RIVER.

Unlike the tributaries of the Mississippi, the streams falling into the Great Lakes or the St. Lawrence River are mostly rapid, and navigable only for a short distance from their mouths.

The following are the principal Rivers that are navigable for any considerable length:

AMERICAN SIDE.		Miles.
St. Louis River, Min................	Superior to Fond du Lac............	20
Fox, or Neenah, Wis.................	Green Bay to Lake Winnebago*.....	36
St. Joseph, Mich....................	St. Joseph to Niles.................	26
Grand River, "	Grand Haven to Grand Rapids......	40
Muskegon, "	Muskegon to Newaygo.............	40
Saginaw "	Saginaw Bay to Upper Saginaw.....	26
Maumee, Ohio	Maumee Bay to Perrysburgh........	18
Genesee, N. Y.....................	Charlotte to Rochester..............	6

CANADIAN SIDE.		Miles.
Thames...........................	Lake St. Clair to Chatham...........	24
Ottawa	La Chine to Carillon...............	40
"	(By means of locks to Ottawa City)†...	70
Richelieu or Sorel....................	Sorel to Lake Champlain (by locks)	75
Saguenay	Tadusac to Chicoutimi.............	70
	(thence to Lake St. John, 50 m.)	

LAKE AND RIVER NAVIGATION,

FROM FOND DU LAC, LAKE SUPERIOR, TO THE GULF OF ST. LAWRENCE.

LAKES, RIVERS, ETC.	Length in miles.	Greatest breadth.	Av. breadth.	Depth in feet.	El. above sea.
Superior......................	460	170	85	900	600 ft.
St. Mary's River...............	60	5	2	10 to 100	
Michigan......................	320	85	58	700	576 "
Green Bay.....................	100	25	18	100	576 "
Strait of Mackinac.............	40	20	10	20 to 200	575 "
Huron.........................	250	100	70	700	574 "
North Channel.................	150	20	10	20 to 200	574 "
Georgian Bay..................	140	55	40	500	574 "
St. Clair River................	38	1½	1	20 to 60	
Lake St. Clair*...............	25	25	18	10 to 20	568 "
Detroit River..................	27	3	1	10 to 60	
Erie..........................	250	70	40	200	564 "
Niagara River.................	35	3	1		
Ontario.......................	180	58	40	600	234 "
St. Lawrence River.............	760	100	2		
Lake St. Francis, foot Long Saut...			4		142 "
Lake St. Louis, foot Cascade Rapids			5		58 "
At Montreal...................			3		13 "
Lake St. Peter.................			12		6 "
Tide-water at Three Rivers.......			1		0 "
At Quebec....................			1		0 "

Total miles navigation........ 2,835

* By means of 17 locks, overcoming an elevation of 170 feet.
† The navigation for steamers extends 150 miles above Ottawa City, by means of portages and locks.
* The *St. Clair Flats*, which have to be passed by all large steamers and sail vessels running from Lake Erie to the Upper Lakes, now affords thirteen feet of water.

ALTITUDE OF VARIOUS POINTS ON THE SHORES OF LAKE SUPERIOR.

LOCALITIES.	Above Lake Superior.	Above the Sea.
Lake Superior................................	000 feet.	600 feet.
Point Iroquois, South Shore..................	350 "	950 "
Gros Cap, C. W., North Shore.................	700 "	1,300 "
Grand Sable, South Shore	345 "	945 "
Pictured Rocks, "	250 "	850 "
Iron·Mountain "	850 "	1,450 "
Huron Mountains "	1,400 "	2,000 "
Mount Houghton, near Keweenaw Point........	1,000 "	1,600 "
Porcupine Mountains, South Shore............	1,380 "	1,980 "
Isle Royale, Michigan.....	300 "	900 "
Minnesota Mountains (estimated)..............	1,200 "	1,800 "
Michipicoten Island, C. W....................	800 "	1,400 "
Pie Island, "	850 "	1,450 "
St. Ignace (estimated) "	1,200 "	1,800 "
McKay's Mountain, "	1,200 "	1,800 "
Thunder Cape, "	1,350 "	1,950 "

TOPOGRAPHY AND METEOROLOGY.

"The mountains of the region along the south shore of Lake Superior, consist of two granite belts in the northwest, the *Huron Mountains* to the southward, a trap range starting from the head of Keweenaw Point, and running west and southwest into Wisconsin, the *Porcupine Mountains*, and the detrital rocks. The Huron Mountains in places attain an elevation of 1,400 feet above the Lake. The highest elevation attained by the Porcupine Mountains is 1,380 feet.

"Meteorological observations were instituted by order of the Government at three military posts in the District, viz.: Forts Wilkins (Copper Harbor), Brady, and Mackinac. From these observations it appears that the mean annual temperature of Fort Brady is about one degree lower than that of Fort Wilkins, although the latter post is nearly a degree further north. This difference arises from the insular · position of Keweenaw Point, which is surrounded on three sides by water. The climate at Fort Brady, during the whole season, corresponds in a remarkable degree with that of St. Petersburg. The temperature of the region is very favorable to the growth of cereals. The annual ratio of fair days at Fort Brady is 168; of cloudy days, 77; rainy days, 71; snowy days, 47.

"The temperature of the water of Lake Superior during the summer, a fathom or two below the surface, is but a few degrees above the freezing point. In the western portion, the water is much colder than in the eastern—the surface flow becoming warmer as it advances toward th outlet. The mirage which frequently oc curs, is occasioned by the difference between the temperature of the air and the Lake. Great difficulties are experienced from this cause in making astronomical observations.

"Auroras, even in midsummer, are of frequent occurrence, and exhibit a brilliancy rarely observed in lower latitudes."
—*Foster & Whitney's Report.*

2

THE UPPER LAKES, OR "INLAND SEA," OF AMERICA.

This appellation applies to Lakes Huron, Michigan, and Superior, including Green Bay, lying within the confines of the United States, and Georgian Bay, which lies entirely in Canada.

These bodies of water embrace an area of about 75,000 square miles, and, as a whole, are deserving of the name of the 'INLAND SEA,' being closely connected by straits or water-courses, navigable for the largest class of steamers or sail vessels. The shores, although not elevated, are bold, and free from marsh or swampy lands, presenting one clean range of coast for about 3,000 miles.

By a late decision of the Supreme Court of the U. States, the Upper Lakes including Lake Erie, with their connecting waters, were declared to be *seas*, commercially and legally. Congress, under this decision, is empowered to improve the harbors of the lakes and the connecting straits, precisely as it has power to do the same on the seaboard. This will probably lead to a vigorous policy in the maintenance of Federal authority, both in improving the harbors, and making provision for the safety of commerce, and protection of life, as well as guarding against foreign invasion. The only fortification of importance that is garrisoned is *Fort Mackinac*, guarding the passage through the Straits of Mackinac.

The islands of these lakes are numerous, particularly in the Straits of Mackinac, and in Georgian Bay, retaining the same bold and virgin appearance as the mainland; most of them are fertile and susceptible of high cultivation, although, as yet, but few are inhabited to any considerable extent.

The dark green waters of the Upper Lakes, when agitated by a storm, or the motion of a passing steamer, presents a brilliancy peculiar only to these transparent waters—they then assume the admixture of white foam, with a lively green tinge, assuming a crystal-like appearance. In this pure water, the *white fish*, and other species of the finny tribe, delight to gambol, affording the sportsman and epicurean untold pleasure, which is well described in the following poem:

THE WHITE FISH.

HENRY R. SCHOOLCRAFT, in his poem. "THE WHITE FISH," says:

" All friends to good living by tureen and dish,
Concur in exulting this prince of a fish;
So fine in a platter, so tempting a fry,
So rich on a gridiron, so sweet in a pie;
That even before it the salmon must fall,
And that mighty *bonne-bouche*, of the land
    beaver's tail.

*   *   *   *

'Tis a morsel alike for the gourmand or faster,
While, white as a tablet of pure alabaster!
Its beauty or flavor no person can doubt,
When seen in the water or tasted without;
And all the dispute that opinion ere makes
Of this king of lake fishes, this ' *deer of the
    lakes*,'*
Regard not its choiceness to ponder or sup,
But the best mode of dressing and serving it up

*   *   *   *

Here too, might a fancy to descant inclined,
Contemplate the love that pertains to the kind,
And bring up the red man, in fanciful strains,
To prove its creation from feminine brains."†

---

* A translation of *Ad-dik-keem-maig*, the Indian name for this fish.
† *Vide* " Indian Tales and Legends."

FISH OF THE UPPER LAKES.

" The numbers, varieties, and excellent quality of lake fish are worthy of notice. It is believed that no fresh waters known can, in any respect, bear comparison. They are, with some exceptions, of the same kind in all the lakes. Those found in Lake Superior and the straits of St. Mary are of the best quality, owing to the cooler temperature of the water. Their quantities are surprising, and apparently so inexhaustible, as to warrant the belief that were a population of millions to inhabit the lake shore, they would furnish an ample supply of this article of food without any sensible diminution. There are several kinds found in Lake Superior, and some of the most delicious quality, that are not found in the lakes below, as the siskowit and muckwaw, which grow to the weight of eight or ten pounds. The salmon and some others are found in Ontario, but not above the Falls of Niagara.

" The following is a very partial list of a few of the prominent varieties: the white fish, Mackinac and salmon-trout, sturgeon, muscalunje, siskowit, pickerel, pike, perch, herring, white, black, and rock bass, cat, pout, eel-pout, bull-head, roach, sun-fish, dace, sucker, carp, mullet, bill-fish, sword-fish, bull-fish, stone-carrier, sheeps-head, gar, &c.

" The lamprey-eel is found in all, but the common eel is found in neither of the lakes, nor in any of their tributaries, except one. The weight to which some of these attain is not exceeded by the fish of any other inland fresh waters, except the Mississippi.   *   *   *   *

" The fish seem to be more numerous some years than others, and likewise of better quality. The kinds best for pickling and export are the white fish, Mackinac and salmon trout, sturgeon, and pickerel. The fisheries at which these are caught are at Mackinac, at several points in each of the four straits, the southeast part of Lake Superior, Thunder Bay, Saginaw Bay, and Fort Gratiot near foot of Lake Huron. The sport of taking the brook trout, which are found in great abundance in the rapids at the Saut Ste Marie, and most all of the streams falling into the Upper Lakes, affords healthful amusement to hundreds of amateur fishermen during the summer and fall months. The modes of taking the different kinds of fish are in seines, dip-nets, and gill-nets, and the trout with hooks.

" Those engaged in catching fish in the Straits of Mackinac, are composed of Americans, Irish, French, half breeds, and Indians. Some are employed by capitalists, others have their own boats and nets. Each one is furnished with a boat, and from fifty to one hundred nets, requiring constantly two or three men for each boat, to run the different gangs of nets. The fish caught are principally white fish, with some trout. The demand for exportation increases every year, and although immense quantities are caught every season, still no diminution in their number is perceived.

" A fleet of two hundred fish-boats are engaged in and about the Straits, embracing, however, all the Beaver group. Each boat will average one barrel of fish per day during the fishing season.
*   *   *   *   *

" Ye, who are fond of sport and fun, who wish for wealth and strength; ye, who love angling; ye, who believe that God has given us a time to pray, a time to dance, &c., &c., go to these fishing-grounds, gain health and strength, and pull out Mackinac trout from 20 to 40 lbs. in weight. One hook and line has, in three to four hours, pulled out enough to fill three to four barrels of fish, without taking the sport into consideration.

" Yours, W. M. J.'

THE INTERNATIONAL BRIDGE.

THIS important work was commenced in May, 1870, and completed in the latter part of 1873, being now open for traffic. It is an iron superstructure built on stone piers in the most substantial manner, extending from Black Rock, Buffalo, across the Niagara River to the Canadian side. The width of the river being broken by Squaw Island, at this point it was found that the main river measured 1,894 feet, and Black Rock Harbor 445 feet, or a total of 2,339 feet to be bridged; while Squaw Island is crossed by an embankment 25 feet in height and 1,328 feet in length. There are two openings for the passage of vessels. The depth of water here varies from 13 to 47 feet at the points for the piers, and the normal current at low water from 2.58 to 5.12 miles per hour.

The *Great Western Railway*, the *Grand Trunk Railway*, and the *Canada Southern Railway* all run westward from the International Bridge, while on the American side of the river all the Railroads diverging from Buffalo have a connection, affording immense facilities for both Eastern and Western traffic.

GRAIN TRADE OF BUFFALO.

In order to show the rapid increase of the carrying trade of Buffalo, by railroad and lake, we insert the following Table showing the shipment of grain and flour since the year 1841.

GRAIN INCLUDING FLOUR AS WHEAT.

	Grain, bush.	Grain, incl'g Flour, bush.		Grain, bush.	Grain, incl'g Flour, bush.
1841	1,852,325	5,592,525	1858	20,202,244	26,812,980
1842	2,015,928	5,687,468	1859	14,429,069	21,530,722
1843	2,055,025	6,642,610	1860	31,441,440	37,053,115
1844	2,335,568	6,910,718	1861	50,662,646	61,460,601
1845	1,848,040	5,581,790	1862	58,642,344	72,872,454
1846	6,493,522	13,386,167	1863	49,845,065	64,735,510
1847	9,868,187	19,153,187	1864	41,044,096	51,177,146
1848	7,396,012	14,641,012	1865	42,473,223	51,415,188
1849	8,628,013	14,665,188	1866	51,820,342	58,388,087
1850	6,618,004	12,059,559	1867	43,079,079	50,168,074
1851	11,449,661	17,740,781	1868	42,573,125	50,197,215
1852	13,392,937	20,390,504	1869	37,456,131	45,489,276
1853	11,078,741	15,956,526	1870	38,208,039	45,477,604
1854	18,553,455	22,252,235	1871	61,319,313	67,529,158
1855	19,788,473	24,472,278	1872	58,703,666	62,550,596
1856	20,123,667	25,753,907	1873	65,498,955	70,962,520
1857	15,348,930	19,578,695	1874	——	——

RAILROAD AND STEAMBOAT ROUTES,

From Buffalo to Niagara Falls, Toronto, Etc.

CROSSCUP & WEST.PHILA.

THE most usual mode of conveyance from Buffalo to the Falls of Niagara, and thence to Lake Ontario, or into Canada, is by the *Erie Railway*, or the *Buffalo, Niagara Falls and Lewiston Railroad*, 28 miles in length. The latter runs through Tonawanda, 11 miles; Niagara Falls, 22 miles; Suspension Bridge, 24 miles, connecting with the Great Western Railway of Canada, and terminates at Lewiston, the head of navigation on Niagara River, 28 miles.

American and Canadian steamers of a large class leave Lewiston several times daily, for different ports on Lake Ontario and the St. Lawrence River.

There is also another very desirable mode of conveyance, by Steamboat, descending the Niagara River, from Buffalo to Chippewa, Can., thence by the *Erie and Ontario Railroad*, 17 miles in length; passing in full view of the Falls, to the Clifton House, three miles below Chippewa; Suspension Bridge, five miles; Queenston, eleven miles, terminating at Niagara, Can., thirty-five miles from Buffalo.

As the steamboat leaves Buffalo, on the latter route, a fine view may be obtained of Lake Erie and both shores of

Niagara River. On the Canada side, the first objects of interest are the ruins of old FORT ERIE, captured by the Americans, July 3d, 1814. It is situated at the foot of the lake, opposite the site of a strong fortress which the United States Government has recently erected for the protection of the river and city of Buffalo.

WATERLOO, Can., three miles below Buffalo and opposite Black Rock (now part of Buffalo), with which it is connected by the new International Bridge, is situated on the west side of Niagara River, which is here about half a mile wide. A Branch of the *Grand Trunk Railway* runs to Goderich, Canada, on Lake Huron; a Branch of the *Great Western Railway of Canada*, and the *Canada Southern Railroad* also commences at the railroad bridge crossing Niagara River to Buffalo; all connecting with Eastern Railroads.

GRAND ISLAND, belonging to the United States, is passed on the right in descending the river.

NAVY ISLAND, belonging to the British, is next passed, lying within gunshot of the mainland. This island obtained great notoriety in the fall and winter of 1837–'38, when it was occupied by the "Patriots," as they were styled, during the troubles in Canada. The Steamer *Caroline* was destroyed December 29th, 1837, while lying at Schlosser's Landing, on the American shore.

CHIPPEWA, 20 miles below Buffalo, and two miles above the Falls, is on the west side of Niagara River, at the mouth of a

creek of the same name, which is navigable to PORT ROBINSON, some eight or ten
miles west; the latter place being on the
line of the Welland Canal. The village of
Chippewa contains a population of about
1,000 souls. Steamboats and lake craft
of a large size are built at this place for
the trade of Lake Erie and the Upper
Lakes. It has obtained a place in history
on account of the bloody battle which
was fought near it in the war of 1812,
between the United States and Great
Britain. The battle was fought on the
5th of July, 1814, on the plains, a short
distance south of the steamboat landing.
The American forces were commanded
by Major-General Jacob Brown, and the
British, by Major-General Riall, who, after an obstinate and sanguinary fight,
was defeated, with considerable loss.

At Chippewa commences the railroad
extending to Niagara, at the mouth of the
river, a distance of 17 miles. Steamboats
continue the line of travel from both ends
of this road, thus furnishing an interesting
and speedy conveyance between Lakes
Erie and Ontario.

On ariving in the vicinity of the FALLS
OF NIAGARA, the cars stop near the *Clifton
House,* situated near the ferry leading to
the American side. The site of this house
was chosen as giving the best view of both
the American and Canadian or Horse-Shoe
Falls, which are seen from the piazzas and
front windows. This is the most interesting approach to the Falls.

In addition to the Falls, there are other
points of attraction on the Canada side of
the river. The collection of curiosities at
the Museum, and the Camera Obscura,
which gives an exact and beautiful, though
miniature image of the Falls, are well worthy of a visit. The *Burning Spring,* two
miles above the Falls, is also much frequented; and the rides to the battle-
grounds in this vicinity makes an exhilarating and very pleasant excursion.

DRUMMONDSVILLE, one mile west of the
Falls, and situated on *Lundy's Lane,* is
celebrated as the scene of another sanguinary engagement between the American and British forces, July 25, 1814.

The following is a brief, though correct
account of the engagement: "On the afternoon of the above day, while the American army was on their march from *Fort
George* toward *Fort Erie,* ascending the
west bank of the river, their rear-guard,
under the immediate command of Gen.
Scott, was attacked by the advanced guard
of the British army, under Gen. Riall, the
British having been reinforced after their
defeat at Chippewa, on the 5th of the same
month. This brought on a general conflict
of the most obstinate and deadly character.
As soon as attacked, Gen. Scott advanced
with his division, amounting to about 3,000
men, to the open ground facing the heights
occupied by the main British army, where,
were planted several heavy pieces of can-
non. Between eight and nine o'clock in
the evening, on the arrival of reinforcements
to both armies, the battle became general
and raged for several hours, with alternate
success on both sides; each army evin-
cing the most determined bravery and resistance. The command of the respective
forces was now assumed by Major Gen.
Brown and Lieut.-Gen. Drummond, each
having under his command a well-disci-
plined army. The brave (American) Col.
Miller was ordered to advance and seize
the artillery of the British, which he
effected at the point of the bayonet in the
most gallant manner. Gen. Riall, of the
English army, was captured, and the possession of the battle-ground contested until near midnight, when 1,700 men being
either killed or wounded, the conflicting
armies, amounting altogether to about
6,000 strong, ceased the deadly conflict,
and for a time the bloody field was left un-
occupied, except by the dead and wounded.

When the British discovered that the
Americans had encamped one or two miles

BROCK'S MONUMENT.—Queenston Heights.

distant, they returned and occupied their former position. Thus ended one of the most bloody conflicts that occurred during the last war; and while each party boasted a victory, altogether too dearly bought, neither was disposed to renew the conflict."

CLIFTON is a new and flourishing village, situated at the western termination of the Great Western Railway, where it connects with the *Suspension Bridge.*

QUEENSTON, situated seven miles below the Falls, and about the same distance above the entrance of Niagara River into Lake Ontario, lies directly opposite the village of Lewiston, with which it is connected by a Suspension Bridge 850 feet in length. It contains about 500 inhabitants, 60 dwelling-houses, one Episcopal, one Scotch Presbyterian, and one Baptist church, four taverns, four stores, and three warehouses. This place is also celebrated as being the scene of a deadly strife between the American and British forces, October 13, 1812. The American troops actually engaged in the fight were commanded by Gen. Solomon Van Rensselaer, and both the troops and their commander greatly distinguished themselves for their bravery, although ultimately overpowered by superior numbers. In attempting to regain their own side of the river many of the Americans perished; the whole loss in killed, wounded, and prisoners amounting to at least 1,000 men.

Major-General BROCK, the British commander, was killed in the middle of the fight, while leading on his men. A new monument stands on the heights, near where he fell, erected to his memory. The first monument was nearly destroyed by gunpowder, April 17, 1840; an infamous act, said to have been perpetrated by a person concerned in the insurrection of 1837–'38.

BROCK'S NEW MONUMENT was commenced in 1853, and finished in 1856; being 185 feet high, ascended on the inside by a spiral staircase of 235 stone steps. The base is 40 feet square and 35 feet in height, surmounted by a tablet 35 feet high, with historical devices on the four sides. The main shaft, about 100 feet, is fluted and surmounted by a Corinthian capital, on which is placed a colossal figure of Major-General Brock, 18 feet in height. This beautiful structure cost £10,000 sterling, being entirely constructed of a cream-colored stone quarried in the vicinity. A massive stone wall, 80 feet square, adorned with military figures and trophies at the corners, 27 feet in height, surrounds the monument, leaving space for a grass-plot and walk on the inside of the enclosure.

The following is the inscription:

Upper Canada
Has dedicated this Monument
to the memory of the late
Major-General Sir ISAAC BROCK, K. B.
Provisional Lieut.-Governor and Commander
of the Forces in this Province,
Whose remains are deposited
in the vault beneath.
Opposing the invading enemy
He fell in action, near the Heights,
on the 13th October, 1812,
In the 43d year of his age,
Revered and lamented by the people
whom he governed, and deplored by
the Sovereign to whose service
His life had been devoted.

The last words of Major-General Brock, when he fell mortally wounded by a musket-shot through the left breast, were, "Never mind, my boys, the death of one man—I have not long to live." Thus departed one of the many noble spirits that were sacrificed on this frontier during the war of 1812.

The village of NIAGARA is advantageously situated on the Canada side, at the entrance of the river into Lake Ontario, directly opposite *Fort Niagara,* on the American side. It contains about 3,000 inhabitants, a court-house and jail; one Episcopal, one Presbyterian, one Methodist, and one Roman Catholic Church; 6 hotels and taverns; and 20 stores of different kinds; also, an extensive locomotive and car factory. This is the most noted place in Canada West for building steamboats and other craft navigating Lake Ontario. Here is a dockyard with a marine railway and foundry attached, capable of making machinery of the largest description, and giving employment to a great number of men. It is owned by the "Niagara Dock Company." Steamers leave daily for Toronto, etc.

FORT GEORGE, situated a short distance south or up-stream from the mouth of the river, is now in ruins. This was the scene of a severe contest in 1813, in which the Americans were victorious. A new fort has been erected on the point of land at the mouth of the river, directly opposite old *Fort Niagara* on the American side. The new fortification is called *Fort Massasauga*.

The whole frontier on the Canada side, from Fort George to Fort Erie, opposite Buffalo, was occupied by the American army in 1814, when occurred a succession of battles of the most determined and brilliant character.

NIAGARA RIVER,

ITS RAPIDS, FALLS, ISLANDS, AND ROMANTIC SCENERY.

" Majestic stream ! what river rivals thee,
Thou child of many lakes, and sire of one—
Lakes that claim kindred with the all-circling
      sea—
Large at thy birth as when thy race is run !
Against what great obstructions has thou won
Thine august way—the rock-formed mountain-
      plain
Has opened at thy bidding, and the steep
Bars not thy passage, for the ledge in vain
Stretches across the channel—thou dost leap
Sublimely down the height, and urge again
Thy rock-embattled course on to the distant
      main."

THIS most remarkable and romantic stream, the outlet of Lake Erie, through which flows all the accumulated waters of the Upper Lakes of North America, very appropriately forms the boundary between two great countries, the British province of Upper Canada on the one side, and the State of New York, the "Empire State" of the Union, on the opposite side. In its whole course, its peculiar character is quite in keeping with the stupendous Cataract from which its principal interest is derived.

The amount of water passing through this channel is immense ; from a computation which has been made at the outlet of Lake Erie, the quantity thus discharged is about twenty millions of cubic feet, or upwards of 600,000 tons per minute, all of which great volume of water, 20 miles below, plunges over the Falls of Niagara.

The Niagara River commences at Bird Island, nearly opposite the mouth of Buffalo harbor, and passes by the site of old Fort Erie and Waterloo on the Canada side. At the later place a steam ferryboat plies across the river to Black Rock, now forming a part of the city of Buffalo. It is here proposed to construct a railroad bridge across the stream, about 1,800 feet in width.

SQUAW ISLAND and STRAWBERRY ISLAND are both small islands lying on the American side of the stream, near the head of Grand Island. The river is here used in part for the Erie Canal, a pier extending from Squaw Island to Bird Island, forming a large basin called Black Rock Harbor.

GRAND ISLAND, attached to Erie Co.,

N. Y., is a large and important body of land, about ten miles long from north to south, and seven miles wide. This island is partly cleared and cultivated, while the larger portion is covered with a large growth of oaks and other forest trees.

The ship or steamboat channel runs along the bank of Grand Island to nearly opposite Chippewa, where the whole stream unites before plunging over the Falls of Niagara, being again separated at the head of Goat Island. From this point the awe-struck traveller can scan the quiet waters above, and the raging rapids below, preparing to plunge over the Cataract.

CAYUGA ISLAND and BUCKHORN ISLAND are small bodies of land belonging to the United States, situated immediately below Grand Island.

NAVY ISLAND, lying opposite the village of Chippewa, 18 miles below the head of the river, is a celebrated island belonging to the Canadians, having been taken possession of by the sympathizing patriots in 1837, when a partial rebellion occurred in Upper and Lower Canada.

TONAWANDA, 11 miles below Buffalo, is situated at the mouth of Tonawanda Creek, opposite Grand Island. The *Erie Canal* here enters the creek, which it follows for several miles on its course toward Lockport. A railroad also runs to Lockport, connecting with the *New York Central Railroad*, extending to Albany. A *ship canal* is proposed to be constructed from Tonawanda to some eligible point on Lake Ontario, thus forming a rival to the Welland Canal of Canada.

SCHLOSSER'S LANDING, two miles above Niagara Falls village, is a noted steamboat landing, opposite Chippewa, from whence the steamer *Caroline* was cut adrift by the British and destroyed, by being precipitated over the Falls during the Canadian rebellion, December 29th, 1837.

THE RAPIDS.—Below Navy Island, between Chippewa and Schlosser, the river is nearly three miles in width, but soon narrows to one mile, when the Rapids commence, and continue for about one mile before reaching the edge of the precipice at the Horse-Shoe Fall.

At the commencement of the Rapids, "the bed of the river declines, the channel contracts, numerous large rocks heave up the rolling surges, and dispute the passage of the now raging and foaming floods. The mighty torrent leaping down successive ledges, dashing over opposing elevations, hurled back by ridges, and repelled from shores and islands—plunging, boiling, roaring—seems a mad wilderness of waters striving against its better fate, and hurried on to destruction by its own blind and reckless impetuosity. Were there no cataract, these Rapids would yet make Niagara the wonder of the world."

IRIS, or GOAT ISLAND, commences near the head of the Rapids, and extends to the precipice, of which it forms a part, separating the American Fall from the Canadian or Horse-Shoe Fall. It is about half a mile in length, eighty rods wide, and contains over sixty acres of arable land, being for the most part covered with a heavy growth of forest trees of a variety of species, and native plants and flowers. A portion of the island, however, has been cleared off, and a garden enclosed, in which are some excellent fruit-trees, and a variety of native and foreign plants and flowers, and a fish-pond. The island is remarkably cool, shady, and pleasant, and is an object of unceasing admiration from year to year. Comfortable seats and arbors are placed at the most interesting points, where the visitor can sit at ease and enjoy the beautiful and sublime views presented to his sight—often entranced by a deafening roar of mighty waters in their descent, accompanied by changing rainbows of the most gorgeous description.

Niagara.

WRITTEN BY LYDIA H. SIGOURNEY.

Flow on forever, in thy glorious robe
Of terror and of beauty; God hath set
His rainbow on thy forehead, and the cloud
Mantles around thy feet, and He doth give
Thy voice of thunder power to speak of Him
Eternally; bidding the lip of man
Keep silence, and upon thy rocky altar
Pour incense of awe-struck praise.

GOAT ISLAND BRIDGE.—The Niagara Falls *Gazette* gives the following description of this new structure:

"This bridge across the east branch of the Niagara River is situated in the Rapids, about sixty rods above the Cataract, on the site of the old wooden bridge. It is 360 feet long, and consists of four arches of ninety feet span each, supported between the abutments of three piers. The piers above water are built of heavy cut stone, and are twenty-two feet long and six feet wide, tapering one foot in the height. The foundations are formed of foot-square oak timber, strongly framed and bolted together in cribs, filled with stone, and covered with timber at the surface of the water. These timber-foundations are protected against wear and injury from ice by heavy plates of iron, and being always covered with water, will be as durable as the stone.

"The superstructure is of iron, on the plan of Whipple's iron-arched bridge. The whole width is twenty-seven feet, affording a double carriage-way of sixteen and a half feet, and two foot-ways of five and a fourth feet each, with iron railings. The arches are of cast iron, and the chords, suspenders, and braces of wrought iron. All the materials used in the construction are of the best quality, and the size and strength of all the parts far beyond what are deemed necessary in bridges exposed to the severest tests.

"This substantial and beautiful structure, spanning a branch of this majestic river in the midst of the rapids, and overlooking the cataract, is worthy of the site it occupies, and affords another instance of the triumph of human ingenuity over the obstacles of nature.

"The islands connected by this bridge with the American shore are the property of Messrs. Porter, and constitute the most interesting features in the scenery surrounding the cataract. This bridge has been erected by them to facilitate communication with these interesting localities not otherwise accessible."

This is a toll-bridge, every foot passenger being charged 25 cents for the season, or single crossing.

There are upward of thirty islands and islets in the Niagara River or Strait, above the cataract. Most of those not described are small, and scarcely worthy of enumeration, although those immediately contiguous to Goat Island form beautiful objects in connection with the rushing and mighty waters by which they are surrounded. *Bath Island, Brig Island, Chapin's Island,* and *Bird Island,* all situated immediately above the American Fall, are reached by bridges.

When on Goat Island, turning to the right toward the Falls, the first object of interest is *Hogg's Back,* a point of land facing the American Fall,—Bridge to Addington Island immediately above the Cave of the Winds, 160 feet below. Sam. Patch's Point is next passed on the right, from which he took a fearful leap some years since. Biddle's Stairs descend to the water's edge below and the Cave of the Winds, which are annually visited by thousands of visitors. Terrapin Bridge and Terrapin Tower afford a grand view of the Canadian or Horse-Shoe Fall and Rapids above the Falls. Three Sister Islands are contiguous to Goat Island, on the American side. Passing around Goat Island toward the south, a grand view is afforded of the river and rapids above the Canadian and American Falls.

THE AMERICAN RAPIDS, FROM THE BRIDGE

THE AMERICAN FALLS BY MOONLIGHT.

Niagara is a word of Indian origin—the orthography, accentuation, and meaning of which are variously given by different authors. It is highly probable that this diversity might be accounted for and explained by tracing the appellation through the dialects of the several tribes of aborigines who formerly inhabited the neighboring country. There is reason to believe, however, that the etymon belongs to the language of the Iroquois, and signifies the "*Thunder of Waters.*"

"When the traveller first arrives at the cataract he stands and gazes, and is lost in admiration. The mighty volume of water which forms the outlet of the great Lakes Superior, Michigan, Huron, and Erie, is here precipitated over a precipice 160 feet high, with a roar like that of thunder, which may be heard, in favorable circumstances, to the distance of fifteen miles, though, at times, the Falls may be nearly approached without perceiving much to indicate a tremendous cataract in the vicinity. In consequence of a bend in the river, the principal weight of water is thrown on the Canadian side, down

what is called the *Horse-Shoe Fall*, which name has become inappropriate, as the edges of the precipice have ceased to be a curve, and form a moderately acute angle. Near the middle of the fall, *Goat Island*, containing 75 acres, extends to the brow of the precipice, dividing the river into two parts; and a small projecting mass of rock at a little distance from it, toward the American shore, again divides the cataract on that side. Goat Island, at the lower end, presents a perpendicular mass of rocks, extending from the bottom to the top of the precipice. A bridge has been constructed from the American shore to Bath Island, and another connects the latter with Goat Island, and a tower is erected on the brow of the Horse-Shoe Fall, approached from Goat Island by a short bridge, on which the spectator seems to stand over the edge of the mighty cataract, and which affords a fine view of this part of it. The distance at the fall from the American shore to Goat Island is 65 rods; across the front of Goat Island is 78 rods; around the Horse-Shoe Fall, on the Canadian side, 144 rods; directly across the Horse-Shoe, 74 rods. The height of the fall near the American shore is 163 feet; near Goat Island, on the same side, 158 feet; near Goat Island, on the Canada side, 154 feet. Table Rock, a shelving projection on the Canadian side, at the edge of the precipice, is 150 feet high. This place is generally thought to present the finest view of the Falls; though, if the spectator will visit the tower on the opposite side on Goat Island, at sunrise, when the whole cavity is enlightened by the sun, and the gorgeous bow trembles in the rising spray, he cannot elsewhere, the world over, enjoy such an incomparable scene. A covered stairway on the American side descends from the top to the bottom of the precipice.

"It has been computed that 100 million tons of water are discharged over the precipice every hour. The Rapids commence about a mile above the Falls, and the water descends 57 feet before it arrives at the cataract. The view from the bridge to Goat Island, of the troubled water dashing tumultuously over the rocks of the American fall, is terrific. While curiosity constitutes an attribute of the human character, these falls will be frequented by admiring and delighted visitors as one of the grandest exhibitions in nature.

"This stupendous Cataract, situated in north latitude 43° 6', and west longitude 2° 6' from Washington, is 22 miles north from the efflux of the river at Lake Erie, and 14 miles south of its outlet into Lake Ontario. The whole length of the river is therefore 36 miles, its general course is a few points to the west of north. Though commonly called a river, this portion of the St. Lawrence is, more properly speaking, a *strait*, connecting, as above mentioned, the Lakes Erie and Ontario, and conducting the superfluous waters of the great seas and streams above, through a broad and divided, and afterward compressed, devious, and irregular channel to the latter lake, into which it empties—the point of union being about 40 miles from the western extremity of Lake Ontario.

"The climate of the Niagara is in the highest degree healthful and invigorating. The atmosphere, constantly acted upon by the rushing water, the noise, and the spray, is kept pure, refreshing, and salutary. There are no stagnant pools or marshes near to send abroad their fetid exhalations and noxious miasmas, poisoning the air and producing disease.

"Sweet-breathing herbs and beautiful wild flowers spring up spontaneously even on the sides, and in the crevices of the giant rocks; and luxuriant clusters of firs and other stately forest trees cover the islands, crown the cliffs, and overhang the banks of Niagara. Here are no mosquitoes to annoy, no reptiles to alarm, and no wild animals to intimidate, yet there is life and vivacity. The many-hued butterfly sips ambrosia from the fresh opened honey-cup; birds carol their lays of love among the spray-starred branches; and the lively squirrel skips chattering from tree to tree. Varieties of water-fowl, at certain seasons of the year, sport among the rapids, the sea-gull plays around the precipice, and the eagle—the banner bird of freedom—hovers above the cataract, plumes his gray pinions in its curling mists, and makes his home among the giant firs of its inaccessible islands.

"No place on the civilized earth offers such attractions and inducements to visitors as Niagara, and they can never be fully known except to those who see and study them, from the utter impossibility of describing such a scene as this wonderful cataract presents. When motion can be expressed by color, there will be some hope of imparting a faint idea of it; but until that can be done, Niagara must remain undescribed."

—————

Cataract of Niagara.

"Shrine of Omnipotence! how vast, how grand,
How awful, yet how beautiful thou art!
Pillar'd around thy everlasting hills,
Robed in the drapery of descending floods,
Crowned by the rainbow, canopied by clouds
That roll in incense up from thy dread base,
Hid by their mantling o'er the vast abyss
Upon whose verge thou standest, whence ascends
The mighty anthem of thy Maker's praise,
Hymn'd in eternal *thunders* !"

Below the Falls, the first objects of interest are the Ferry Stairs and Point View on the American side; while on the opposite side is a ferry-house and landing, where carriages are usually to be found to convey passengers to the Clifton House, Table Rock, and other places of great interest.

The new *Suspension Bridge* is erected near the Falls immediately below the Clifton House, on the Canada side. Here the American and Canadian, or Horse-Shoe Falls may be seen to advantage.

The SUSPENSION BRIDGE, the greatest artificial curiosity in America, is situated two miles and a half below the Falls, where has recently sprung into existence *Niagara City*, or better known as the *Suspension Bridge*, on the American side, and *Clifton* on the Canadian side of the river, here being about 800 feet in width, with perpendicular banks of 325 feet.

The *Whirlpool* and *Rapids*, one mile below the Bridge, are terrific sights of great interest, and well worthy a visit.

The *Devil's Hole*, one mile farther down, is also a point of great attraction, together with the *Bloody Run*, a small stream where a detachment of English soldiers were precipitated in their flight from an attack by Indians during the old French war in 1759. An amphitheatre of high ground spreads around and perfectly encloses the valley of the Devil's Hole, with the exception of a narrow ravine formed by Bloody Run—from which, against a large force, there is no escape, except over the precipice. The *Ice Cave* is another object of interest connected with the Devil's Hole.

The *Rapids* below the Whirlpool are the next object of attraction; then Queenston. Heights and Brock's Monument on the Canadian side, and the *Suspension Bridge* at Lewiston; altogether forming objects of interest sufficient to fill a well-sized volume.

The Niagara River is navigable from Lewiston to its mouth at Fort Niagara, a farther distance of seven miles, or fourteen below the Falls of Niagara.

The village of NIAGARA FALLS, Niagara Co., N. Y., is situated on the east side of Niagara River, in the immediate vicinity of the grand Cataract, 22 miles from Buffalo and 303 miles from Albany by railroad route. No place in the Union exceeds this favored spot as a fashionable place of resort during the summer and fall months, when hundreds of visitors may be seen every day flocking to Goat Island, or points contiguous to the Rapids and Falls. The village contains several large hotels for the accommodation of visitors, the most noted of which are the Cataract House and the International Hotel; the Monteagle Hotel, situated two miles below the Falls, near the Suspension Bridge, and the Clifton House, on the Canada side, are all alike popular and well-kept hotels; there are five churches of different denominations; 15 stores, in many of which are kept for sale Indian curiosities and fancy work of different kinds. The water-power here afforded by the descending stream, east of Goat Island, is illimitable. A paper-mill, a flouring-mill, two saw-mills, a woollen factory, a furnace and machine shop, together with other manufacturing establishments, here use the water-power so bountifully supplied. The population is about 3,500.

The railroads centring at the Falls are the *Buffalo, Niagara Falls and Lewiston Railroad*, and the *New York Central Railroad*; also, the *New York and Erie Railway*, forming with other roads a direct route to New York, Philadelphia, Baltimore, and Washington.

An *Omnibus Line* and hacks run from the village of Niagara Falls to Niagara City, or Suspension Bridge, during the summer months, and thence to the Clifton House and Table Rock on Canada side.

NIAGARA CITY, situated two miles below the Falls, at the *Suspension Bridge*, is a new and flourishing place containing about 1,500 inhabitants. Here are situated two or three public houses.

SUSPENSION BRIDGE
AND THE
Cataract and Rapids of Niagara.

To give the reader some idea of the grandeur of this triumph of engineering skill—THE SUSPENSION BRIDGE— we copy the following article from a late Buffalo paper:

AN ENGINEER'S MONUMENT.

Spanning the chasm of the Niagara River, uniting the territories of two different Governments, and sustaining the uninterrupted railroad traffic of the Provinces of Canada with the United States, 250 feet above a flood of water which man has never been able to ferry, stands the monument of JOHN A. ROEBLING. The *Niagara Railway Suspension Bridge*, is the grandest and the most distinguishing achievement of Art in this world. It is the proudest, it is the most beautiful, and will prove to be the most enduring monument anywhere set up on this continent.

Regard this wonderful product of engineering skill. Its span is 822 feet. Yet an engine, tender and passenger car, loaded with men, and weighing altogether 47 tons, depress the long floor in the centre but 5½ inches. The Bridge, loaded with a loaded freight train, covering its whole length, and weighing 326 tons, is deflected in the middle only 10 inches. This extreme depression is perceptible only to practised eyes. The slighter changes of level require to be ascertained with instruments. Delicate as lace work, and seemingly light and airy, it hangs there high between heaven and the boiling flood below, more solid than the earthbeds of the adjacent railways. The concussions of fast moving trains are sensibly felt miles off through solid rocky soil. In cities lo· comotives shake entire blocks of stone dwellings. The waters of the Cayuga Lake tremble under the wheels of the express trains, a mile away from the bridge. But a freight train traversing JOHN A. ROEBLING'S Monument, at the speed of five miles an hour, communicates no jar to passengers walking upon the carriage way below. The land cables of the bridge do not tremble under it—the slight concussions of the superstructure do not go over the summits of the towers. This last fact in the stiffness of the great work is of much importance. It furnishes a guarantee of the durability of the masonry. Fast anchored with stone and grouted in solid rock cut down to the depth of twenty-five feet, the great cables are immovable by any mechanical force incidental to the use of the bridge, or the natural influences it will be subject to. The ultimate strength of these cables is 12,400 tons. The total weight of the material of the bridge, and of the traffic to which it will ordinarily be subjected is 2,262 tons, to sustain which the Engineer has provided in his beautiful and scientific structure, a strength of 12,400 tons. He demonstrates, too, that while the strength of the cables is nearly six times as great as their ordinary tension, THAT STRENGTH WILL NEVER BE IMPAIRED BY VIBRATION. This was the question raised by THE DEMOCRACY, a year ago, which excited such general, and in instances such angry discussion. ROEBLING treated our doubts with a cool reason and the stores of an extensive engineering experience, which gave us to believe that Art had at last attained to a method of suspending Iron Bridges for Railroad use, that should en tirely obviate the objections to them felt by most of the Iron-Masters of the United States. He has since that demonstrated it in a most wonderful structure.

There are in the bridge 624 "suspenders," each capable of sustaining 30 tons—and all of sustaining 18,720 tons. The weight they have ordinarily to support is only 1,000 tons. But the Engineer has skilfully distributed the weight of the burdens, by the means of "girders" and "trusses." These spread the 34 tons heft of a locomotive and tender over a length of 200 feet. How ample is this provision made for defective iron or sudden strains!

The Anchor Chains are composed of 9 links, each 7 feet long, save the last, which is 10 feet. The lowest link is made of 7 bars of iron, 7 inch by 1½. It is secured to a cast iron anchor plate 3½ inches thick, and 6 feet 6 inches square. The other links are equally strong. The iron used was all made from Pennsylvania charcoal, Ulster county, N. Y., and Salisbury Pig, and can be depended upon for a strength of 64,000 pounds to the square inch. The central portions of the anchor plates, through which the links pass is 12 inches thick. The excavations in the solid rock were not vertical. They inclined from the river. The rock upon which the work may rely on the New York side of the chasm is 100 feet long, 70 feet wide, and 20 feet deep. It weighs 160 pounds to the cubic foot, and presents a resistance of 14,000 tons, exclusive of the weight of the superincumbent masonry and embankment.

The TOWERS are each 15 feet square at the base, 60 feet high above the arch, and 8 feet square at the top. The limestone of which they are built will support a pressure of 500 tons on each square foot without crushing. While the greatest weight that can fall upon the tower will rarely exceed 600 tons, a pressure of 32,000 tons will be required to crush the top course. There are 4,000 tons' weight in each of the towers on the New York side.

The cables are 4 in number, 10 inches in diameter, and composed each of 3,640 small No. 9 wires. Sixty wires form one square inch of solid section, making the solid section of the entire cable 60.40 square inches, wrapping not included. These immense masses of wire are put together so that each individual wire performs its duty, and in a strain all work together. On this, Mr. ROEBLING, who is a moderate as well as a modest man, feels justified in speaking with the word PERFECT. Each of the large cables is composed of four smaller ones, called "strands." Each strand has 520 wires. One is placed in the centre. The rest are placed around that. These strands were manufactured nearly in the same position the cables now occupy. The preparatory labors, such as oiling, straightening, splicing, and reeling, were done in a long shed on the Canada side. Two strands were made at the same time, one for each of the two cables under process of construction. On the completion of one set, temporary wire bands were laid on, about nine inches apart, for the purpose of keeping the wires closely united, and securing their relative position. They were then lowered to occupy their permanent position in the cable, On completion of the seven pairs of strands, two platform carriages were mounted upon the cables, for laying on a continuous wrapping, by means of ROEBLING'S patent wrapping machines. During this process the whole mass of wire was again saturated with oil and paint, which, together with the wrapping, will protect them effectually against all oxidation. Five hundred tons of this wire is English. American manufacturers did not put in proposals. That used was remarkably uniform, and most carefully made.

The law deduced from large use of wire rope in Pennsylvania, is, that its durability depends upon its usage. It will last much longer under heavy strains moving slowly, than it will under light strains moving rapidly. This law was borne constantly in mind by the Engineer of the Niagara Railway Bridge. The cables and suspenders are, so to speak, at rest. They are so well protected, too, from rust, that they may be regarded as eternally durable.

Among the interesting characteristics of this splendid architecture, is its elasticity. The depression under a load commences at the end, of course, and goes regularly across. After the passage of a train, the equilibrium is perfectly restored. The elasticity of the cables is fully equal to this task, and WILL NEVER BE LOST.

The equilibrium of the Bridge is less affected in cold weather than in warm. If a change of temperature of 100 degrees should take place, the difference in the level of the floor would be 2 feet 3 inches.

So solid is this Bridge in its weight, its stiffness, and its staying, that not the slightest motion is communicated to it by the severest gales of wind that blow up through the narrow gorge which it spans.

Next to violent winds, suspension bridge builders dread the trotting of cattle across their structures. Mr. ROEBLING says that a heavy train running 20 miles an hour across his Bridge, would do less injury to it than would 20 steers passing on a trot. It is the severest test, next to that of troops marching in time, to which bridges, iron or wooden, suspension or tubular, can be subjected. Strict regulations are enforced for the passage of hogs, horses, and oxen, in small bodies, and always on a walk.

This great work cost only $500,000. The same structure in England (if it could possibly have been built there) would have cost $4,000,000. It is unquestionably the most admirable work of art on this continent, and will make an imperishable monument to the memory of its Engineer, JOHN A. ROEBLING.

We append a Table of Quantities for the convenience of our readers, and the more easy comprehension of the character of the structure:

Length of span from centre to centre of Towers	822 feet
Height of Tower above rock on American side	88 feet
Height of Tower above rock, Canada side	78 feet
Height of Tower above floor of Railway	60 feet
Number of Wire Cables	4
Diameter of each Cable	10 inches
Number of No. 9 wires in each Cable	3,569
Ultimate aggregate strength of Cables,	12,400 tons
Weight of Superstructure	750 tons
Weight of Superstructure and maximum loads	1,250 tons
Ultimate supporting strength	730 tons
Height of Track above water	250 feet
Base of Towers	16 feet square
Top of Towers	8 " "
Length of each Upper Cable	1,256¼ feet
" " Lower Cable	1,190 feet
Depth of Anchor Pits below surface of Rock	30 feet
Number of Suspenders	624
Ultimate strength of Suspenders	18,720 tons
Number of Overfloor Stays	64
Aggregate strength of Stays	1,920 tons
Number of River Stays	56
Aggregate strength of Stays	1,680 tons
Elevation of Railway Track above middle stage of River	245 feet
Total length of Wires	4,000 miles

The weights of the materials in the bridge are as follows:

	LBS.
Timber	919,130
Wrought Iron and Suspenders	113,120
Castings	44,332
Rails	66,740
Cables (between towers)	535,400
Total	1,678,722

The GREAT WESTERN RAILWAY OF CANADA, which unites with the *New York Central Railroad*, terminating on the American side of the river, here commences and extends westward through Hamilton, London, and Chatham to Windsor, opposite Detroit, Mich., forming one of the great through lines of travel from Boston and New York to Detroit, Chicago, and the Far West.

This road also furnishes a speedy route of travel to Toronto, Montreal, etc.

Fort Niagara.—Mouth Niagara River.

N. ORR-CO.

TO INVALIDS AND SEEKERS OF PLEASURE.

Mountains, Lakes and *Rivers* are the elements, in the physical world, that go to purify the atmosphere, fertilize the soil, and beautify the landscape for the abode of man. So intimately are they connected in the economy of Nature that they are indispensable one to the other. From the lakes, rivers and ocean—the latter the recipient of the former—arises the vapor and moisture that forms the clouds,—these returning their distilled contents, made pure by natures alchemy,—overshadow the mountain top and the plain, and descend in rain or snow, giving vigor and life to animated nature far and wide; these influences being modified by that portion of the earth's surface on which they descend. Hence, the *Tropics*, which have the greatest degree of heat and moisture—and where there are but two seasons, the *wet* and the *dry*—produce rank vegetation, reptiles, and animals of a ferocious character, while *man*, for the most part, is indolent and unused to labor,—an abundant nature supplying most all his wants for clothing and subsistence.

In the *Temperate climate,* where there are *four* seasons, about equally divided, there falls a less amount of rain, owing, no doubt, to the sun's rays not absorbing so much moisture as near the equator. This is the region for pure lakes, lovely streams, and rich valleys; where *man* loves to congregate, and where the highest state of civilized society is found to exist. This favored region, however, is greatly modified by soil and altitude, as well as by isothermal lines, moderately increasing in temperature on the southern limit: vegetation often assumes a sub-tropical appearance, and the human race are to a certain extent unfitted for labor during the warm season. In the middle of the temperate zone *man* attains his greatest perfection,—here the cereals, the grasses, and the fruit, yield an abundant supply for the animal kingdom, and a surplus to spare for less favored portions of the earth. On the northern limit of the temperate zone, while the cold increases and vegetation becomes more stinted, the waters are more pure and alive with fish of various kinds. Here a hardy race of men abound, and health generally prevails. To the north-west, along the chain of the Great Lakes of America, reaching to the base of the Rocky Mountains, is the great health-restoring region of North America.

During the Summer and early Fall months the temperature usually ranges from 60 to 80 degress Fahrenheit, giving strength and vitality to the human frame,—most perceptibly felt by those visiting this region from a more southern latitude. Here all those seeking health and pleasure should yearly resort, as most astonishing benefit has been found by those laboring under respiratory diseases as well as general debility. The Upper Peninsula of Michigan, Northern Wisconsin, and the whole of Minnesota, are favorably situated as regards health, being a most delightful resort, free from the heat of the Summer months.

The more northern, and less favored portion of the earth's surface, lying to the north of the temperate zone, is given up to cold, too intense for vegetation that goes to sustain the human race,— hence, to the north of the Upper Lakes, in Canada, settlements cease to exist, except where is found a Hudson Bay Company's Post, or a tribe of roving Indians. So abrupt is this civilization, that, for fifteen hundred miles from the head waters of the Saguenay river to the head of Lake Superior, along the North Shore, no villages are now found to exist. Here, however, sportsmen resort for hunting and fishing during the summer months, the air being pure and invigorating. To the west of Lake Superior, on the Upper Mississippi and the Red River of the North, the climate is modified by a *Siberian Summer*, affording three months of warm weather, which brings cereals and vegetables to great perfection, rendering this portion of North America capable of sustaining a dense population.

This region of country will, no doubt, soon become a great resort for invalids and sportsmen during the Summer months.

MEDICAL INFLUENCE OF CLIMATE.

"The Influence of Climate on human life is now so universally allowed, that it is quite unnecessary for us to say a word respecting its beneficial action on the animal economy; the benefit resulting from the change from a cold, humid atmosphere, to a warm, dry one, is also as well understood, and as marked in its effects, as a change of treatment from an ignorant to a scientific system is satisfactory and apparent. The influence exercised on the respiratory organs and the skin by a bland atmosphere is not only immediate but apparent—not merely confined to those organs, but, by the improved condition of the blood, resulting from such a change, *reciprocating the benefit acquired on the brain*, by the quicker and lively state of the imagination—on the nutritive system, by a fuller condition of the body from a perfect digestion; and on the nervous temperament, by the more regular and natural performance of all the functions of the body—the best indication at all times of sound physical health."—*Zell's Encyclopedia.*

"*Seek Nature's perfect cure.*"

"*Throw physic to the dogs.*"

ADVICE TO PLEASURE TRAVELLERS.

TOURISTS, in search of health or pleasure, who intend to visit the region of the Great Lakes of America, if starting from New York or any of the cities of the eastern or middle States, are advised to take the most direct route for NIAGARA FALLS, where may be seen the magnitude of the accumulated waters of the "Inland Seas," as exhibited by viewing the American and Canadian, or Horse-Shoe Fall of this mighty Cataract. The Suspension Bridges, Rapids, and Islands, with other objects combined, form attractions that will profitably employ several days sojourn at this fashionable resort.

Here are several well-kept Hotels, both on the American and Canadian sides of the river, from whence delightful drives are afforded in almost every direction, while bringing into view new objects of interest, either on ascending on descending the banks of this majestic stream—this whole section of country, above and below the Falls, being historic ground. The battle-fields of Chippewa, Lundy's Lane, Queenstown, and old Fort George, opposite Fort Niagara, on the American side, all deserve a visit. (For further description, see page 21.)

On leaving Niagara Falls the tourist can proceed westward, *via* the *Great Western Railway of Canada*, to Detroit, 230 miles, passing through an interesting section of Canada; or, proceed to Buffalo, by rail, 22 miles.

GRAND PLEASURE EXCURSION, THROUGH LAKES HURON AND SUPERIOR.

Steamers of a large class leave *Buffalo*, during the season of navigation, every alternate day for *Erie, Cleveland* and *Detroit*, proceeding on their way to the *Saut Ste. Marie* and *Duluth*, Lake Superior, a distance of about 1,200 miles.

Passengers taking the Round Trip can stop, to suit their convenience, at any of the Lake ports, before arriving at Detroit. The City of Erie, 90 miles from Buffalo, is a place of growing importance, where terminates the *Philadelphia and Erie Railroad*, forming a direct and speedy communication with the cities of New York, Philadelphia and Baltimore. This is a favorite line of travel, crossing the Alleghany range and connecting with the Great Lakes. The City of Cleveland, 95 miles further, is fast becoming a great mart of trade, and a stopping-place for pleasure travellers. The railroad lines, in connection with its shipping facilities, afford this port great commercial advantages—no city on the Lakes exceeding it in natural advantages as regards a healthy climate, lovely situation, beautiful avenues, and delightful drives. Steamers run from this place to Put-in-Bay, Kelley's Island, Sandusky and Toledo, as well as direct to Detroit, Mich., each affording pleasant summer excursions.

On leaving Detroit, if bound for Lake Superior, commences the Grand Excursion—passing through Lake St. Clair and St. Clair river, forming the boundary between the United States and Canada. The steamer usually stops at Sarnia, Can., or Port Huron, Mich. to land and receive passengers. Immediately, after leaving the latter port, Point Edward and Fort Gratiot are passed, and the steamer enters the broad waters of Lake Huron. Here is experienced during warm weather the most delightful change imaginable. The upward bound vessels usually keep near the Michigan shore, on the left, while on the right nothing but the broad waters are visible for some two hundred miles.

Point au Barque and Light are reached 70 miles above Port Huron, and Saginaw Bay entered, here presenting a most magnificent expanse of waters, which, in stormy weather is dreaded by the mariner. The sight of land is usually lost to view until Thunder Bay and Light are sighted, 75 miles from Point au Barque. The steamer now runs direct for the De Tour passage, 85 miles further, when the grand and lovely St. Mary's river is entered, presenting a succession of islands, lakes or expansions—affording a view of river scenery of the most enchanting character, before arriving at the Saut Ste. Marie, the gate-way to Lake Superior.

SAUT STE. MARIE, in connection with the Ship Canal, Fort Brady, and the fisheries below the rapids—the rapids themselves having a descent of 20 feet—and the Hudson Bay Company's post on the Canada side, present great and varied attractions. Here

fishing parties are fitted out for long excursions along the Canada or North Shore of Lake Superior,—often proceeding as far as the Nepigon river, where brook trout, of a large size, are taken in great quantities.

On the American or South Shore of the Lake lies Grand Island Harbor, where are two or three settlements, situated near the Pictured Rocks, 120 miles above the Saut. Here are many points of attraction, which, no doubt, is destined to become a fashionable resort.

MARQUETTE, 170 miles above the Saut, is one of the largest and most frequented resorts for invalids and seekers of pleasure that the Lake region affords. Comfortable hotels and boarding houses are here wanted in order to make this embryo city the "Newport" of the Upper Lakes.

HOUGHTON, COPPER HARBOR, EAGLE HARBOR, and EAGLE RIVER, situated on Keeweenaw Point are all places of great attraction.

ONTONAGON, BAYFIELD and LA POINTE, situated on one of the Twelve Apostle Islands are old and favorite resorts.

DULUTH and SUPERIOR CITY, situated at the head of the Lake, where enters the St. Louis river, have become places of great attraction, in a commercial point of view. Here parties are fitted out who desire to explore the North Shore either for fishing or seeking health and pleasure on the pure waters of this Inland Sea.

Tourists seeking health and pleasure may safely start on this excursion in June, and remain in the Lake Region or Upper Mississippi Valley until October.

APPROACHES TO LAKE SUPERIOR.

There are now *six Great Routes* of Travel open to Tourists to and from the Lake Superior country.

The *first* is by the LAKE SUPERIOR LINE STEAMERS. Starting from *Buffalo* and stopping at Erie and Cleveland, they pass through Lake Erie and enter the Detroit River, stopping at *Detroit* to land and receive passengers — cross Lake St. Clair, and ascend the St. Clair River to Port Huron, Mich., stopping at Fort Gratiot, where the *Grand Trunk Railway of Canada* crosses the river near Sarnia. The broad waters of Lake Huron are next crossed — passing Saginaw Bay — then St. Mary's River is entered at Point de Tour, passing upwards to the Saut Ste. Marie, and through the Ship Canal to Lake Superior; a distance of about 400 miles from Detroit.

The *second* is by the Canadian route, starting from Toronto and proceeding by *Northern Railway of Canada* to Collingwood, 94 miles; then crossing Georgian Bay and passing through the North Channel and St. Mary's River to Saut Ste. Marie, entering Lake Superior and running along the North Shore. This route affords some of the grandest lake and river scenery imaginable.

The *third* is by the Chicago and Milwaukee Line of Steamers, passing through Lake Michigan and the Straits of Mackinac for a distance of about 400 miles, when the far-famed Island of Mackinac is reached; from thence the steamers run to the mouth of the St. Mary's River, ascending this beautiful stream to Lake Superior; a total distance of 500 miles from Chicago.

The *fourth* is via the *Chicago and North-western Railroad*, running to Green Bay and Escanaba, Mich., and from thence by the *Peninsula Railroad* to Marquette, situated on the South Shore of Lake Superior; a total distance of 431 miles. This route is direct and speedy, passing through an interesting section of country for most of the distance.

The *fifth* is via ST. PAUL, passing over the *Lake Superior and Mississippi Railroad* to DULUTH, 155 miles. This route affords an easy access to the Lake Superior region from the South; passing up the noble Mississippi to the head of navigation,—uniting the *"Imperial Lakes with the Father of Waters."*

The *sixth* is via the *Northern Pacific Railroad*, now completed from Duluth to the Upper Missouri River, a distance of 450 miles. This important railroad, when finished, will extend to Puget Sound, Washington Territory, with a branch running to Portland, Oregon. It now affords a direct line to travel to FORT GARRY, Manitoba. The favorable features of this extended route Across the Continent, in a commercial and climatic point of view, cannot be over-estimated, which will afford a speedy and desirable route *"Around the World."*

Two other Lines of Railroad will soon be completed, affording additional means of reaching Lake Superior, viz.: the *Grand Rapids* and *Indiana Railroad*, extending from Fort Wayne, Ind., to Old Mackinac, Mich., and the *Wisconsin Central Railroad*, running from Menasha and Portage City, Wisconsin, to Ashland, on the South Shore of Lake Superior.

NOTE. — The numerous Lines of Railroad, on the East and South, which connect with the above Through Lines of Travel to Lake Superior, make this whole region of country easily accessible to the pleasure traveller or man of business.

HINTS TO PLEASURE TRAVELLERS.

1. Purchase through tickets previously to entering the cars.

2. Attend to checking your baggage in person before taking your seat in the car.

3. Select a seat on the shady side of the car.

4. When you leave your seat, place a parcel, coat, or something belonging to you on it, which is an evidence of the seat being engaged.

5. Have the exact change to pay your fare on the cars, or you are subjected to be ejected from the cars—it has been decided by law that a conductor is not obliged to make change for a passenger.

6. Railroad CHECKS are good only for the train for which they are used; passengers cannot lay over for another train without making arrangements with the conductor.

7. LADIES without escort in travelling should be very particular with whom they become acquainted.

" If your lips would save from slips,
   Five things observe with care:
Of whom you speak—to whom you speak,
   And how—and when—and where."

8. If you see a lady unaccompanied, do not obtrude yourself upon her notice.

9. If she needs your services, tender them as though they were due to her, without unnecessary forwardness or undue empressment.

10. Such services do not entitle you to after recognition, unless by permission of the lady.

11. Ladies travelling with children should invariably have a basket of eatables, a tumbler or a goblet, for the children to drink from, and keep the children in their seats.

12. Keep your head and arms inside the car windows.

13. Never talk on politics in the cars —it is usually disagreeable to some of your fellow-travellers.

14. Never talk loudly while the train is in motion; it may not annoy any one, but it will injure your lungs.

15. A gentleman should not occupy more than one seat at a time.

16. Gentlemen should not spit tobacco juice in the cars where there are ladies; it soils their skirts and dresses.

17. Always show your ticket (without getting into a bad humor,) whenever the conductor asks for it. Observe this rule and it will pay.

18. Never smoke in a car where there are ladies. No gentlemen would be guilty of such an act.

19. Never use profane language in a railroad car.

20. If you cannot sleep yourself, don't prevent others from doing so, by whistling or loud talking.

21. Make a bargain with the hackman before getting into his carriage.

22. Look out for pickpockets.

23. Remember, that unless you pay for two seats you are entitled to but one, and every gentlemen and lady too, will respect the rights of others, and be mindful especially of the weak, the aged, and the infirm.

24. Provide yourself with sleeping berths before starting—you may then have a choice—the double lower birth is preferable.

25. Always be at the railroad station in good time to take the train. Better be an hour too early than a minute too late.

NOTE.—Many of the above rules are as applicable to Steamboat travelling as when travelling on Railroads. Often much comfort can be obtained by writing or sending a telegraph in order to secure state rooms, &c.

TABLE OF DISTANCES,

FROM BUFFALO, ERIE, CLEVELAND AND DETROIT, TO THE SAUT STE.

MARIE AND DULUTH, MINN.

PASSING THROUGH LAKES ERIE, HURON AND SUPERIOR.

MILES.	PORTS, &C.	MILES.	MILES.	PORTS, &C.	MILES.
660	**BUFFALO,** N. Y...	0	14	CHURCH'S LANDING....36	646
629	Silver Creek...............	31	0	**Saut Ste. Marie.**14	660
620	DUNKIRK....................9	40			
592	State Line...................28	68		*(Ship Canal and Rapids.)*	
572	**Erie,** Pa...................20	88			
557	Girard.......................15	103	570	**Saut Ste. Marie.**	660
544	Conneaut, Ohio...........13	116	564	POINT AUX PINS, Can. 6	666
531	ASHTABULA...............13	129	555	Pt. Iroquois & Light ⎱ 9	675
505	Painesville 26	155		—Gros Cap, Can. ⎰	
475	**Cleveland**...............30	185	530	White Fish Point......25	700
415	Point Pelée Is. & Light...60	245			
375	MALDEN, Can..............40	285		*(Lake Superior.)*	
373	Grosse Isle, Mich......... 2	287			
365	Wyandotte, " 8	295	470	Grand Sauble............60	760
356	WINDSOR, Can............. 9	304	452	Cascade Falls............18	778
355	**Detroit,** Mich......... 1	305	446	Pictured R'ks—Cha-⎱ 6	784
348	Lake St. Clair.............. 7	312		pel, Arch. Rock, &c ⎰	
315	Algonac, Mich.............33	345	436	Grand Is. and Harbor.10	794
309	NEWPORT, " 6	351	400	**Marquette,** Mich..36	830
299	ST. CLAIR, " 10	361	350	Huron Is. and Light...50	880
282	**Port Huron,** ⎱ 17	378	320	Portage Entry...........30	910
	SARNIA, Can. ⎰				
280	FORT GRATIOT............ 2	380		(HOUGHTON & HANCOCK, 14 Miles.)	
			270	Keweenaw Point.......50	960
	(Lake Huron.)		255	COPPER HARBOR.......15	975
			239	EAGLE HARBOR.......16	991
269	Lakeport, Mich...........11	391	229	EAGLE RIVER............19	1,001
258	Lexington, " 11	402	209	Entrance Ship Canal*.20	1,021
246	Port Sanilac " 12	414	169	ONTONAGON, Mich.....40	1,061
232	Forrestville " 14	428	89	LA POINTE, Wis.........80	1,141
216	Port Hope, Mich..........16	444	86	BAYFIELD " 3	1,144
210	Point au Barque—⎱ 6	450			
	off Saginaw Bay, ⎰			*(Twelve Apostle Islands.)*	
135	Thunder Bay Island.....75	525			
50	POINT DE TOUR—⎱ ...85	610	6	SUPERIOR CITY, Wis..80	1,224
	St. Mary's River, ⎰		0	**Duluth,** Minn....... 6	1,230

* A distance of 85 miles is saved by passing through the Ship Canal.

TO SEEKERS OF HEALTH AND PLEASURE.

Grand Pleasure Excursion for the Season of 1874

—FROM—

BUFFALO, ERIE, CLEVELAND AND DETROIT,

TO DULUTH AND ST. PAUL,

PASSING THROUGH

LAKES HURON AND SUPERIOR.

To CONTINUE DURING THE SUMMER MONTHS.

A Daily Line of STEAMERS will run from Buffalo, Erie, &c., to Saut Ste. Marie, Marquette and Duluth,—Connecting with Cars on the Lake Superior & Mississippi Railroad, running to St. Paul, Minn.

FROM St. Paul Steamers run Daily on the Mississippi River, during the season of Navigation, to La Crosse, Prairie du Chien, Dubuque and St. Louis,—Connecting with the Lines of Railroad running to Milwaukee, Chicago and Detroit,—thus furnishing a ROUND TRIP of over *two thousand miles*, by land and water, through one of the most healthy and interesting regions on the Continent.

DULUTH TO BISMARCK, DAKOTA,

VIA

NORTHERN PACIFIC RAILROAD

This new and HEALTH-RESTORING LINE OF TRAVEL, by means of steamers on the UPPER LAKES OF AMERICA affords an extended EXCURSION of 1,650 miles from BUFFALO TO BISMARCK, Dakota—connecting with Steamers on the Red River of the North, and on the Upper Missouri, extending for 1,200 miles. further to FORT BENTON, Montana—forming altogether the

GRANDEST EXCURSION IN THE WORLD.

TRIP THROUGH THE LAKES,

GIVING A DESCRIPTION OF CITIES, TOWNS, &c.

Buffalo, "QUEEN CITY OF THE LAKES," possessing commanding advantages, being 22 miles above Niagara Falls, is distant from Albany 298 miles by Railroad, about 350 miles by the line of the Erie Canal, and 443 miles N. W. of the City of New York; in N. lat. 42° 53′, W. long. 78° 55′ from Greenwich. It is favorably situated for commerce at the head of Niagara River, the outlet of Lake Erie, and at the foot of the great chain of Upper Lakes, and is the point where the vast trade of these Inland Seas is concentrated. The harbor, formed of Buffalo Creek, lies nearly east and west across the southern part of the city, and is separated from the waters of Lake Erie by a peninsula between the creek and lake. This harbor is a very secure one, and is of such capacity, that although steamboats, propellers, barques, schooners, and other lake craft, and canal-boats, to the number, in all, of from three to four hundred, have sometimes been assembled there for the

transaction of the business of the lakes, yet not one-half part of the water accommodations has ever yet been occupied by the vast business of the great and growing West. The harbor of Buffalo is the most capacious, and really the easiest and safest of access on our inland waters. Improvements are annually made by dredging, by the construction of new piers, wharves, warehouses, and elevators, which extend its facilities, and render the discharge and transshipment of cargoes more rapid and convenient; and in this latter respect it is without an equal.

Buffalo was first settled by the whites in 1801. In 1832 it was chartered as a city, being now governed by a mayor, recorder, and board of aldermen. Its Population in 1830, according to the United States Census, was 8,668; in 1840, 18,213; and in 1850, 42,261. Since the latter period the limits of the city have been enlarged by taking in the town of Black Rock; it is now divided into thirteen wards, and according to the Census of 1860, contained 81,130 inhabitants, in 1870, 117,715, being now the third city in point of size in the State. The public buildings are numerous, and many of them fine specimens of architecture while the private buildings, particularly those for business purposes, are of the most durable construction and modern style. The manufacturing establishments, including several extensive ship-yards for the building and repairing of lake craft, are also numerous, and conducted on a large scale, producing manufactured articles for the American and Canadian markets.

The principal public buildings are an United States Custom-House and Post-Office, State Insane Asylum, City Hall, Court House, and Jail; two Theatres, and sixty Churches of different denominations. Here are also eight Banking Houses, six Savings Banks, several Fire and Marine Insurance Companies, and thirty large Elevators, with a capacity of storing upwards of 8,000,000 bushels of grain. The *Buffalo Park* is known as one of the most famous parks in the country.

The Lines of Steamers and Railroads diverging from Buffalo tend to make it one of the greatest thoroughfares in the Union. Steamers and Propellers run to Cleveland, Sandusky, Toledo, Detroit, Mackinac, Green Bay, Milwaukee, and Chicago; also, to Sault Ste. Marie and ports on Lake Superior.

During the Season of 1874 two Lines of Steamers, of a large class, will run from Buffalo, stopping at Erie, Cleveland and Detroit, to Duluth.

RAILROADS RUNNING FROM BUFFALO.

1. *New York Central*, to Albany and Troy, 298 miles.
2. *Buffalo, Niagara Falls and Lewiston*, 28 miles.
3. *Erie Railway*, running to the City of New York, 443 miles.
4. *Lake Shore Railroad*, to Cleveland, Ohio, 183 miles, and thence to Chicago.
5. *Buffalo, Corry and Pittsburgh*, running to Corry, Penn., 92 miles.
6. *Buffalo and Jamestown Railroad.*
7. *Buffalo, New York* and *Philadel-*

phia, running to Emporium, Penn., 121 miles.

8. *Great Western* (Canada) *Railway* (Suspension Bridge to Detroit, Mich.) 230 miles; also, from International Bridge.
9. *Grand Trunk Railway*, running from International Bridge to Goderich, Can., 160 miles.
10. *Canada Southern Railway*, running from Buffalo, via International Bridge, to Amherstburg, Can., Detroit, Toledo, &c.

There are also four Lines of City Railroads running to different points within the limits of Buffalo.

The principal Hotels are the *Tifft House*, and *Mansion House*, on Main Street. A new Public House with all the modern improvements is much needed in Buffalo.

The *Erie Canal* terminating at Buffalo, and the numerous Lines of Propellers running on Lake Erie and the Upper Lakes, in connection with the Lines of Railroad, transport an immense amount of agricultural products and merchandise to and from the Eastern markets.

The completion of the *International Bridge* at Black Rock affords great facilities for travel, passing through Buffalo and Canada to the Far West.

"The climate of Buffalo is, without doubt, of a more even temperature than any other city in the same parallel of latitude from the Mississippi to the Atlantic Coast. Observations have shown that the thermometer never ranges as low in winter, nor as high in summer, as at points in Massachusetts, the eastern and central portions of this State, the northern and southern shores of Lake Erie in Michigan, northern Illinois, and Wisconsin.

"Buffalo, with its broad, well-paved streets and shade-trees; its comparatively mild winters; its cool summers; its pleasant drives and picturesque suburbs, and its proximity to the '*Falls*,' combine to render it one of the most desirable residences on the continent."

STEAMBOAT ROUTE FROM BUFFALO TO CLEVELAND, TOLEDO, DETROIT, &c.

STEAMERS and Propellers of a large class leave Buffalo daily, during the season of navigation, for the different ports on the American or south shore of Lake Erie, connecting with Railroad cars at Erie, Cleveland, Sandusky, Toledo, and Detroit, for the East, South and West.

On leaving Buffalo harbor, which is formed by the mouth of Buffalo Creek —where is erected a breakwater by the United States government—a fine view is afforded of the City of Buffalo, the Canada shore, and Lake Erie stretching off in the distance, with here and there a steamer or sail vessel in sight. As the steamer proceeds westward, through the middle of the lake, the landscape fades in the distance, until nothing is visible but a broad expanse of green waters.

STURGEON POINT, 20 miles from Buffalo, is passed on the south shore, when the lake immediately widens, by the land receding on both shores. During the prevalence of storms, when the full blast of the wind sweeps through this lake, its force is now felt in its full power, driving the angry waves forward with the velocity of the race-horse, often causing the waters to rise at the lower end of the lake to a great height, so as to overflow its banks and forcing its surplus waters into the Niagara river, which causes the only perceptible rise and increase of the rush of waters at the Falls.

Dunkirk, N. Y., 40 miles from Buffalo, is advantageously situated on the shore of Lake Erie, where terminates the *New York and Erie Railway*, 460 miles in length. Here is a good and secure harbor, affording about twelve feet of water over the bar. A light-house, a beacon-light, and breakwater, the latter in a dilapidated state, have here been erected by the United States government. As an anchorage and port of refuge this harbor is extremely valuable, and is much resorted to for that purpose by steamers and sail vessels during the prevalence of storms.

The village was incorporated in 1837, and now contains about seven thousand inhabitants, eight hundred dwelling-houses, five churches, a bank, three hotels, and thirty stores of different kinds, besides several extensive storehouses and manufacturing establishments.

FREDONIA, three miles from Dunkirk, with which it is connected by a plank-road, is handsomely situated, being elevated about 100 feet above Lake Erie. It contains about twenty-five hundred inhabitants, three hundred dwelling-houses, five churches, one bank, an incorporated academy, four taverns, twenty stores, besides some mills and manufacturing establishments situated on Canadoway Creek, which here affords good water-power. In the village, near the bed of the Creek, is an inflammable spring, from which escapes a sufficient quantity of gas to light the village.

WESTFIELD, N. Y., 17 miles from Dunkirk, is a handsome village on the line of the Railroad.

RIPLEY, 65 miles, and STATE LINE, 68 miles, are small settlements that are passed before entering the State of Pennsylvania.

NORTH EAST, Erie Co., Pa., is a small village, 73 miles from Buffalo and 15 miles from the City of Erie.

HARBOR CREEK, 7 miles further, is another small village situated near Lake Erie.

Erie, "THE LAKE CITY OF PENNSYLVANIA," distant 451 miles from Philadelphia by railroad, 90 miles from Buffalo, and 95 miles from Cleveland, is beautifully situated on a bluff, affording a prospect of Presque Isle Bay and the Lake beyond. It has one of the largest and best harbors on Lake Erie, from whence sailed Commodore Perry's fleet during the war of 1812. The most of the vessels were here built, being finished in seventy days from the time the trees were felled; and here the gallant victor returned with his prizes after the naval battle of Lake Erie, which took place off Put-in-Bay, September 10, 1813. The remains of his flag-ship, the *Lawrence*, lie in the harbor, from which visitors are allowed to cut pieces as relics. On the high bank, a little distance from the town, are the ruins of the old French fort, Presque Isle.

The City contains many fine public buildings and private residences; a court house and city hall, 20 churches, 4 banks, 3 savings banks, 2 large hotels, the *Reed House* and *Ellsworth House*, besides several other public houses; a ship yard, 3 grain elevators,

Erie car works, a blast furnace, 5 iron works, gas works and water works, besides several extensive manufacturing establishments, and about 28,000 inhabitants. It is the greatest market for bituminous coal on the Lakes, the coal being supplied from the rich Shenango Valley, the receipts exceeding half a million tons annually. Lake Superior iron ore is also shipped in large quantities, *via* Erie, to the numerous furnaces in the Shenango Valley, and east of the Alleghany Mountains; anthracite coal, and the lumber trade, *via* the Pennsylvania and Erie Railroad, are fast increasing in importance. In addition to the *Lake Shore Railroad*, the *Philadelphia and Erie Railroad*, and the *Erie and Pittsburgh Railroad* terminates at this place, affording a direct communication with New York, Philadelphia, Baltimore, and other Eastern cities. One of the natural advantages which Erie possesses are the veins of *inflammable gas*, which are struck at the depth of from one to five hundred feet from the surface; 30 or 40 gas wells are now employed for propelling machinery, as well as affording heat and light for private residences.

"The City of Erie is not surpassed in healthiness by any of the Lake towns, and, except in winter, very few pleasanter places for a residence can be found. The purity of the atmosphere, kept in constant motion by the invigorating Lake breezes, is such that malaria, with its attendant evils, is unknown. The commercial interests of Erie are varied, and in the aggregate of considerable extent; but are in small proportion to what the advantages of geographical position as a railroad centre and a good lake-port would justify. There is no place on the Great Lakes better designed by nature for a commercial mart; and, in spite of opposition or lethargy, it *must* eventually become one of the largest manufacturing cities in the Union."

Erie is the key or outlet to the large iron and coal district of Western Pennsylvania, through which is distributed the Lake Superior iron ore, supplying numerous blast furnaces in Western Pennsylvania, as well as the anthracite furnaces of the Susquehanna region, and the Lake market for the coal of that valley, which is the best iron making bituminous coal, as well as gas coal, in the United States. It is here the rich ores of the Lake Superior and St. Lawrence region meet the best iron smelting anthracite and bituminous coals known to exist in America.

PRESQUE ISLE BAY is a lovely sheet of water, protected by an island projecting into Lake Erie. There is a light-house on the west side of the entrance to the Bay, in lat., 42° 8′ N; it shows a fixed light, elevated 128 feet above the surface of the Lake, and visible for a distance of 19 miles. The beacon shows a fixed light, elevated 38 feet, and is visible for 13 miles.

CONNEAUT, Ohio, 117 miles from Buffalo and 68 from Cleveland, situated in the northeast corner of the State, stands on a creek of the same name, near its entrance into Lake Erie. It exports large quantities of lumber, grain, pork, beef, butter, cheeese, etc., being surrounded by a rich agricultural section of country. The village contains about 1,500 inhabitants. The harbor of Conneaut lies 2 miles from the village, where is a light-house, a pier, and several warehouses.

ASHTABULA, Ohio, 14 miles farther west, stands on a stream of the same name, near its entrance into the Lake.

This is a thriving place, inhabited by an intelligent population estimated at 3,000. The harbor of Ashtabula is 2½ miles from the village, at the mouth of the river, where is a light-house.

FAIRPORT stands on the east side of Grand river, 155 miles from Buffalo. It has a good harbor for lake vessels, and is a port of considerable trade. This harbor is so well defended from winds, and easy of access, that vessels run in when they cannot easily make other ports. Here is a light-house and a beacon to guide the mariner.

PAINESVILLE, Ohio, 3 miles from Fairport and 30 miles from Cleveland, is a beautiful and flourishing town, being surrounded by a fine section of country. It is the county-seat for Lake County, and contains a court-house, 5 churches, a bank, 20 stores, a number of beautiful residences, and about 3,700 inhabitants.

Cleveland, "THE FOREST CITY," Cuyahoga County, Ohio, is situated on a plain, elevated 80 feet above the waters of Lake Erie, at the mouth of the Cuyahoga river, which forms a secure harbor for vessels of a large class; being in N. lat. 41° 30′, W. long. 81° 42′. The bluff on which it is built rises abruptly from the Lake level, where stands a light-house, near the entrance into the harbor, from which an extensive and magnificent view is obtained, overlooking the City, the meandering of the Cuyahoga, the line of railroads, the shipping in the harbor, and the vessels passing on the Lake.

The City is regularly and beautifully laid out, ornamented with numerous shade-trees, from which it takes the name of "Forest City." Near its centre is a large public square, in

the British fleet on Lake Erie, September 10, 1813. Cleveland is the mart of one of the greatest grain-growing States in the Union, and has a ready communication by railroad with New York, Baltimore and Philadelphia on the east, while continuous lines of railroads run south and west; Propellers run to Ogdensburg and the Upper Lakes. It is distant 185 miles from Buffalo, 135 miles from Columbus, 107 miles from Toledo, and 150 miles from Pittsburgh, by railroad route; 120 miles from Detroit, by steamboat route.

It contains a county court-house and jail; city hall, U. S. custom-house and post office building; 1 theatre; a library association with a public reading room; 2 medical colleges, 2 orphan asylums, 60 churches of different denominations; 7 banks, 2 savings banks, and 6 insurance companies; also, numerous large manufacturing companies, embracing iron and copper works, ship-building, &c.; gas works, water works, and six city railroad companies. The stores and warehouses are numerous, and many of them well built. It now boasts of 100,000 inhabitants, and is rapidly increasing in numbers and wealth. The Lake Superior trade is a source of great advantage and profit, while the other Lake traffic together with the facilities afforded by railroads and canals, makes Cleveland one of the most favored cities on the Inland Seas of America.

The principal hotels are the *Kennard House, Weddell House* and *Forest City House.*

RAILROADS DIVERGING FROM CLEVELAND.

1. *Lake Shore,* to Buffalo, 183 miles. Toledo Division, 113 miles.

2. *Cleveland, Columbus, Cincinnati and Indianapolis,* to Cincinnati, 258

Perry Monument, Erected Sept. 10, 1860.

which stands a beautiful marble statue of Commodore OLIVER H. PERRY, which was inaugurated September 10, 1860, in the presence of more than 100,000 people. It commemorates the glorious achievement of the capture of

miles. Indiana Division, to Indianapolis, 282 miles.

3. *Cleveland and Pittsburgh,* connecting with Wheeling, W. Va., 150 miles.

4. *Cleveland, Mount Vernon and Columbus,* running to Columbus, via Hudson, 145 miles.

5. *Cleveland* and Mahoning Railroad.

6. *Atlantic and Great Western,* connecting with Erie Railway.

STEAMERS and PROPELLERS of a large class leave Cleveland daily, during the season of navigation, for Buffalo, Toledo, Detroit, Mackinac, Green Bay, Milwaukee, Chicago, the Saut Ste. Marie, Marquette and Duluth, stopping at ports on the Upper Lakes; altogether, transporting an immense amount of merchandise, grain, lumber, coal, iron and copper ore.

BLACK RIVER, 28 miles from Cleveland, is a small village with a good harbor, where is a ship-yard and other manufacturing establishments.

VERMILION, 10 miles farther, on the line of the Cleveland and Toledo Railroad, is a place of considerable trade, situated at the mouth of the river of the same name.

HURON, Ohio, 50 miles from Cleveland and 10 miles from Sandusky, is situated at the mouth of Huron river, which affords a good harbor. It contains several churches, 15 or 20 stores, several warehouses, and about 1,500 inhabitants.

Sandusky, "THE BAY CITY," capital of Erie Co., Ohio, is a port of entry and a place of considerable trade. It is advantageously situated on Sandusky Bay, three miles from Lake Erie, in N. lat. 41° 27′, W. long. 82° 45′. The Bay is about twenty miles long, and five or six miles in width, forming a capacious and excellent harbor, into which steamers and vessels of all sizes can enter with safety. The average depth of water is from ten to twelve feet. The City is built on a bed of limestone, producing a good building material. It contains about 15,000 inhabitants, a court-house and jail, 8 churches, 2 banks, several well-kept hotels, and a number of large stores and manufacturing establishments of different kinds. This is the terminus of the *Sandusky, Dayton and Cincinnati Railroad,* 153 miles to Dayton; the *Sandusky, Mansfield and Newark Railroad,* 116 miles in length—now known as the Lake Erie Division of the Baltimore and Ohio Railroad—forms, in part, a direct line of travel to Baltimore and Washington, D. C.

KELLEY'S ISLAND AND BASS ISLANDS.—LAKE ERIE.

THESE important Islands, forming a group in connection with some smaller islands, lie on the southwest end of Lake Erie, in N. lat. 41° 30′. There are other islands attached to Canada, situated immediately to the North, the most important of which is *Point Pelée Island.* These islands in a measure partake of the same favorable climatic influence that pervades the American islands, they being celebrated for the luxurious growth of grapes and other kinds of fruit, as well as for health-restoring influences.

The largest of the American Islands are *Kelley's Island, South Bass or Put-in-Bay Island, Middle Bass Island,* and *North Bass Island.* The smaller islands or islets are *Ballast Island, Gibraltar Island, Sugar Island, Rattlesnake Island,*

Green Island and Light, and *West Sister Island* and Light,—the two latter belonging to the United States government.

The Canada islands are *Point Pelée Island, Middle Island, East Sister, Middle Sister,* and the *Old Hen and Chickens.* All the above islands are celebrated as possessing fine fishing grounds; the Bass Islands taking their names from the great quantities of bass taken during the Spring and Fall months. Other fish, of a fine flavor, are also taken in the waters contiguous to the above islands.

KELLEY'S ISLAND, the largest of the American group, contains about 2,800 acres of land, being mostly under a high state of cultivation, about one-third of which is devoted to the culture of grapes. Here is a population of about 1,000. On the island there are 2 hotels and several private boarding-houses, affording good accommodations. The *Island House,* situated near the steamboat landing, is the principal hotel; it has long been a favorite resort for those seeking health and pleasure. On this island is located Kelley's Island Wine Company, occupying a large building over a cellar cut out of solid limestone, with a capacity of holding 100,000 gallons of wine. It is an object of great interest to visitors. Here are immense limestone quarries, from which are annually shipped an enormous amount of stone for building, paving, and smelting purposes.

SOUTH BASS ISLAND, the next in size, contains about 1,500 acres of superior land, being underlaid by limestone; there are some 500 acres where grapes are raised, mostly of the Catawba and Delaware species. Apples, plums, cherries, peaches, pears and other fruit here flourishes and are raised in considerable quantities.

PUT-IN-BAY, favorably situated on the northwest shore of this island, is a good harbor and safe refuge for vessels in all seasons during navigation. Here steamers arrive and depart daily, running to Detroit, Sandusky, Toledo, Cleveland, and other ports on Lake Erie. The great attraction, however, of this favored island is its delightful climate, beautiful harbor, and the ample accommodations afforded for those seeking health or pleasure. Here can be enjoyed bathing, boating and fishing, besides indoor amusements. The *Put-in-Bay House* and the *Beebee House* are two well-kept hotels, being usually thronged with visitors during the Summer months. No place on the Great Lakes exceeds this favored place as a fashionable resort.

Sailing or rowing in the Bay and around the Islands afford delightful amusement and healthy exercise, while a moonlight excursion over the waters of the Lake are a feature of enjoyment offered, which, once participated in, are never forgotten.

PUT-IN-BAY, now containing about 600 inhabitants, received its name and was made celebrated by being the rendezvous of Commodore Perry's flotilla before and after the decisive naval battle of Lake Erie, September 10, 1813, which resulted in the capture of the entire British fleet. Here were interred, under a willow tree, the officers of the American fleet, killed in the above battle, their remains occupy a small space, inclosed by posts and chains lying near the steamboat landing.

The PERRY CAVE, situated near the centre of the island, is annually visited by thousands of admirers. It has recently been put in order by enlarging the entrance, gravelling the bottom, and lighting the cave with numerous coal oil lamps, making it as bright as daylight. The scenery in the cave is most interesting and beautiful. A miniature lake with tiny boats floating on its bosom heightens its beauty. The depth of the cave, over the mouth of which a building is erected, is about 50 feet; length, 15 rods; roof elevated from 5 to 10 feet, spreading out into a large chamber with a small body of water at its base. When lighted it presents a grand appearance. The thermometer in the chamber stands at 48° Fahrenheit, with little or no variation throughout the year.

NAVAL BATTLE ON LAKE ERIE.

September 10th, 1813, the hostile fleets of England and the United States, on Lake Erie, met near the head of the Lake, and a sanguinary battle ensued. The fleet, bearing the "red cross" of England, consisted of six vessels, carrying 64 guns, under command of the veteran Commodore Barclay; and the fleet, bearing the "broad stripes and bright stars" of the United States, consisted of 9 vessels, carrying 54 guns, under command of the young and inexperienced, but brave, Commodore Oliver H. Perry. The result of this important conflict was made known to the world in the following laconic dispatch, written at 4 P. M. of that day:—

"DEAR GENERAL: *We have met the enemy, and they are ours: Two ships, two brigs, one schooner, and one sloop.*
With esteem, etc. O. H. PERRY.
"Gen. William H. Harrison."

GIBRALTAR ISLAND, lying in Put-in-Bay, is a small island mostly covered with forest trees, rising beautifully from the water's edge. It is owned and occupied by JAY COOKE, Esq., as a country residence.

MIDDLE BASS ISLAND is another lovely piece of ground, containing about 800 acres of land, which is mostly devoted to the culture of grapes and fruit of different kinds. Here is a steamboat landing and a number of fine residences.

NORTH BASS ISLAND, containing about 700 acres, is also equally favorably situated and productive; producing grapes and almost every variety of fruit in abundance.

These islands, the favorite abode of an intelligent and thriving population, are also rich in geological interest, affording to the naturalist a broad field for investigation. Put-in-Bay Island, itself, is noted for its many and curious subterranean caverns, some of which are very large, with smooth floors, high walls, and clear, cool miniature lakes—apparently designed by nature to refresh and please the weary visitor.

On the Peninsula lying north-west of Sandusky Bay, and on the mainland, further inland, in Ohio, grapes are now being successfully cultivated, and native wine of a fine flavor is produced—equalling in reputation the wines of southern Ohio or California. The grapes raised for wine purposes, which arrive to perfection in September, sell for about five dollars a hundred pounds, and those packed in boxes for the Eastern market for about twice the above price.

The BASS ISLANDS and KELLEY'S ISLAND are easily reached by steamer running from Detroit to Sandusky, Ohio; distance 80 miles; stopping at the above islands. A steamer also runs from Sandusky to Kelley's Island and the Bass Islands, twice daily, affording a most delightful excursion.

Put-in-Bay.—Our Nearest Watering Place.—*Copied from a Detroit paper.*

"This place of summer resort offers unusual attractions to the pleasure-seeker, the tourist or the excursionist. The natural beauties of its location, and the historic interest that attaches to it, have been so often described as to render any repetition thereof needless, if not tiresome. The pleasures of either a visit to or a sojourn at the Bay, have, however, been greatly enhanced by the decided improvements just made in both the extent of the accommodation it can afford, and the general convenience of the appointments pertaining thereto.

"The Put-in-Bay Island is easily accessible from Detroit, *via* the steamer Jay Cooke—a handsome, fast, well-officered, and in every respect first-class boat. Her time-table is so arranged as to place a trip to the Bay within the reach of almost any one. An absence of a day thus gives any one a magnificent ride down our river, over the Lake, and among the islands; leaves him with the afternoon and evening in which to enjoy the pleasures of watering-place life, and lands him in our city, after a night's rest, early on the following day.

"The *Put-in-Bay House* has been greatly enlarged since last summer, and is now under the proprietorship of Messrs. West & Sweeny. A long wing has been added, the dining room has been much enlarged, and all the various features of the establishment have been amplified and improved. "*Stacey's*" remains still in full blast, with its various means of amusement and refreshment, and the stables connected with this house can now boast of fine turn-outs and good animals. Among the employés of the hotel the material for an excellent band for dancing music has been found, and hops are the regular order of the night. The *Perry House* has been transformed into the "Beebe House," and has also been greatly enlarged and improved. Its location on the bank of the Bay is delightful, and it is kept on home-like principles. In addition to a decided increase in its accommodating capacity, a new building with bowling alleys, billiard room, bar, dancing hall, and dressing-rooms, etc., has also been erected adjoining it, and it is thus more on a par with its larger rival in its resources of amusement."

Toledo, the capital of Lucas Co., Ohio, one of the most favored cities of the Lakes, is situated on the Maumee River, 4 miles from its mouth, and 10 miles from Turtle Island Light, at the outlet of the Maumee Bay into Lake Erie. The harbor is good, and the navigable channel from Toledo of sufficient depth for all steamers or sail vessels navigating the lakes. Toledo is the eastern terminus of the *Wabash and Erie Canal,* running through Maumee and Wabash valleys, and communicating with the Ohio river at Evansville, Ind., a distance of 474 miles. The *Miami and Erie Canal* branches from the above canal 68 miles west of Toledo, and runs southerly, communicating with the Ohio river at Cincinnati.

The City contains a court house and

jail; a United States custom-house and post office building; 30 churches, many of them costly edifices; an extensive high-school edifice, and 7 large brick ward school-houses; a board of trade building; 6 banks, 4 savings banks; several fire, marine and life insurance companies; 6 hotels, 8 grain elevators, and a great number of wholesale stores and warehouses. The manufacturing interest is immense, embracing mills and machine shops of almost every variety. The shipping interest is rapidly increasing, here being transshipped, annually, a large amount of grain—exceeded only by Chicago—and and other kinds of agricultural products of the Great West brought here by canal and railroad. This City is destined, like Chicago, to export direct to European ports.

Toledo is the nearest point for the immense country traversed by the above canals and railroads terminating at this point, where a transfer can be made of freight to the more cheap transportation by the Lakes, and thence through the Erie Canal, Welland Canal, or Oswego Canal to the Seaboard.

Toledo now ranks as the third city in the State, and the third on the shores of Lake Erie, being only exceeded by Cleveland and Buffalo, and is rapidly increasing in wealth and numbers. The population in 1850, was 3,829; in 1860, 13,768; in 1865, 19,500; in 1870, 31,584. At this time there are in process of erection, in Toledo, many handsome dwellings and numerous handsome stores, which when completed can be classed among the most elegant structures. The principal hotels are the new Boody House, Island House, Oliver House and St. Charles Hotel.

RAILROADS AND BRANCHES DIVERGING FROM TOLEDO.

	MILES.
LAKE SHORE & MICHIGAN SOU'N..	
Michigan Sou'n Div. to Chicago.	244
Toledo Division to Cleveland.	113
Air Line to Elkhart, Ind	127
Detroit and Toledo Division	65
DAYTON AND MICHIGAN	202
TOLEDO, WABASH & WESTERN to	
Quincy, Ill.	476
Hannibal Branch	48
Keokuk Branch.	42
St. Louis Division.	104

RAILROAD ROUTE AROUND LAKE ERIE.

This important body of water being encompassed by a band of iron, we subjoin the following *Table of Distances:*

Buffalo to Paris, Can., via *Buffalo and Lake Huron Railroad,* 84 miles; Paris to Windsor or Detroit, via *Great Western Railway,* 158 miles; Detroit to Toledo, Ohio, via *Detroit and Toledo Railroad,* 65 miles; Toledo to Cleveland, via *Cleveland and Toledo Railroad,* 113 miles; Cleveland to Erie, Pa., via *Lake Shore Railroad,* 95 miles; Erie to Buffalo, N. Y., via *Lake Shore Railroad,* 88 miles. Total miles, 603.

Comparative Growth of Lake Cities.

CITIES.	1860	1870	INCREASE.
BUFFALO, N. Y.	81,129	117,714	36,585
CHICAGO, Ill.	109,260	298,977	189,717
CLEVELAND, Ohio	43,417	92,829	49,412
DETROIT, Mich.	45,619	79,577	33,958
DULUTH, Minn.		3,131	
ERIE, Penn.	9,419	19,646	10,227
GREEN BAY, Wis.	2,275	4,666	2,391
MARQUETTE, Mich.	1,647	4,000	2,353
MILWAUKEE, Wis.	45,246	71,440	26,194
MICHIGAN CITY, Ind.	3,320	3,985	665
MONROE, Mich.	3,892	5,086	1,194
OSWEGO, N. Y.	16,816	20,910	4,094
PORT HURON, Mich.	4,371	5,973	1,620
RACINE, Wis.	7,822	9,880	2,058
SANDUSKY, Ohio.	8,408	13,000	4,592
TOLEDO, Ohio.	13,768	31,584	17,816

Climatology of Ohio—By LORIN BLODGET. "The climate of Ohio is, on the whole, remarkably favorable both in a sanitary point of view, and in its capacity for the most abundant agricultural production. It has less injurious extremes of heat or cold than any other State in or near the same latitude, and it suffers less from extremes of excessive rains, or want of rain. The northern border of the State is greatly favored by the presence of the waters of Lake Erie, which modify the heat of summer, and the cold of winter, very sensibly. No finer climate can be found in the United States for delicate vegetable growths than that of the portions of Ohio bordering on the west end of Lake Erie; and this is abundantly attested by the recent extensive development there of the grape culture. The finest fruit, and the best wines, are now obtained there in great abundance; and this partially is due to the softening influence of the Lake on both the winter and summer climate. The average for the central part of the State is 72° Fahr. for the summer. The range of the mean annual temperature of Ohio extends from 50° to 54° Fahr. from north to south; while the fall of moisture or rain averages from 32 inches on the north, in the vicinity of Lake Erie, to 48 inches on the south, on the borders of the Ohio River, the fall of rain in the centre of the State being 42 inches. These results give a remarkable favorable climatic influence to this portion of the United States."

The City of MONROE, capital of Monroe Co., Mich., is situated on both sides of the River Raisin, three miles above its entrance into Lake Erie, 25 miles from Toledo, and 40 miles from Detroit. It is connected with the Lake by a ship canal, and is a terminus of the *Michigan Southern Railroad*, which extends west, in connection with the Northern Indiana Railroad, to Chicago, Ill. The town contains about 5,000 inhabitants, a court house and jail, a United States land office, 8 churches, several public houses, and a number of large stores of different kinds. Here are two extensive piers, forming an outport at the mouth of the river; the railroad track running to the landing. A plank-road also runs from the outport to the City, which is an old and interesting locality, being formerly called *Frenchtown*, which was known as the scene of the battle and massacre of River Raisin in the war of 1812. The *Detroit, Monroe, and Toledo Railroad* passes through this city. Steamers run from Detroit to Toledo, stopping at Monroe.

TRENTON, situated on the west bank of Detroit river, is a steamboat landing and a place of considerable trade. Population, 1,000.

WYANDOTTE, ten miles below Detroit, is a new and flourishing manufacturing village, where are located the most extensive iron works in Michigan. The iron used at this establishment comes mostly from Lake Superior, and is considered equal in quantity to any in the world. The village contains about 2,000 inhabitants.

Fort Wayne.—The United States Government has recently made extensive improvements at *Fort Wayne*, below Detroit, which will render it one of the strongest fortifications in the country, and almost impregnable against a land assault. The site of the fort, as is well known, is in Springwells, about three miles below the city of Detroit. Its location is admirable, being on a slight eminence, completely commanding the river, which at that point is narrower than in any other place of its entire length. Guns properly placed there could effectually blockade the river against ordinary vessels, and, with the aid of a few gunboats, could repulse any hostile fleet which might present itself.

The Steamers from Cleveland bound for Detroit river, usually pass to the north side of Point Pelée Island, and run across *Pigeon Bay* towards *Bar Point*, situated at the mouth of Detroit river. Several small islands are passed on the south, called *East Sister, Middle Sister, and West Sister;* also, in the distance, may be seen the BASS ISLANDS, known as the "North Bass," "Middle Bass," and "South Bass."

POINT PELÉE ISLAND, belonging to Canada, is about seven miles long, and two or three miles in width. It is inhabited by a few settlers. The island is said to abound with red cedar, and possesses a fine limestone quarry. A lighthouse is situated on the east side.

DETROIT RIVER, forming one of the links between the Upper and Lower Lakes, is next approached, near the mouth of which may be seen a light on the Michigan shore called *Gibraltar Light*, and another light on an island attached to Canada, the steamers usually entering the river, through the east or *British Channel* of the river, although vessels often pass through the west or *American Channel*.

AMHERSTBURGH, CAN., eighteen miles below Detroit, is an old and important town. The situation is good; the banks of the river, both above and below the village,—but particularly the latter, where the river emerges into Lake Erie,—are very beautiful. The *Chicago* and *Canada Southern Railroad* pass through this place, running from Buffalo to Chicago.

FORT MALDEN, capable of accommodating a regiment of troops, is situated about half a mile above Amherstburgh, on the east bank of the river, the channel of which it here commands.

At BROWNSTOWN, situated on the opposite side of the river, in Michigan, is the *battle ground* where the Americans, under disadvantageous circumstances, and with a slight loss, routed the British forces, which lay in ambush, as the former were on their way to relieve the fort at Frenchtown; which event occurred, August 5, 1812.

SANDWICH, CAN., is beautifully situated on the river, two miles below Detroit, and nine miles below Lake St. Clair. It stands on a gently sloping bank a short distance from the river, which is here about a mile wide. This is one of the oldest settlements in Canada West. The town contains 3,133 inhabitants. Here is a *Sulphur Spring* of great celebrity, the water rising 16 feet above the surface, presenting a fine appearance. Here is erected an hotel and bathing-houses.

WINDSOR, CAN., situated in the township of Sandwich, is a village directly opposite Detroit, with which it is connected by three steam ferries. It was laid out in 1834, and is now a place of considerable business, having a population of about 4,000 inhabitants. Here terminates the *Great Western Railway* of Canada, which extends from Niagara Falls or Suspension Bridge, *via* Hamilton and London, to opposite Detroit.

8

Detroit, "THE CITY OF THE STRAITS," a port of entry, and the great commercial mart of the State, is favorably situated in N. lat. 42° 20', W. long. 82° 58', on a river or strait of the same name, elevated some 30 or 40 feet above its surface, being seven miles below the outlet of Lake St. Clair and twenty above the mouth of the river, where it enters into Lake Erie. It extends for the distance of upward of a mile upon the southwest bank of the river, where the stream is three-fourths of a mile in width. The principal public and private offices and wholesale stores are located on Jefferson and Woodward avenues, which cross each other at right angles, the latter running to the water's edge. There may usually be seen a great number of steamboats, propellers, and sail vessels of a large class, loading or unloading their rich cargoes, destined for Eastern markets or for the *Great West*, giving an animated appearance to this place, which is aptly called the *City of the Straits*. It was incorporated in 1815, being now divided into ten wards, and governed by a mayor, recorder, and board of aldermen. The new City Hall, on Campus Martius, is a magnificent edifice, from the dome of which a fine view is obtained of the City and vicinity. The other public buildings are a United States custom-house and post office, and United States lake survey office; an opera house and theatre, masonic hall, fire-

men's hall, mechanic's hall, odd-fellows' hall, the young men's society building, 2 market buildings, 52 churches of different denominations, 6 hotels, besides a number of taverns; 6 banks, 4 savings banks, 2 orphan asylums, gas works, water works, 5 grain elevators, iron foundries, copper smelting works, 5 steam grist-mills, and several saw-mills, besides a great number of other manufacturing establishments. There are also several extensive ship-yards and machine-shops, where are built and repaired vessels of almost every description. The population in 1860 was 45,619. In 1870, 79,530. In 1874, estimated 100,000.

The principal hotels are the *Biddle House*, and *Michigan Exchange*, on Jefferson avenue, and the *Russell House*, on Woodward avenue, facing *Campus Martius*, an open square near the centre of the City.

Detroit may be regarded as one of the most favored of all the Western cities of the Union. It was first settled by the French explorers as early as 1701, as a military and fur trading post. It changed its garrison and military government in 1760 for a British military commander and troops; enduring, under the latter *regime*, a series of Indian sieges, assaults, and petty but vigilant and harassing warfare, conducted against the English garrison by the celebrated Indian warrior Pontiac. Detroit subsequently passed into possession of the American revolutionists; but on the 16th of August, 1812, it was surrendered by Gen. Hull, of the United States army, to Gen. Brock, commander of the British forces. In 1813 it was again surrendered to the Americans, under Gen. Harrison.

The following Railroad lines diverge from Detroit:—

1. The *Detroit, Monroe and Toledo Railroad,* 65 miles in length, connecting with the Michigan Southern Railroad at Monroe, and with other roads at Toledo.

2. The *Michigan Central Railroad,* 284 miles in length, extends to Chicago, Ill. This important road, running across the State from east to west, connects at Michigan City, Ind., with the New Albany and Salem Railroad

3. The *Detroit and Bay City R. R.,* 110 miles in length.

4. The *Detroit, Eel River and Illinois Railroad.*

5. The *Detroit, Lansing and Michigan Railroad.*

6. The *Detroit and Milwaukee Railroad* runs to Grand Haven, on Lake Michigan, opposite Milwaukee, Wis., connecting with a Line of Steamers.

7. The *Grand Trunk Railway* runs from Detroit to Port Huron, Mich., 62 miles, extending from Sarnia, Canada, to Montreal, Quebec, and Portland, Me.

8. The *Great Western Railway* of Canada, 230 miles in length, has its terminus at Windsor, opposite Detroit, the two places being connected by three steam ferries—thus affording a speedy line of travel through Canada, and thence to Eastern cities of the U. States.

There are also three City Railroads.

Steamers of a large class run from Detroit to Buffalo, Cleveland, Toledo, and other ports on Lake Erie; others run to Port Huron, Saginaw, Goderich, Ont., and other ports on Lake Huron.

The *Lake Superior* Line of Steamers running from Buffalo, Cleveland and Detroit direct for the Sault Ste. Marie, and all the principal ports on Lake Superior, are of a large class, carrying passengers and freight. This has become one of the most fashionable and healthy excursions on the continent.

———

The DETROIT RIVER, or *Strait,* is a noble stream, through which flows the surplus waters of the Upper Lakes into Lake Erie. It is twenty-seven miles in length, and from half a mile to two miles in width, forming the boundary between the United States and Canada, commencing at the foot of Lake St. Clair and emptying into Lake Erie. It has a perceptible current, and is navigable for vessels of the largest class. Large quantities of fish are annually taken in the river, and the sportsman usually finds an abundance of wild ducks, which breed in great numbers in the marshes bordering some of the islands and harbors of the coast.

There are altogether 17 islands in the river. The names of these are:— *Clay, Celeron, Hickory, Sugar, Bois Blanc, Ella, Fox, Rock, Grosse Isle, Stoney, Fighting, Turkey, Mammy, Judy, Grassy, Mud, Belle* or *Hog,* and *Ile la Pêche.* The two latter are situated a few miles above Detroit, near the entrance to Lake St. Clair, where large quantities of white fish are annually taken.

ILE LA PÊCHE, attached to Canada, was the home of the celebrated Indian chief *Pontiac.* Parkman, in his "History of the Conspiracy of Pontiac," says:—"Pontiac, the Satan of this forest-paradise, was accustomed to spend the early part of the summer upon a small island at the opening of Lake St. Clair." Another author says: —"The king and lord of all this country lived in no royal state. His cabin was a small, oven-shaped structure of bark and rushes. Here he dwelt with his squaws and children; and here, doubtless, he might often have been seen carelessly reclining his naked form on a rush-mat or a bear-skin, like an ordinary Indian warrior."

The other fifteen islands, most of them small, are situated below Detroit, within the first twelve miles of the river after entering it from Lake Erie, the largest of which is GROSSE ISLE, attached to Michigan, on which are a number of extensive and well-culti-vated farms. This island has become a very popular retreat for citizens of Detroit during the heat of summer,

there being here located good public houses for the accommodation of visi-tors.

Father Hennepin, who was a pas-senger on the "Griffin," the first vessel that crossed Lake Erie, in 1679, in his description of the scenery along the route says: "The islands are the finest in the world; the strait is finer than Niagara; the banks are vast meadows, and the prospect is terminated with some hills covered with vineyards, trees bearing good fruit, groves and forests so well disposed that one would think that nature alone could not have made, without the help of art, so charming a prospect."

TABLE OF DISTANCES

FROM DETROIT TO THE SAULT STE. MARIE AND LAKE SUPERIOR, PASSING THROUGH LAKE HURON.

MILES.	PORTS, &C.	MILES.	MILES.	PORTS, &C.	MILES.
370	**Detroit,** Mich.......... 0		150	Thunder Bay and Light	75 220
363	Lake St. Clair............. 7			(ALPENA, Mich.)	
345	St. Clair Flats and Canal 18	25	120	Presque Isle Light.......30	250
335	Algonac......................10	35	65	*Point de Tour*......... 55	305
327	Marine City 8	43		Entrance to St. Mary's Riv.	
317	St. Clair 10	53		(Is. OF MACKINAC, 35 miles West.)	
301	**Port Huron** .... ⎫ 16	69	57	St. Joseph's Island, Can. 8	313
	SARNIA, Can .... ⎭		49	Mud Lake................... 8	321
300	FORT GRATIOT......... ⎫ 1	70	45	Sailor's Encampment Is. 4	325
	Point Edward, Can.... ⎭		38	Nebish Rapids............. 7	332
	(*Lake Huron,* 235 miles.)		35	Lake George 3	335
289	Lakeport...................11	81	29	*Church's Landing*.........: 6	341
278	Lexington 11	92	26	Garden River Set., Can.. 5	344
266	Port Sanilac.................12	104	15	**Sault Ste. Marie**...11	355
256	Richmondville..............10	114		(Usual time, 36 hours.)	
251	Forestville 5	119	14	Ship Canal 1	356
238	Sand Beach 13	132	8	*Point au-Pins,* Can 6	362
232	Port Hope 6	138	0	Pt. Iroquois & Gros Cap 8	370
225	Point aux Barque and Light...................... 7	145		(Head of St. Mary's Riv.)	
	(*Off Saginaw Bay.*)			*White Fish Point*..........25	395
				Entrance to Lake Superior	

Canal through St. Clair Flats.

The length of the Canal is 8,200 feet; width between dykes, 300 feet; depth of water, 13 feet below lowest stage known during navigation. This work is constructed by the U. S. Government, and will be completed during the season of 1871.

St. Mary's Ship Canal.

Since the completion of this Canal in 1855, the commerce of Lake Superior has augmented so very much as now to require the capacity of the Canal to be increased. The following are the proposed enlargements:—

.1. Deepen the existing locks to a depth of 16 feet on the miter sill. There is now, in the low stage of navigation, only $10\frac{2}{3}$ ft.

2. Deepen the canal to 17 feet.

3. Construct a prolongation of the upper end of the North Canal bank.

4. Construct another lock, overcoming the fall, (20 feet) with one lift, alongside the present two locks. Estimated cost, $1,000,000. The work to be completed by the United States Government, and finished in 1874.

TRIP FROM DETROIT TO MACKINAC, SAUT STE. MARIE.

During the season of navigation propellers of a large class, with good accommodations for passengers, leave Detroit daily direct for Mackinac, Green Bay, Milwaukee, and Chicago, situated on Lake Michigan.

Steamers of a large class, carrying passengers and freight, also leave Detroit, almost daily for the Saut Ste Marie, from thence passing through the *Ship Canal* into Lake Superior—forming delightful excursions during the summer and early autumn months.

For further information of steamboat routes, see *Advertisements.*

On leaving Detroit the steamers run in a northerly direction, passing *Bell* or *Hog Island*, two miles distant, which is about three miles long and one mile broad, presenting a handsome appearance. The Canadian shore on the right is studded with dwellings and well cultivated farms.

PECHE ISLAND is a small body of land attached to Canada, lying at the mouth of Detroit River, opposite which, on the Michigan shore, is *Wind-Mill Point* and light-house.

LAKE ST. CLAIR commences seven miles above Detroit; it may be said to be 20 miles long and 25 miles wide, measuring its length from the outlet of St. Clair River to the head of Detroit River. Compared with the other lakes it is very shallow, having a depth of only from 8 to 24 feet as indicated by Bayfield's chart. It receives the waters of the Upper Lakes from the St. Clair Strait by several channels forming islands, and discharges them into the Detroit River or Strait. In the upper portion of the lake are several extensive islands, the largest of which is *Walpole Island ;* it belongs to Canada, and is inhabited mostly by Indians. All the islands to the west of Walpole Island belong to Michigan. The Walpole, or "Old Ship Channel," forms the boundary between the United States and Canada. The main channel, now used by the larger class of vessels, is called the "North Channel." Here are passed the "*St Clair Flats,*" a great impediment to navigation, for the removal of which Congress will no doubt make ample appropriation sooner or later. The northeastern channel, separating Walpole Island from the main Canada shore, is called " *Chenail Ecarte.*" Besides the waters passing through the Strait of St. Clair, Lake St. Clair receives the river Thames from the Canada side, which is navigable to Chatham, some 24 miles; also the waters of Clinton River from the west or American side, the latter being navigable to Mt. Clemens, Michigan. Several other streams flow into the lake from Canada, the principal of which is the River Sydenham. Much of the land bordering on the lake is low and marshy, as well as the islands; and in places there are large plains which are used for grazing cattle.

ASHLEY, or NEW BALTIMORE, situated on the N. W. side of Lake St. Clair, 30 miles from Detroit, is a new and flourishing place, and has a fine section of country in the rear. It contains three steam saw-mills, several other manufactories, and about 1,000 inhabitants. A steamboat runs from this place to Detroit.

MT. CLEMENS, Macomb Co., Mich., is situated on Clinton River, six miles above its entrance into Lake St. Clair, and about 30 miles from Detroit by lake and river. A steamer plies daily to and from Detroit during the season of navigation. Mt. Clemens contains the county buildings, several churches, three hotels, and a number of stores and manufacturing establishments, and about 2,000 inhabitants. Detroit is distant by plank road only 20 miles.

CHATHAM, ONT., 46 miles from Detroit by railroad route, and about 24 miles above the mouth of the river Thames, which enters into Lake St. Clair, is a port of entry and thriving place of business, where have been built a large number of steamers and sail-vessels.

ALGONAC, Mich., situated near the foot of St. Clair River, 40 miles from Detroit, contains a church, two or three saw-mills, a grist-mill, woollen factory, and about 700 inhabitants.

MARINE CITY, Mich. 7 miles farther

north, is noted for steamboat building, there being extensive ship-yards, where are annually employed a large number of workmen. Here are four steam saw-mills, machine shops, etc. Population about 1,400. Belle River here enters the St. Clair from the west.

ST. CLAIR STRAIT connects Lake Huron with Lake St. Clair, and discharges the surplus waters of Lakes Superior, Michigan, and Huron. It flows in a southerly direction, and enters Lake St. Clair by six channels, the north one of which, on the Michigan side, is the only one at present navigated by large vessels in ascending and descending the river. It receives several tributaries from the west, or Michigan; the principal of which are Black River, Pine River, and Belle River, and several rivers flow into it from the east, or Canadian side. It has several flourishing villages on its banks. It is 48 miles long, from a half to a mile wide, and has an average depth of from 40 to 60 feet, with a current of three miles an hour, and an entire descent of about 15 feet. Its waters are clear and transparent, the navigation easy, and the scenery varied and beautiful—forming for its entire length, the boundary between the United States and Canada. The banks of the upper portion are high; those of the lower portion are low and in parts inclined to be marshy. Both banks of the river are generally well settled, and many of the farms are beautifully situated. There are several wharves constructed on the Canada side, for the convenience of supplying the numerous steamboats passing and repassing with wood. There is also a settlement of the Chippewa Indians in the township of Sarnia, Canada; the Indians reside in small log or bark houses of their own erection.

The CITY OF ST. CLAIR, Mich., is pleasantly situated on the west side of St. Clair Strait, 56 miles from Detroit and 14 miles from Lake Huron. This is a thriving place, with many fine buildings, and is a great lumber depôt. It contains a number of fine residences, a bank, several churches and hotels, two flouring mills and six steam saw-mills, besides other manufacturing establishments, and about 3,000 inhabitants. St. Clair has an active business in the construction of steamers and other lake craft. The site of old *Fort St. Clair*, now in ruins, is on the border of the town.

SOUTHERLAND, CAN., is a small village on the Canada shore, opposite St. Clair. It was laid out in 1833 by a Scotch gentleman of the same name, who here erected an Episcopal church, and made other valuable improvements.

MOORE, is a small village ten miles below Sarnia on the Canada side.

FROMEFIELD, or TALFOURD'S, CAN., is another small village, handsomely situated four and a half miles below Sarnia. Here is an Episcopal church, a windmill, and a cluster of dwellings.

The city of PORT HURON, St. Clair Co., Mich., a port of entry, is advantageously situated on the west bank of St. Clair River, at the mouth of Black River, two miles below Lake Huron. It contains the county buildings for St. Clair Co., one Congregational, one Episcopal, one Baptist, one Methodist, and one Roman Catholic Church; six hotels and public houses, forty stores, and several warehouses; one steam flouring-mill, eight steam saw-mills, producing annually a large amount of lumber, the logs being rafted down Black River, running through an extensive pine region; here are also, two yards for building of lake craft, two refineries of petroleum oil, one iron foundry, and several other manufacturing establishments. Population in 1860, 4,300; in 1870, 5,973.

During the season of navigation, there is daily intercourse by steamboats with Detroit, Saginaw, and ports on the Upper Lakes. A steam ferry-boat also plies between Port Huron and Sarnia, C. W., the St. Clair River here being about one mile in width. A branch of the Grand Trunk Railway runs from Fort Gratiot, one mile and a half above Port Huron, to Detroit, a distance of 62 miles, affording altogether speedy modes of conveyance. A railroad is also proposed to run from Port Huron, to intersect with the Detroit and Milwaukee Railroad, at Owasso, Michigan.

FORT GRATIOT, one and a half miles north of Port Huron, lies directly opposite Point Henry, C. W., both situated at the foot of Lake Huron, where commences St. Clair River. It has become an important point since the completion of the Grand Trunk Railway of Canada, finished in 1859, which road terminates by a branch at Detroit, Mich., thus forming a direct railroad communication from Lake Huron, eastward, to Montreal, Quebec, and Portland, Maine.

The village stands contiguous to the site of Fort Gratiot, and contains besides the railroad buildings, which are extensive, one church, five public houses, the Gratiot House being a well-kept hotel; two stores, one oil refinery, and about 400 inhabitants. A steam ferry-boat plies across the St. Clair River, to accommodate passengers and freight; the river here being about 1,000 feet wide, and running with considerable velocity, having a depth of from 20 to 60 feet.

In a military and commercial point of view, this place attracts great attention, no doubt, being destined to increase in population and importance. The Fort was built in 1814, at the close of the war with Great Britain, and consists of a stockade, including a magazine, barracks, and other accommodations for a garrison of one battalion. It fully commands the entrance to Lake Huron from the American shore, and is an interesting landmark to the mariner.

SARNIA, ONT., situated on the east bank of St. Clair River, two miles below

Lake Huron and 68 above Detroit, is a port of entry and a place of considerable trade; two lines of railroad terminate at this point, and it is closely connected with Port Huron on the American shore by means of a steam ferry. The town contains a court-house and jail, county registers office and town hall; one Episcopal, one Methodist, one Congregational, one Baptist, one Roman Catholic, and one Free Church; seven public houses, the principal being the *Alexander House* and the *Western Hotel;* twenty stores and several groceries; two grain elevators; two steam saw-mills; one steam gristmill, one large barrel factory, one steam cabinet factory, one steam iron foundry, and one refinery of petroleum oil, besides other manufacturing establishments. Population, 2,000.

The Grand Trunk Railway of Canada terminates at Point Edward, 2 miles from Sarnia, extending eastward to Montreal, Quebec and Portland, Me.; a branch of the Great Western Railway also terminates at Sarnia, affording a direct communication with Niagara Falls, Boston, and New York. Steamers run from Sarnia to Goderich and Saugeen, Ont.; also to and from Detroit, and ports on the Upper Lakes.

Canadian Steamers running to and from Detroit on their way to the different ports on the east shore of Lake Huron, hug the Canada side, leaving the broad waters of the lake to the westward.

POINT EDWARD, 2 miles above Sarnia, lies at the foot of Lake Huron, opposite Fort Gratiot, where are erected a large depôt and warehouses connected with the *Grand Trunk Railway* of Canada. Here terminates the grand railroad connection extending from the Atlantic ocean to the Upper Lakes. It also commands the entrance into Lake Huron and is an important military position although at present unfortified. In the vicinity is an excellent fishery, from whence large quantities of fish are annually exported.

BAYFIELD, Ont., 108 miles from Detroit, is a new and flourishing place, situated at the mouth of a river of the same name.

GODERICH, 120 miles north of Detroit, is situated on elevated ground at the mouth of Maitland River, where is a good harbor. This is a very important and growing place, where terminates the *Buffalo and Huron Railroad,* 160 miles in length.

KINCARDINE, 30 miles from Goderich, is another port on the Canadian side of Lake Huron, where the British steamers land and receive passengers on their trips to Saugeen.

SAUGEEN, Ont., is situated at the mouth of a river of the same name, where is a good harbor for steamers and lake craft.

SOUTHAMPTON is a port situated on the east side of Lake Huron, where terminates the *Wellington, Grey and Bruce Railroad,* connecting with the Great Western Railway of Canada.

Steamers leave Southampton for Lake Superior, forming a through line, starting from Sarnia, Can. This Steamboat Line forms the most direct route from Buffalo and Niagara Falls, passing over the *Great Western Railway of Canada.*

Steamboat Route from Port Huron to Saginaw City, etc.

On leaving the wharf at Port Huron, the steamers pass Fort Gratiot and enter the broad waters of Lake Huron, one of the Great Upper Lakes, all alike celebrated for the sparkling purity of their waters. The shores are for the most part low, being covered by a heavy growth of forest trees.

LAKEPORT, 10 miles from Port Huron, is a small village lying on the lake shore.

LEXINGTON, 12 miles further, is the capital of Sanilac County, Michigan, where is a good steamboat landing and a flourishing village of 1,000.

PORT SANILAC, 34 miles above Port Huron, is another small village.

FORRESTVILLE, Mich., 47 miles from Port Huron, and 120 miles north of Detroit, situated on the west side of Lake Huron, is a new settlement, where is erected an extensive steam saw-mill. It has some five or six hundred inhabitants, mostly engaged in the lumber trade. Several other small settlements are situated on the west shore of Lake Huron, which can be seen from the ascending steamer, before reaching Point aux Barques, about seventy-five miles above Port Huron.

SAGINAW BAY is next entered, presenting a wide expanse of waters; Lake Huron here attaining its greatest width, where the mariner often encounters fierce storms, which are prevalent on all of the Upper Lakes. To the eastward lies the Georgian Bay of Canada, with its innumerable islands.

BAY CITY, favorably situated near the mouth of Saginaw River, is a flourishing town with a population of about 9,000. Here is a good harbor, from whence a large amount of lumber is annually exported. It has fifteen saw-mills, and other manufacturing establishments.

Steamers run daily to Detroit and other ports.

EAST SAGINAW, situated on the right bank of the river, about one mile below Saginaw City, is a new and flourishing place, and bids fair to be one of the most important cities of the state. It is largely engaged in the lumber trade, and in the manufacture of salt of a superior quality. There are several large steam saw-mills, many with gangs of saws, and capable of sawing from four to five million feet of lumber annually; grist and flouring-mills, with four run of stones, planing-mills, foundries, machine shops, breweries, a ship-yard, and other manufacturing establishments, giving employment to a great number of workmen. Here is a well-kept hotel, and several churches; a banking office and a number of large stores and warehouses. Coal of a good quality is abundant, being found near the river, and the recent discovery of *salt springs* in the neighborhood is of incalculable value, the manufacture of salt being carried on very extensively. Population in 1870, 11,350, now 14,000.

Several lines of steamers, and one of propellers, sail from this port regularly for Detroit and other lake ports. It is near the head of navigation for lake craft, where five rivers unite with the Saginaw, giving several hundred miles of water communication for river rafting and the floating of saw-logs. The surrounding country is rich in pine, oak, cherry, blackwalnut, and other valuable timber. A railroad is finished from this place to *Holly*, connecting with the Detroit and Milwaukee Railroad.

SAGINAW CITY, Saginaw County, Mich., is handsomely situated on the left bank of the river, 23 miles above its mouth. It contains a court-house and jail, several churches, two hotels, fifteen stores, two warehouses, and six steam saw-mills. Population about 10,000. There is a fine section of country in the rear of Saginaw much of which is heavily timbered; the soil produces grain in abundance, while the streams afford means of easy transportation to market. Steamers run daily from Saginaw City and East Saginaw to Detroit, Chicago, &c., and other ports on the lakes, during the season of navigation.

LAKE HURON.

The waters of Lake Huron, lying between 43° and 46° north latitude, are surrounded by low shores on every side. The most prominent features are Saginaw Bay on the southwest, and the Georgian Bay on the northeast; the latter large body of water being entirely in the limits of Canada. The lake proper, may be said to be 100 miles in width, from east to west, and 250 miles in length, from south to north, terminating at the Straits of Mackinac. It is nearly destitute of islands, presenting one broad expanse of waters. It possesses several good harbors on its western shores, although as yet but little frequented. Point aux Barques, Thunder Bay, and Thunder Bay Islands, are prominent points to the mariner.

TAWAS, or OTTAWA BAY, lying on the northwest side of Saginaw Bay, affords a good harbor and refuge during storms, as well as THUNDER BAY, lying farther to the north. Off Saginaw Bay, the widest part of the lake, rough weather is often experienced, rendering it necessary for steamers and sail vessels to run for a harbor or place of safety.

In addition to the surplus waters which Lake Huron receives through the Straits of Mackinac and the St. Mary's River from the north, it receives the waters of Saginaw River, and several other small streams from the west. This lake drains but a very small section of country compared to its magnitude, while its depth is a matter of astonishment, being from 100 to 750 feet, according to recent surveys;

altitude above the ocean, 574 feet, being 26 feet below the surface of Lake Superior. Its outlet, the St. Clair River, does not seem to be much larger than the St. Mary's River, its principal inlet, thus leaving nearly all its other waters falling in the *basin*, to pass off by evaporation. On entering the *St. Clair River*, at Fort Gratiot, after passing over the Upper Lakes, the beholder is surprised to find all these accumulated waters compressed down to a width of about 1,000 feet, the depth varying from 20 to 60 feet, with a strong downward current.

The *Straits of Mackinac*, connecting Lakes Huron and Michigan, is a highly interesting body of water, embosoming several picturesque islands, with beautiful headlands along its shores. It varies in width from 5 to 30 miles, from mainland to mainland, and may be said to be from 30 to 40 miles in length. Here are good fishing grounds, as well as at several other points on Lake Huron and Georgian Bay.

The climate of Lake Huron and its shores is perceptibly warmer than Lake Superior during the spring, summer, and autumn months, while the winter season is usually rendered extremely cold from the prevalence of northerly winds passing over its exposed surface. On the 30th of July, 1860, at 8 A. M., the temperature of the air near the middle of Lake Huron, was 64° Fahr., the water on the surface, 52°, and at the bottom, 50 fathoms (300 feet) 42° Fahr.

THE LOWER PENINSULA OF MICHIGAN.

THE *Lower Peninsula of Michigan* is nearly surrounded by the waters of the Great Lakes, and, in this respect, its situation is naturally more favorable for all the purposes of trade and commerce than any other of the Western States.

The numerous streams which penetrate every portion of the Peninsula, some of which are navigable for steamboats a considerable distance from the lake, being natural outlets for the products of the interior, render this whole region desirable for purposes of settlement and cultivation. Even as far north as the Strait of Mackinac, the soil and climate, together with the valuable timber, offer great inducements to settlers; and if the proposed railroads, under the recent grant of large portions of these lands by Congress, are constructed from and to the different points indicated, this extensive and heavily timbered region will speedily be reclaimed, and become one of the most substantial and prosperous agricultural portions of the West.

It is well that in the system of compensation, which seems to be a great law of the universe, the vast prairies which comprise so large a portion of this great Western domain are provided so well with corresponding regions of timber, affording the necessary supply of lumber for the demand of the increasing population which is so rapidly pouring into these Western States.

The State of Michigan—all the waters of which flow into the Basin of the St. Lawrence—Northern Wisconsin, and Minnesota are the sources from which the States of Ohio, Indiana, Illinois, and Iowa, and a large portion of the prairie country west of the Mississippi, must derive their supply of this important article (lumber). The supply in the West is now equal to the demand, but the consumption is so great, and the demand so constantly increasing with the development and settlement of the country, that of necessity, within comparatively a very few years, these vast forests will be exhausted. But as the timber is exhausted the soil is prepared for cultivation, and a large portion of the northern part of the southern Peninsula of Michigan will be settled and cultivated, as it is the most reliable wheat-growing portion of the Union.

Besides the ports and towns already described, there are on *Lake Huron*, after leaving *Saginaw Bay*, going north, several settlements and lumber establishments, fisheries, &c.

TAWAS CITY, the county seat of Iosco county, is a flourishing place of about 900 inhabitants, situated at the mouth of Tawas river, being about 180 miles north of Detroit. It contains, besides the county buildings, a church, 2 hotels, 3 steam saw-mills and 12 or 15 stores.

ALPENA, capital of Alpena county, is favorably situated on the shore of Thunder Bay, Lake Huron, 220 miles above Detroit, lying in N. lat. 45° 05'; W. long. 83° 30'. It is reached during the season of navigation by steamers from Detroit and Bay City. Alpena was incorporated as a city in 1871, and now contains about 3,500 inhabitants. An important mineral spring is here located possessing rare medicinal qualities and is much frequented by strangers. A commodious Hotel has recently been erected for the accommodation of guests.

OLD MACKINAC, lying on the mainland, is one of the most interesting points, being celebrated both in French and English history when those two great powers contended for the possession of this vast Lake Region. It is proposed to build a railroad from Old Mackinac to Saginaw, and one to the southern confines of the State, while another line of road will extend northwestward to Lake Superior, crossing the straits by a steam ferry. A town plot has been surveyed, and preparations made for settlement.

Passing around the western extremity of the Peninsula, at the *Waugoshance* Light and Island, the next point is *Little Traverse Bay,* a most beautiful sheet of water.

About fifteen miles southwesterly from Little Traverse we enter GRAND TRAVERSE BAY, a large and beautiful arm of the lake, extending about thirty miles inland. This bay is divided into two parts by a point of land, from two to four miles wide, extending from the head of the bay about eighteen miles toward the lake. The country around this bay is exceedingly picturesque, and embraces one of the finest agricultural portions of the State. The climate is mild, and fruit and grain of all kinds suitable to a northern latitude are produced, with less liability to injury from frost than in some of the southern portions of the State.

GRAND TRAVERSE CITY is located at the head of the west arm of the bay, and is the terminus of the proposed railroad from Grand Rapids, a distance of about 140 miles.

Passing out of the bay and around the point dividing the west arm from the lake, we first arrive at the river *Aux Becs Sceis.* There is here a natural harbor, capable of accommodating the larger class of vessels and steamboats. A town named FRANK-FORT has been commenced at this place, and with its natural advantages, and the enterprise of parties who now contemplate making further improvements, it will soon become a very desirable and convenient point for the accommodation of navigators.

The islands comprising the Beavers, the Manitous, and Fox Isles should here be noticed. The *Beavers* lie a little south of west from the entrance to the Strait of Mackinac, the Manitous a little south of these, and the Foxes still farther down the lake. These are all valuable for fishing purposes, and for wood and lumber. Lying in the route of all the steamboat lines from Chicago to Buffalo and the Upper Lakes, the harbors on these islands are stopping-points for the boats, and a profitable trade is conducted in furnishing the necessary supplies of wood, etc.

We next arrive at MANISTEE, a small but important settlement at the mouth of the Manistee River. The harbor is a natural one, but requires some improvement. A large trade is carried on with Chicago in lumber.

The next point of importance is the mouth of the *Père Marquette* River. Here is the terminus of the proposed railroad from Flint, in Genesee County, connecting with Detroit by the Detroit and Milwaukee Railway, a distance of about 180 miles.

The harbor is very superior, and the country in the vicinity is well adapted for settlement. About 16 miles in the interior is situated one of the most compact and extensive tracts of pine timber on the western coast.

About forty miles south of this, in the county of Oceana, a small village is located at the mouth of *White River.* The harbor here is also a natural one, and the region is settled to considerable extent by farmers. Lumber is, however, the principal commodity, and the trade is principally with the Chicago market.

The next point, MUSKEGON, at the mouth of the *Muskegon River,* is supported principally by the large lumber region of the interior. Numerous steam saw-mills are now in active operation here, giving the place an air of life and activity.

The harbor is one of the best on the lake, and is at present accessible for all the vessels trading between Muskegon and Chicago.

GRAND HAVEN, Ottawa Co., Mich., is situated on both sides of Grand River, at its entrance into Lake Michigan, here eighty-five miles wide; on the opposite side lies Milwaukee, Wis.

DIRECT STEAMBOAT ROUTE FROM DETROIT TO GREEN BAY, CHICAGO, &c.

Sailing direct through Lake Huron to Mackinac, or to the De Tour entrance to St. Mary's River, a distance of about 330 miles, the steamer often runs out of sight of land on crossing Saginaw Bay.

Thunder Bay Light is first sighted and passed, and then Presque Isle Light, when the lake narrows and the Strait of Mackinac is soon entered, where lies the romantic Island of Mackinac. The Strait of Mackinac, with the approaches thereto from Lakes Huron and Michigan, will always command attention from the passing traveller. Through this channel will pass, for ages to come, a great current of commerce, and its shores will be enlivened with civilized life.

In this great commercial route, Lake Huron is traversed its entire length, often affording the traveller a taste of sea-sickness and its consequent evils. Yet there often are times when Lake Huron is hardly ruffled, and the timid passenger enjoys the voyage with as much zest as the more experienced mariner.

MACKINAC, crowned by a fortress, where wave the *Stars and the Stripes*, the gem of the Upper Lake islands, may vie with any other locality for the salubrity of its climate, for its picturesque beauties, and for its vicinity to fine fishing-grounds. Here the invalid, the seeker of pleasure, as well as the sportsman and angler, can find enjoyment to their heart's content during warm weather. *For description, see p. 88.*

On leaving Mackinac for Green Bay, the steamer generally runs a west course for the mouth of the bay, passing the Beaver Islands in Lake Michigan before entering the waters of Green Bay, about 150 miles.

SUMMER ISLAND lies on the north side and ROCK ISLAND lies on the south side of the entrance to Green Bay, forming a charming view from the deck of a steamer.

WASHINGTON or POTAWATOMEE ISLAND. CHAMBERS' ISLAND, and other small islands are next passed on the upward trip toward the head of the bay.

WASHINGTON HARBOR, situated at the north end of Washington Island, is a picturesque fishing station, affording a good steamboat-landing and safe anchorage.

GREEN BAY, about 100 miles long and from 20 to 30 miles wide, is a splendid sheet of water, destined no doubt to be enlivened with commerce and pleasure excursions.

Ports of Lake Michigan— East and South Shores.

Michigan City, Ind., situated at the extreme south end of Lake Michigan, is distant 45 miles from Chicago by water and 228 miles from Detroit by railroad route. The *New Albany & Salem Railroad*, 228 miles in length, terminates at this place, connecting with the Michigan Central Railroad. Several plank roads also terminate here, affording facilities for crossing the extensive prairies lying in the rear. Here are several large store-houses, situated at the mouth of Trail Creek, intended for the storage and shipment of wheat and other produce; 20 or 30 stores of different kinds, several hotels, and a branch of the State Bank of Indiana. It now contains 5,000 inhabitants, and is steadily increasing in wealth and numbers

The harbor of Michigan City, which has been closed for a number of years to large vessels by the accumulation of sand in the channel is again open, with a good depth of water. Vessels loaded with iron ore and with lumber enter and discharge cargoes without difficulty. An efficient dredge is at work, and Michigan City will soon become a lake port of importance.

NEW BUFFALO, Mich., lying 50 miles east of Chicago by steamboat route, is situated on the line of the Michigan Central Railroad, 218 miles west of Detroit. Here have been erected a light-house and pier, the latter affording a good landing for steamers and lake craft. The settlement contains two or three hundred inhabitants, and several stores and storehouses. It is surrounded by a light, sandy soil, which abounds all along the east and south shores of Lake Michigan.

ST. JOSEPH, Berrien Co., Mich., is advantageously situated on the east shore of Lake Michigan, at the mouth of St. Joseph River, 194 miles west of Detroit. Here is a good harbor, affording about 10 feet of water. The village contains about 1,000 inhabitants, and a number of stores and storehouses. An active trade in lumber, grain, and fruit is carried on at this place, mostly with the Chicago market, it being distant about 70 miles by water. Steamers of a small class run from St. Joseph to Niles and Constantine, a distance of 120 miles, to which place the St. Joseph River is navigable.

St. Joseph River rises in the southern portion of Michigan and Northern Indiana, and is about 250 miles long. Its general course is nearly westward; is very serpentine, with an equable current, and flowing through a fertile section of country, celebrated alike for the raising of grain and different kinds of fruit. There are to be found several flourishing villages on its banks. The principal are Constantine, Elkhart, South Bend, and Niles.

NILES, situated on St. Joseph River, is 26 miles above its mouth by land, and 191 miles from Detroit by railroad route. This is a flourishing village, containing about 3,000 inhabitants, five churches, three hotels, several large stores and flouring mills; the country around producing large quantities of wheat and other kinds of grain. A small class of steamers run to St. Joseph below and other places

above, on the river, affording great facilities to trade in this section of country.

SOUTH HAVEN, Van Buren Co., lies at the mouth of Black River.

NAPLES, Allegan Co., lies on the east side of Lake Michigan, near the mouth of the Kalamazoo River.

AMSTERDAM, Ottawa Co., is a small village lying near the Lake shore, about 20 miles south of Grand Haven.

HOLLAND, situated on *Black Lake*, a few miles above Amsterdam, is a thriving town, settled mostly by Hollanders. Here is a good and spacious harbor.

The counties of Berrien, Cass, Van Buren, Kalamazoo, Allegan, Kent, and Ottawa are all celebrated as a fruit-bearing region.

The Ports extending from Grand Haven to Saginaw Bay are fully described in another portion of this work, as well as the bays and rivers falling into Lakes Michigan and Huron.

Steamers on Lake Michigan.

STEAMERS leave Chicago daily, during the season of navigation, for Grand Haven and Muskegon, connecting with steamers running to Pere Marquette, Manistee, and other ports on the Michigan side of the lake.

Daily lines of steamers also leave Chicago for Racine, Milwaukee, Port Washington, Sheboygan, Manitowoc, Two Rivers, and other ports on the Wisconsin side of the lake.

Lines of propellers also leave Chicago daily for all the principal ports on the west side of Lake Michigan, stopping at Mackinac, Port Huron, Detroit, etc., affording a speedy and cheap mode of conveyance from Chicago and Milwaukee to all the principal ports on Lakes Erie and Ontario.

CHICAGO,

"THE GARDEN CITY," the largest city of Illinois, is advantageously situated on the south-western shore of Lake Michigan, at the mouth of Chicago river, in N. lat., 41° 52', and W. long., from Greenwich, 87° 35'; being elevated eight to ten feet above the lake, the level of which great body of water is 578 feet above the Atlantic Ocean. This city has within forty years risen from a small settlement around an old fort (Dearborn), to a place of great commercial importance, being now one of the largest interior cities in the United States, exhibiting a rapidity of growth and wealth never before known in the annals of the country. The harbor and river has a depth of from 12 to 14 feet of water, which makes it a commodious and safe haven; and it has been much improved artificially by the construction of piers, which extend on each side of the entrance of the river, for some distance into the lake, to prevent the accumulation of sand upon the bar. The light-house is on the south side of the harbor, and shows a fixed light on a tower 40 feet above the surface of the lake; there is also a beacon light on the end of the pier. In a naval and military point of view, this is one of the most important ports on the Upper Lakes, and should be strongly defended. Along the river and its branches, for several miles, are immense grain warehouses, some of which are capable of storing upward of 1,000,000 bushels of grain—and alongside of which vessels can be loaded within a few hours. The whole capacity for storage of grain exceeds 10,000,000 bushels. There are also immense storehouses for the storage of flour, beef, pork, whisky, and other merchandise, and capacious docks and yards for lumber, wood, coal, &c., Chicago now being one of the greatest grain, provision, and lumber markets in the world; the shipment of flour and grain alone, in 1873, being nearly 100,000,000 bushels.

The city of Chicago is laid out at right angles, the streets run from the lake westward, intersected by others, all of which are about 80 feet wide; it extends along the lake, north and south, about 8 miles, there being a gradual rise in the ground, affording a good drainage into the river and lake. The business portion of the city is mostly built of brick, and a fine quality of stone, sometimes called "Athens marble." This stone is found in the vicinity of the city, and is highly prized as a building material. The dwelling-houses are mostly constructed of wood, except costly residences, which are of brick, or stone and marble.

The city contains a United States custom-house and post-office building, a court-house and jail, the county buildings, a Marine Hospital, Rush Medical College, and Chicago Medical College; the Chamber of Commerce, a new edifice, built of Athens stone; a new opera house, academy of music, and other places of amusement; market houses; several large hotels; 160 churches of different denominations, many of which are costly edifices; 20 banks; 15 marine and fire insurance companies; gas works and water works. The manufacturing establishments of Chicago are numerous and extensive, consisting of iron founderies and machine shops, railroad car manufactory, steam saw, planing, and flouring mills, manufactories of agricultural implements, breweries, distilleries, &c. Numerous steamers, propellers, and sailing vessels ply between this place and the ports on Lake Michigan and Green Bay; also, to the Lake Superior ports, Collingwood and

Goderich, Can., Detroit, Cleveland, Dunkirk, Buffalo, and to the ports on Lake Ontario, passing through the Welland Canal; vessels occasionally sailing to and from European ports, *via* the St. Lawrence river.

The *Illinois and Michigan Canal*, connecting Lake Michigan with Illinois river, which is 60 feet wide at the top, 6 feet deep, and 107 miles in length, including 5 miles of river navigation, terminates here, through which is brought a large amount of produce from the south and southwest; and the numerous Railroads radiating from Chicago, add to the vast accumulation which is here shipped for the Atlantic sea-board. Chicago being within a short distance of the most extensive coal-fields to be found in Illinois, and the pineries of Michigan and Wisconsin, as well as surrounded by the finest grain region on the face of the globe, makes it the natural outlet for the varied and rich produce of an immense section of fertile country.

The establishment of the great *Union Stock Yard* renders Chicago more attractive than ever as a cattle market.

The *Lake Tunnel*, extending under Lake Michigan, supplies the City with pure and wholesome water. Two *Artesian Wells* are also in operation, situated three miles west of the lake, yielding 1,200,000 gallons of pure water daily. The *City Railroads* extend to the limits of the City in every direction, affording a cheap and speedy mode of conveyance, while from the numerous Railroad depôts, passengers are conveyed to remote points, east, west, north and south.

Two *Tunnels* are completed under the Chicago river, for free passage of pedestrians and wheeled vehicles.

POPULATION OF CHICAGO AT DIFFERENT PERIODS.

United States Census, 1840....	4,853
State Census, 1845..............	12,088
United States Census, 1850....	29,963
State Census, 1855..............	80,000
United States Census, 1860....	109,260
State Census, 1865..............	178,900
United States Census, 1870....	298,977

GREAT FIRE OF 1871.—Chicago was visited with the most destructive fire on record on the 8th and 9th of October, 1871. Property to an immense amount was destroyed, including most of the public buildings, hotels and stores in the heart of the city. The fire extended over about 2,000 acres, including a large portion of the city limits facing on Lake Michigan. The duration of the fire was about 30 hours, destroying upwards of 20,000 buildings. Estimated loss $200,000,000, of which amount about one half was insured. Notwithstanding the great destruction of property and the derangement of business, the city has improved with wonderful rapidity. The business of 1873 is said to exceed any previous year. The grain, lumber and provision business is enormous, exceeding any other place on the continent.

OBJECTS OF INTEREST. — Chicago can boast of several PUBLIC PARKS, situated in different parts of the City. *Lake Park* and *Lincoln Park* lie on the shores of Lake Michigan. *Union Park*, lying on the west side of the City, is of large size, and beautifully ornamented with a small lake in its centre, being crossed by three bridges. The other parks are,—*Jefferson Park, Washington Park, Dearborn Park, Ellis Park,* and *Vernon Park.*

AVENUES AND STREETS.—There are 37 miles of Avenues and Streets, paved with wooden block pavements, not including the graveled or macadamized streets. Formerly Chicago was the subject of complaint, owing to poor thoroughfares, there are to-day few cities in the Union presenting more drives, of miles in length, of smoothly paved streets.

TUNNELS AND BRIDGES.— *Washington* and *La Salle Street Tunnels*, the latter, just finished, affords great facilities for passing under the Chicago river, while numerous bridges span the stream at the intersection of the different streets running north and west. *Chicago River* itself, crowded with Steam Propellers and sail vessels, presents great attraction.

HOTELS OF CHICAGO.

The principal Hotels erected since the great fire, are the Grand Pacific, Palmer House, Sherman House, Tremont House, Gardiner House, Briggs House, Honore Hotel, Anderson's European Hotel and several others, varying in charges from $2 to $5 per day. Besides the above there are numerous other Hotels and Restaurants.

RAILROADS.—The numerous Railroads that diverge from Chicago, running south, north, east and west, carrying an immense number of passengers and great quantities of freight, is the wonder of the world.

IRON PROPELLER.

MONTHLY TEMPERATURES.

The following Table is of interest, as showing the relative Temperatures of the air and of the water at the crib, at its entrance into the Tunnel.

MONTHS.	AIR.—Fahr.			WATER.—Fahr.		
	HIGHEST.	MEAN.	LOWEST.	HIGHEST.	MEAN.	LOWEST.
April, 1869.	61°	44°	31°	44°	38°	33°
May, "	69°	51°	41°	51°	46°	43°
June, "	74°	63°	53°	60°	53°	50°
July, "	80°	71°	62°	67°	63°	60°
August, "	88°	75°	66°	72°	67°	64°
September, "	76°	65°	53°	72°	63°	57°
October, "	59°	45°	28°	57°	47°	40°
November, "	50°	35°	28°	40°	36°	33°
December, "	38°	30°	18°	33°	32°	32°
January, 1870	40°	28°	10°	32°	32°	32°
February, "	39°	30°	9°	33°	32°	32°
March, "	41°	33°	12°	33°	32°	32°
MEAN ANN. TEMP.	47½° Fahr.			45° Fahr.		

STEAMBOAT ROUTE,
FROM CHICAGO AND MILWAUKEE TO MACKINAC, SAUT STE. MARIE AND DULUTH.

MILES.	PORTS, &C.	MILES.	MILES.	PORTS, &C.	MILES.
	(*Lake Michigan.*)		715	Beaver Island, Mich.. 70	345
1,060	**Chicago,** Ill	0	684	Point Waugoshance... 31	376
1,025	Waukegan, Ill............	35		(*Straits of Mackinac.*)	
1,009	Kenosha, Wis............16	31			
998	RACINE, Wis............11	62	668	OLD MACKINAC....... 16	392
975	**Milwaukee**, Wis..23	85	660	**Mackinac**.......... 8	400
950	Port Washington, Wis.25	110	624	DE TOUR.............. 36	436
925	SHEBOYGAN, Wis.......25	135		(*St. Mary's River.*)	
895	Manitowoc, Wis.........30	165			
888	Two Rivers, Wis........ 7	172	584	Church's Landing..... 40	476
866	Kewaunee, Wis 22	194	570	**Saut Ste. Marie.** 14	490
855	Annapee, Wis............11	205		(*Ship Canal.*)	
820	Bayley's Harbor, Wis..35	240	530	White Fish Point...... 40	530
800	Death's Door, Wis......20	260		(*Lake Superior.*)	
	(To GREEN BAY, 80 Miles.)		400	**Marquette**.........130	660
785	Washington Harbor....15	275		Keeweenaw Point.......130	790
	(To ESCANABA, 30 Miles.)		0	**Duluth**..............270	1,060

NOTE.—This distance is shortened about 90 Miles, by passing through Portage Lake and the Ship Canal.

LAKE MICHIGAN.

GOODRICH'S STEAMERS

Leave CHICAGO for Racine, Milwaukee, etc., daily (Sundays excepted,) 9 A. M.

Grand Haven, Muskegon, etc., daily (Sundays excepted,) 7 P. M.

St. Joseph, Tuesday, Thurs. and Saturday, 11 P. M.

Manistee and Ludington, Tues. and Thurs. 9 A. M.

Green Bay and intermediate ports, Tuesday and Friday, 7 P. M.

Kewaunee and Ahnapee, Friday, 9 'A. M., and Tuesday and Friday, 7 P. M.

Office and Docks, foot Michigan Av., Chicago, Ill.

ENGELMANN TRANSPORTATION COMPANY.

STEAMERS LEAVE

68 West Water Street, Milwaukee,

DAILY, } For Gd. Haven, Whitehall, Muskegon,
9 P. M. } Ludington, Manistee, Pentwater, etc.

DAILY, } For Grand Haven, Saginaw, Detroit,
9 P. M. } AND ALL POINTS EAST.

Tickets to Manistee, etc., good via Grand Haven.

ENGELMANN TRANSPORTATION CO.

ROUTE FROM CHICAGO TO MACKINAC AND SAUT STE. MARIE.

On starting from the steamboat wharf near the mouth of the Chicago River, the Marine Hospital and depot of the Illinois Central Railroad are passed on the right, while the Lake House and lumber-yards are seen on the left or north side of the stream. The government piers, long wooden structures, afford a good entrance to the harbor; a light-house has been constructed on the outer end of the north pier, to guide vessels to the port.

The basin completed by the Illinois Central Railroad to facilitate commerce is a substantial work, extending southward for nearly half a mile. It affords ample accommodation for loading and unloading vessels, and transferring the freight to and from the railroad cars.

The number of steamers, propellers, and sailing vessels annually arriving and departing from the harbor of Chicago is very great; the carrying trade being destined to increase in proportionate ratio with the population and wealth pouring into this favored section of the Union.

On reaching the green waters of Lake Michigan, the city of Chicago is seen stretching along the shore for four or five miles, presenting a fine appearance from the deck of the steamer. The entrance to the harbor at the bar is about 200 feet wide. The bar has from ten to twelve feet water, the lake being subject to about two feet rise and fall. The steamers bound for Milwaukee and the northern ports usually run along the west shore of the lake within sight of land, the banks rising from thirty to fifty feet above the water.

LAKE MICHIGAN is about seventy miles average width, and 340 miles in extent from Michigan City, Ind., on the south, to the Strait of Mackinac on the north; it presents a great expanse of water, now traversed by steamers and other vessels of a large class, running to the Saut Ste. Marie and Lake Superior; to Collingwood and Goderich, Can.; to Detroit, Mich.; to Cleveland, Ohio, and to Buffalo, N. Y. From Chicago to Buffalo the distance is about 1,000 miles by water; while from

Chicago to Superior City, at the head of Lake Superior, or Fond du Lac, the distance is about the same, thus affording two excursions of 1,000 miles each, over three of the great lakes or inland seas of America, in steamers of from 1,000 to 2,000 tons burden. During the summer and early autumn months the waters of this lake are comparatively calm, affording safe navigation. But late in the year, and during the winter and early spring months, the navigation of this and the other great lakes is very dangerous.

WAUKEGAN, Lake Co., Ill., 36 miles north of Chicago, is handsomely situated on elevated ground, gradually rising to 50 or 60 feet above the water. Here are two piers, a light-house, several large storehouses, and a neat and thriving town containing about 5,000 inhabitants, six churches, a bank, several well-kept hotels, thirty stores, and two steam-flouring mills.

KENOSHA, Wis., 52 miles from Chicago, is elevated 30 or 40 feet above the lake. Here are a small harbor, a light-house, storehouses, mills, etc. The town has a population of about 4,310 inhabitants, surrounded by a fine back country. Here is a good hotel, a bank, several churches, and a number of stores and manufacturing establishments doing a large amount of business. The *Kenosha and Rockford Railroad*, 73 miles, connects at the latter place with a railroad running to Madison, the capital of the State, and also to the Mississippi River.

The City of RACINE, Wis., 62 miles from Chicago and 23 miles south of Milwaukee, is built on an elevation some forty or fifty feet above the surface of the lake. It is a beautiful and flourishing place. Here are a light-house, piers, storehouses, etc., situated near the water, while the city contains some fine public buildings and private residences. The population is about 9,880, and is rapidly increasing. Racine is the second city in the State in commerce and population, and possesses a fine harbor. Here are located the county buildings, fourteen churches, several hotels, *Congress Hall* being the largest; elevators, warehouses, and numerous stores of different kinds.

The *Racine and Mississippi Railroad* extends from this place to the Mississippi River at Savanna, 142 miles. The Chicago and Milwaukee Railroad also runs through the town, near the Lake Shore.

MILWAUKEE HARBOR.

Milwaukee, "THE CREAM CITY," 85 miles from Chicago, by railroad and steamboat route, is handsomely situated on rising ground on both sides of the Milwaukee River, at its entrance into Lake Michigan. In front of the city is a bay or indentation of the lake, affording a good harbor, except in strong easterly gales. The harbor is now being improved, and will doubtless be rendered secure at all times of the season. The river affords an extensive water-power, capable of giving motion to machinery of almost any required amount. The city is built upon beautiful slopes, descending toward the river and lake. It has a United States Custom House and Post-Office building; a court house, city hall, a United States land-office, the University Institute, a college for females, three academies, three orphan asylums, forty-five churches, several well-kept hotels, the *Newhall House* and the *Walker House* being the most frequented; seven banks, six insurance companies, a Chamber of Commerce, elevators, extensive ranges of stores, and several large manufacturing establishments. The city is lighted with gas, and well supplied with good water. Its exports of lumber, agricultural produce, etc. are immense, giving profitable employment to a large number of steamers and other lake craft, running to different ports on the Upper Lakes, Detroit, Buffalo, etc. The growth of this city has been astonishing; twenty years since its site was a wilderness; now it contains 1870, 71,464 inhabitants, and of a class inferior to no section of the Union for intelligence, sobriety, and industry.

The future of Milwaukee it is hard to predict; here are centring numerous railroads finished and in course of construction, extending south to Chicago, west to the Mississippi River, and north to Lake Superior, which, in connection with the Detroit and Milwaukee Railroad, terminating at Grand Haven, 85 miles distant by water, and the lines of steamers running to this port, will altogether give an impetus to this favored city, blessed with a good climate and soil, which the future alone can reveal.

During the past few years an unusual number of fine buildings have been erected, and the commerce of the port has amounted to $60,000,000. The bay of Milwaukee offers the best advantages for the construction of a harbor of refuge of any point on Lake Michigan. The city has expended over $100,000 in the construction of a harbor; this needs extension and completion, which will no doubt be effected.

The approach to Milwaukee harbor by water is very imposing, lying between two headlands covered with rich foliage, and dotted with residences indicating comfort and refinement not to be exceeded on the banks of the Hudson or any other body of water in the land. This city, no doubt, is destined to become the favored residence of opulent families, who are fond of congregating in favored localities.

THE GRANARIES OF MINNESOTA AND WISCONSIN.—The La Crosse *Democrat* speaks as follows of the great strides of agriculture in a region which ten years ago was a wilderness. It says: "We begin to think that the granaries of Minnesota and Northwestern Wisconsin will never give out; there is no end to the amount, judging from the heavy loads the steamers continually land at the depot of the La Crosse and Milwaukee Railroad. Where does it all come from? is the frequent inquiry of people. We can hardly tell. It seems impossible that there can be much more left, yet steamboat men tell us that the grain is not near all hauled to the shipping points on the river. What will this country be ten years hence, at this rate? Imagine the amount of transportation that will become necessary to carry the produce of the upper country to market. It is hard to state what will be the amount of shipments of grain this season (1863), but it will be well into the millions."

RAILROADS RUNNING FROM MILWAUKEE.

Detroit and Milwaukee (Grand Haven to Detroit, 189 miles), connecting with steamers on Lake Michigan.

La Crosse and Milwaukee, 200 miles, connecting with steamers on the Upper Mississippi.

Milwaukee and Prairie du Chien, 192 miles, connecting with steamers on the Mississippi and railroad to St. Paul.

Milwaukee and Horicon, 93 miles.

Milwaukee and Western, 71 miles.

Milwaukee and Chicago, 85 miles; also, the River and Lake Shore City Railway, running from the entrance of the harbor to different parts of the city.

PORT WASHINGTON, Ozaukee Co., Wis., 25 miles north of Milwaukee, is a flourishing place, and capital of the county. The village contains, besides the public buildings, several churches and hotels, twelve stores, three mills, an iron foundry, two breweries, and other manufactories. The population is about 2,500. Here is a good steamboat landing, from which large quantities of produce are annually shipped to Chicago and other lake ports.

SHEBOYGAN, Wis., 50 miles north of Milwaukee and 130 miles from Chicago, is a thriving place, containing 1870, 5,310 inhabitants. Here are seven churches, several public-houses and stores, together with a light-house and piers ; the harbor being improved by government works. Large quantities of lumber and agricultural products are shipped from this port. The country in the interior is fast settling with agriculturists, the soil and climate being good. A railroad nearly completed runs from this place to FOND DU LAC, 42 miles west, lying at the head of Lake Winnebago.

MANITOUWOC, Wis., 70 miles north of Milwaukee and 33 miles east from Green Bay, is an important shipping port. It contains 1870, 5,168 inhabitants; five churches, several public-houses, twelve stores, besides several storehouses ; three steam saw-mills, two ship-yards, light-house, and pier. Large quantities of lumber are annually shipped from this port. The harbor is being improved so as to afford a refuge for vessels during stormy weather.

"Manitouwoc is the most northern of the harbors of Lake Michigan improved by the United States Government. It derives additional importance from the fact that, when completed, it will afford the first point of refuge from storms for shipping bound from any of the other great lakes to this, or to the most southern ports of Lake Michigan."

TWO RIVERS, Wis., seven miles north from Manitouwoc, is a new and thriving place at the entrance of the conjoined streams (from which the place takes its name) into Lake Michigan. Two piers are here erected, one on each side of the river ; also a ship-yard, an extensive leather manufacturing company, chair and pail factory, and three steam saw-mills. The village contains about 2,000 inhabitants.

KEWAUNEE, Wis., 25 miles north of Two Rivers and 102 miles from Milwaukee, is a small shipping town, where are situated several saw-mills and lumber establishments. Green Bay is situated about 25 miles due west from this place.

AHNEEPEE, 12 miles north of Kewaunee, is a lumbering village, situated at the mouth of Ahneepee, containing about 1,000 inhabitants. The back country here assumes a wild appearance, the forest trees being mostly pine and hemlock.

GIBRALTAR, or BAILEY'S HARBOR, is a good natural port of refuge for sailing craft when overtaken by storms. Here is a settlement of some 400 or 500 inhabitants, mostly being engaged in fishing and lumbering.

PORT DES MORTS or DEATH'S DOOR, the entrance to Green Bay, is passed 20 miles north of Bailey's Harbor, *Detroit Island* lying to the northward.

POTTOWATOMEE, or WASHINGTON ISLAND, is a fine body of land attached to the State of Michigan ; also, Rock Island, situated a short distance to the north. (*See route to Green Bay, &c.*).

On leaving *Two Rivers*, the steamers passing through the Straits usually run for the Manitou Islands, Mich., a distance of about 100 miles. Soon after the last vestige of land sinks below the horizon on the west shore, the vision catches the dim outline of coast on the east or Michigan shore at *Point aux Bec Scies*, which is about 30 miles south of the Big Manitou Island. From this point, passing northward by *Sleeping Bear Point*, a singular shaped headland looms up to the view.

LITTLE, OR SOUTH MANITOU ISLAND, 260 miles from Chicago, and 110 miles from Mackinac, lies on the Michigan side of the lake, and is the first island encountered on proceeding northward from Chicago. It rises abruptly on the west shore 2 or 300 feet from the water's edge, sloping toward the east shore, on which is a light-house and a fine harbor. Here steamers stop for wood. BIG OR NORTH MANITOU is nearly twice as large as the former island, and contains about 14,000 acres of land. Both islands are settled by a few families, whose principal occupation is fishing and cutting wood for steamers and sailing vessels.

FOX ISLANDS, 50 miles north from South Manitou, consist of three small islands lying near the middle of Lake Michigan, which is here about 60 miles wide. On the west is the entrance to Green Bay, on the east is the entrance to Grand Traverse Bay, and immediately to the north is the entrance to little Traverse Bay.

GREAT and LITTLE BEAVER Islands, lying about midway between the Manitou Islands and Mackinac, are large and fertile bodies of land, formerly occupied by Mormons, who had here their most eastern settlement.

GARDEN and HOG Islands are next passed before reaching the Strait of Mackinac, which, opposite old Fort Mackinac, is about six miles in width.

STRAITS OF MACKINAC.—The Straits of Mackinac, where stands Mackinac City on the site of Old Fort Mackinac, have been the theatre of interesting and exciting events from the earliest times down to the present. While the whole southern portion of Michigan was yet a wilderness which no white man had ever penetrated, Mackinac was the home of the missionary, the trader and the warrior, and the centre of a valuable and important traffic with the Indians of the Northwest.

These are significant facts. The early French Jesuits and traders fixed upon Mackinac as a basis of their missionary and commercial operations, not by mere chance, but because of its natural advantages. Nature alone has given it its advantages and made it what it has been in history. For a series of years, however, its natural seemed to be overlooked, and the surging wave of population rolled across southern Michigan and so on to the westward. Yet it has never been quite forgotten, and at the present time we believe it to be rapidly rising into favor, owing to the fact that it is appreciated as a railroad terminus, or connection with lines of travel across the continent.

GROSSE ISLE ST. MARTIN and Isle St. Martin lie within the waters of the Strait, eight or ten miles north of the island of Mackinac. In the neighborhood of these different islands are the favorite fishing-grounds both of the indian and the "pale face."

MACKINAC CITY, lying on the Straits of Mackinac, opposite point St. Ignace, on the north main shore, is an embryo settlement at the most northern point of the Lower Peninsula of Michigan and where formerly stood old *Fort Mackinac* of Indian fame.

Mackinac.—This important town and fortress is situated in N. lat. 45° 54', W. lon. 84° 30' from Greenwich, being seven degrees thirty minutes west from Washington. It is 350 miles north from Chicago, 100 miles south of Saut Ste. Marie by the steamboat route, and about 300 miles northwest from Detroit. *Fort Mackinac,* garrisoned by U. States troops, stands on elevated ground, about 200 feet above the water, overlooking the picturesque town and harbor below. In the rear, about half a mile distant stand the ruins of old *Fort Holmes,* situated on the highest point of land, at an elevation of 320 feet above the water, affording an extensive view.

The town contains two churches, five hotels, ten or twelve stores, 100 dwelling-houses, and about 700 inhabitants. The climate is remarkably healthy and delightful during the summer months, when this favored retreat is usually thronged with visitors from different parts of the Union, while the Indian warriors, their squaws and their children, are seen lingering around this their favorite island and fishing-ground.

The Island of MACKINAC, lying in the Strait of Mackinac, is about three miles long and two miles wide. It contains many deeply interesting points of attraction in addition to the village and fortress; the principal natural curiosities are known as the Arched Rock, Sugar Loaf, Lover's Leap, Devil's Kitchen, Robinson's Folly, and other objects of interest well worthy the attention of the tourist. The *Mission House* and *Island House* are the principal hotels, while there are several other good public-houses for the accommodation of visitors.

ISLAND OF MACKINAC.—The view given represents the Island, approaching from the eastward. "A cliff of limestone, white and weather-beaten, with a narrow alluvial plain skirting its base, is the first thing which commands attention;" but, on nearing the harbor, the village (2), with its many picturesque dwellings, and the fortress (3), perched near the summit of the Island, are gazed at with wonder and delight. The promontory on the left is called the "Lover's Leap" (1), skirted by a pebbly beach, extending to the village. On the right is seen a bold rocky precipice, called "*Robinson's Folly*" (5), while in the same direction is a singular peak of nature called the "*Sugar Loaf.*" Still farther onward, the "*Arched Rock,*" and other interesting sights, meet the eye of the explorer, affording pleasure and delight, particularly to the scientific traveller and lover of nature. On the highest ground, elevated 320 feet above the waters of the Strait, is the signal station (4), situated near the ruins of old *Fort Holmes.*

The settlement of this Island was commenced in 1764. In 1793 it was surrendered to the American government; taken by the British in 1812; but restored by the treaty of Ghent, signed in Nov., 1814

FORT MACKINAC, Mackinac County, Mich.

As this important Military Post is attracting the attention of the Government, as well as pleasure-seekers, where is to be reserved a National Park, we insert a late report of the Assistant Surgeon of the United States Army.

FORT MACKINAC is situated on a bluff on the south-eastern portion of the island of Mackinac, near the Straits of the same name, which connect Lakes Huron and Michigan, latitude 45° 51' north, longitude 84° 41' west; height above the the lake, 155 feet; above the sea, 728 feet. The nearest post is Fort Brady, 60 miles to the northeast. The island was first occupied by the English as a military post, soon after the destruction of old Fort Mackinac, (8 miles southwest,) and its garrison on the mainland by the French in 1763, on account of its security from attacks from Indians. About 1795, it was turned over to the United States Government by treaty, as a part of the results of the revolutionary war, but in 1812, (after war was declared) it was again occupied by the English. The island is about nine miles in circumference, and rises on its eastern and southern shore ,in abrupt rocky cliffs, the highest point being 250 feet above the water, Fort Mackinac being situated on the south-side, near the lake, situated on the highest point of the island, and about half a mile to the rear of the fortress is "Fort Holmes," which was built by the English during the occupancy of the island in 1812-'13-'14, and called by them "Fort George." It was upon this point that the United States forces were making an attack when Major Holmes, of the United States army, was killed, which circumstance subsequently gave the present name to the work.

Geologically the island is made up of the Onondaga salt group of the upper Silurian system, and the upper Helderberg limestone group of the Devonian System. The former is 25 feet in thickness, forming the base, and the latter is about 275 feet in depth, forming the body and cap. The face of the south end of the island is most plainly terraced. Beginning with the top of Fort Holmes, more than 200 feet above the present level of the lake, there are four distinctly marked tables or terraces before we come to the water, each bearing the undulating line of aqueous formation. Another proof of the existence of wave action, which must have been in process for a long period of time, is the fact that from the base of Fort Holmes to the present beach, worn, rounded pebbles, similar to those on the beach, are found upon digging two or three feet into the earth at any point on the line indicated; all arranged and sorted according to size, just as they are on the beach at the present time. The existence of the island is therefore evidently due to no sudden uplift, but to the gradual subsidence of the waters of the lakes, consuming thousands of years of time.

The timber on the island is mostly small, probably owing to its having been cut down at not a very remote period. It is composed of beach, maple, oak, and poplar, principally, with a liberal supply of the *coniferæ*, viz.:—pine, spruce, hemlock, cedar, tamarack, &c.

The reservation contains a little over

137

two square miles. The surface is regular, but there is very little soil covering the underlying rock. The climate is agreeable, the presence of a large body of water preventing extremes of temperature. The extremes are 9° Fahr., and 83° Fahr., the average about 40° Fahr.

The water supply for the fort is from the lake, by water carts, and from cisterns. The natural drainage is good, and is the only form in use. The general sanitary condition of the post is good and there are no prevaling diseases.

———

CHEBOYGAN, lying 18 miles southeast of the Island of Mackinac, on the lower Peninsula of Michigan, is a flourishing commercial town, situated on Duncan Bay at the mouth of the Cheboygan River. It is attracting the attention of business men and pleasure-seekers. Six miles in the interior is Mullet's Lake, some 12 miles in length by 6 in breadth. Still further back is Burt Lake, nearly as large, also Cheboygan Lake and other lakes of smaller dimensions, abounding in fish and wild game of different kinds, affording sportsmen the largest scope for enjoyment. The lakes and several of the streams are navigable for a small class of steamers.

Cheboygan has a population of about two thousand, 2 churches, several good hotels, 12 or 15 stores, grist mill, and several large saw mills. No more desirable or satisfactory pleasure-trip could be made than to this place and up the beautiful chain of lakes and rivers here entering into Lake Huron.

———

The time is not distant when the lovely Island of Mackinac with its National Park, will be the centre for the congregation of fashion and refinement, from whence pleasure boats and steamers will run to all the lovely and charming resorts in and about the Straits of Mackinac; also to the Saut Ste. Marie, passing through the lovely St. Mary's River, with its lakes and rapids, where trouts and white fish abound.

———◆———

" *Beauteous Isle!* I sing of thee,
    Mackinac, my Mackinac.
Thy lake-bound shores I love to see,
    Mackinac, my Mackinac.
From Arch Rock's bright and shelving
        steep
To western cliffs and Lover's Leap,
Where memories of the lost one sleep,
    Mackinac, my Mackinac.

"Thy Northern shore-trod British foe,
    Mackinac, my Mackinac.
That day saw gallant Holmes laid low,
    Mackinac, my Mackinac.
Now freedom's flag above thee waves,
And guards the rest of fallen braves,
Their requiem sung by Huron's waves,
    Mackinac, my Mackinac."

The Lover's Leap.—MACKINAC ISLAND.

The huge rock called the "Lover's Leap," is situated about one mile west of the village of Mackinac. It is a high perpendicular bluff, 150 to 200 feet in height, rising boldly from the shore of the Lake. A solitary pine-tree formerly stood upon its brow, which some Vandal has cut down.

Long before the pale faces profaned this island home of the Genii, Me-che-ne-mock-e-nung-o-qua, a young Ojibway girl, just maturing into womanhood, often wandered there, and gazed from its dizzy heights and witnessed the receding canoes of the large war parties of the combined bands of the Ojibways and Ottawas, speeding South, seeking for fame and scalps.

It was there she often sat, mused, and hummed the songs Ge-niw-e-gwon loved; this spot was endeared to her, for it was there that she and Ge-niw-e-gwon first met and exchanged words of love, and found an affinity of souls or spirits existing between them. It was there she often sat and sang the Ojibway love song—

> " Mong-e-do-gwain, in-de-nain-dum,
> Mong-e-do-gwain, in-de-nain-dum;
> Wain-shung-ish-ween, neen-e-mo-shane,
> Wain-shung-ish-ween, neen-e-mo-shane,
> A-nee-wau-wau-sau-bo-a-zode,
> A-nee-wau-wau-sau-bo-a-zode."

I give but one verse, which may be translated as follows:

> A loon, I thought was looming,
> A loon, I thought was looming;
> Why! it is he, my lover,
> Why! it is he, my lover.
> His paddle, in the waters gleaming,
> His paddle in the waters gleaming.

From this bluff she often watched and listened for the return of the war parties, for amongst them she knew was Ge-niw-e-gwon; his head decorated with war-eagle plumes, which none but a brave could sport. The west wind often wafted far in advance the shouts of victory and death, as they shouted and sang upon leaving Pe-quot-e-nong (old Mackinac), to make the traverse to the Spirit, or Fairie Island.

One season, when the war party returned, she could not distinguish his familiar and loved war-shout. Her thinking spirit, or soul (presentiment) told her that he had gone to the Spirit Land of the west. It was so, an enemy's arrow had pierced his breast, and after his body was placed leaning against a tree, his face fronting his enemies he died; but ere he died he wished the mourning warriors to remember him to the sweet maid of his heart. Thus he died far away from home and the friends he loved.

Me-che-ne-mock-e-nung-o-qua's heart hushed its beatings, and all the warm emotions of that heart were chilled and dead. The moving, living spirit or soul of her beloved Ge-niw-e-gwon she witnessed, continually beckoning her to follow him to the happy hunting grounds of spirits in the west—he appeared to her in human shape, but was invisible to others of his tribe.

One morning her body was found mangled at the foot of the bluff. The soul had thrown aside its covering of earth, and had gone to join the spirit of her beloved Ge-niw-e-gwon, to travel together to the land of spirits, realizing the glories and bliss of a future, eternal existence.

Yours, &c.,

WM. M. J * * * * * *

ALTITUDE OF VARIOUS POINTS ON ISLAND OF MACKINAC.

Localities.	Above Lake Huron.	Above the Sea.
Lake Huron.................	000 feet.	574 feet
Fort Mackinac	150 "	724 "
Old Fort Holmes	315 "	889 "
Robinson's Folly......... .	128 "	702 "
Chimney Rock...........	131 "	705 "
Top of Arched Rock.......	140 "	714 "
Lover's Leap..............	145 "	719 "
Summit of Sugar Loaf... .	284 "	858 "
Principal Plateau of Mackinac Island............ }	160 "	734 "
Upper Plateau............	800 "	874 "
La Cloche Mountain, north side Lake Huron, C. W. }	1,200 "	1,774 "

The whole Island of Mackinac is deeply interesting to the scientific explorer, as well as to the seeker of health and pleasure. The following extract, illustrated by an engraving, is copied from "FOSTER and WHITNEY'S *Geological Report*" of that region:

"As particular examples of denuding action on the island, we would mention the 'Arched Rock' and the 'Sugar Loaf.' The former, situated on the eastern shore, is a feature of great interest. The cliffs here attain a height of nearly one hundred feet, while at the base are strewn numerous fragments which have fallen from above. The *Arched Rock* has been excavated in a projecting angle of the limestone cliff, and the top of the span is about ninety feet above the lake-level, surmounted by about ten feet of rock. At the base of a projecting angle, which rises up like a buttress, there is a small opening, through which an explorer may pass to the main arch, where, after clambering over the steep slope of debris and the projecting edges of the strata, he reaches the brow of the cliff.

"The beds forming the summit of the arch are cut off from direct connection with the main rock by a narrow gorge of no great depth. The portion supporting the arch on the north side, and the curve of the arch itself, are comparatively fragile, and cannot, for a long period, resist the action of rains and frosts, which, in this latitude, and on a rock thus constituted, produce great ravages every season. The arch, which on one side now connects this abutment with the main cliff, will soon be destroyed, as well as the abutment itself, and the whole be precipitated into the lake.

"It is evident that the denuding action roducing such an opening, with other attendant phenomena, could only have operated while near the level of a large body of water like the great lake itself; and we find a striking similarity between the denuding action of the water here in time past, and the same action as now manifested in the range of the *Pictured Rocks* on the shores of Lake Superior. As an interesting point in the scenery of this island, the Arched Rock attracts much attention, and in every respect is worthy of examination." (*See Engraving.*)

Other picturesque objects of great interest, besides those enumerated above, occur at every turn on roving about this enchanting island, where the pure, bracing air and clear waters afford a pleasurable sensation, difficult to be described unless visited and enjoyed.

The bathing in the pure waters of the Strait at this place is truly delightful, affording health and vigor to the human frame.

The Island of Mackinac.

ROMANTIC AND PICTURESQUE APPEARANCE OF THE ISLAND AND SURROUNDING COUNTRY—ITS PURITY OF ATMOSPHERE—A MOONLIGHT EXCURSION, &c., &c.

——"From whose rocky turrets battled high,
Prospect immense spread out on all sides round;
Lost now between the welkin and the main,
Now walled with hills that slept above the storm,
Most fits such a place for musing men;
Happiest, sometimes, when musing without aim."
[POLLOK.

In this Northern region, Nature has at last fully resumed her green dress. Flowers wild, but still beautiful, bloom and disappear in succession. Birds of various hues have returned to our groves, and welcome us as we trace these shady walks. "In all my wand'rings round this world of care," I have found no place wherein the climate, throughout the summer season, seems to exercise on the human constitution a more beneficial influence than on this Island. In other parts of this country and in Europe, the places of *Resort* are beautiful, indeed; but a certain oppressiveness there at times pervades the

ARCHED ROCK.—Mackinac.

air, that a person even with the best health in the world, feels a lassitude creeping through his frame. Here, we seldom, if ever, experience such a feeling from this cause. For the western breeze even in the hottest days passing over this island, keeps the air cool, and, especially if proper exercise be taken by walking or riding, ne feels a bracing up, a certain buoyancy of spirits that is truly astonishing.

Ye inhabitants of warm latitudes, who pant in cities for a breath of cool air, fly to this isle for comfort. Ye invalid, this is the place in which to renovate your shattered constitution. The lovers of beautiful scenery or the curious in nature, and the artist, whose magic pencil delights to trace nature's lineaments, need not sigh for the sunny clime of Italy for subjects on which to feed the taste and imagination.

This island is intersected by fine carriage roads, shaded here and there by a young growth of beech, maple, and other trees. On the highest part of it, about 300 feet, are the ruins of Old Fort Holmes. From this point of elevation, the scenery around is extensive and beautiful. In sight, are some localities connected with "the tales of the times of old," both of the savage and the civilized. Looking westwardly, and at the distance of about four miles across an arm of Lake Huron, is Point St. Ignace, which is the southernmost point of land, of the greater portion of the Upper Peninsula. Immediately south of it are the "Straits of Mackinac," which separating the Northern and Southern Peninsulas from each other, are about four miles wide. On the south shore, may still be seen traces of Old Fort Mackinac, which is well known in history as having been destroyed by Indians, in 1763, at the instigation of Pontiac, an Indian Chief. Turning our gaze southeastwardly, we see the picturesque "Round Island," as it were at our feet. And further on, is "Bois-Blanc Island," stretching away

with its winding shores, far into Lake Huron. Look to the east, and there stands this inland sea, apparently "boundless and deep," and "pure as th' expanse of heaven." Directly north from our place of observation, are the "Islands of St. Martin;" while beyond them in the Bay, are two large rivers—the Pine, and Carp Rivers. And lastly, casting our eyes towards the northwest, we see on the main land the two "Sitting Rabbits;" being two singular looking hills or rocks, and so called by the Indians from some resemblance at a distance to rabbits in a sitting posture. As a whole, this scenery presents, hills, points of land jutting into the lake, and "straits," bays, and islands. Here, the lake contracts itself into narrow channels, or straits, which at times are whitened by numerous sails of commerce; and there, it spreads itself away as far as the eye can reach. And, while contemplating this scene, perhaps a dark column of smoke, like the Genii in the Arabian Tales, may be seen rising slowly out of the bosom of Lake Huron, announcing the approach of the Genii of modern days, the Steamboat! Let us descend to the shore.

It is evening! The sun, with all his glory has disappeared in the west; but the moon sits in turn the arbitress of heaven. And now—

"How sweet the moonlight sleeps upon this bank; Here will we sit, and let the sounds of music Creep in our ears; soft stillness and the night, Becomes the touches of sweet harmony."

Such a moonlight night I once enjoyed. The hum of day-life had gradually subsided, and there was naught to disturb the stillness of the hour, save the occasional laughter of those who lingered out in the open air. In the direction of the moon, and on the Lake before me, there was a broad road of light trembling upon its bosom. A few moments more, two small boats with sails up to catch the gentle breeze, were seen passing and re-passing this broad road of light. Then the vocal song was raised on the waters, and woman's voice was borne on moonlight beam to the listening ear in the remotest shades. The voices became clearer and stronger as the boats approached nearer; then, again, dying away in the distance, seemed to be merged with the mellow rays of the moon. But let us leave poetry and fancy aside, and come to matters of fact, matters of accommodation, prepared for those who may favor our island with their visits this summer.

There are several large hotels, with attentive hosts, ever ready to contribute towards the comforts of their visitors. Walking, riding, fishing, shooting, and sailing can be here pursued with great benefit to health. We have billiard-rooms and bowling-alleys; in the stores are found Indian curiosities; and, perhaps, the Indians themselves, who resort to this island on business, may be curiosities to those who have never seen them; they are the true "native Americans," the *citizens* of this North American Republic.

ROUND ISLAND is a small body of land lying a short distance southeast of Mackinac, while BOIS BLANC ISLAND is a large body of land lying still farther in the distance, in the Straits of Mackinac.

ST. MARTIN'S BAY, and the waters contiguous, lying north of Mackinac, afford fine fishing grounds, and are much resorted to by visitors fond of aquatic sports. *Great St. Martin's* and *Little St. Martin's Islands* are passed before entering the bay, and present a beautiful appearance.

CARP and PINE rivers are two small streams entering into St. Martin's Bay, affording an abundance of brook trout of a large size. From the head of the above bay to the foot of Lake Superior, is only about 30 miles in a northerly direction, passing through a wilderness section of country, sparsely inhabited by Indians, who have long made this region their favored hunting and fishing grounds.

POINT DE TOUR, 36 miles east from Mackinac, is the site of a light-house and settlement, at the entrance of St. Mary's River, which is here about half a mile in width; this passage is also called the West Channel. At a distance of about two miles above the Point is a new settlement, where have been erected a steamboat pier, a hotel, and several dwellings.

DRUMMOND ISLAND, a large and important body of land belonging to the United States, is passed on the right, where are to be seen the ruins of an old fort erected by the British. On the left is the mainland of Northern Michigan. Ascending St. Mary's River, next is passed ROUND or PIPE ISLAND, and other smaller islands on the right, presenting a beautiful appearance, most of them belonging to the United States.

ST. JOSEPH ISLAND, 10 miles above Point de Tour, is a large and fertile island belonging to Canada. It is about 20 miles long from east to west, and about 15 miles broad, covered in part with a heavy growth of forest-trees. Here are seen the ruins of an old fort erected by the British, on a point of land commanding the channel of the river.

CARLTONVILLE is a small settlement on the Michigan side of the river, 12 miles above the De Tour. Here is a steam saw-mill and a few dwelling-houses.

LIME ISLAND is a small body of land belonging to the United States, lying in the main channel of the river, about 12 miles from its mouth. The channel here forms the boundary between the United States and Canada.

MUD LAKE, as it is called, owing to its waters being easily riled, is an expansion of the river, about five miles wide and ten miles long, but not accurately delineated on any of the modern maps, which appear to be very deficient in regard to St. Mary's River and its many islands—presenting at several points most beautiful river scenery. In the St. Mary's River there are about fifty islands belonging to the United States, besides several attached to Canada.

NEBISH ISLAND, and *Sailor's Encampment,* situated about half way from the Point to the Saut, are passed on the left while sailing through the main channel.

SUGAR ISLAND, a large body of fertile land belonging to the United States, is reached about 30 miles above Point de Tour, situated near the head of St. Joseph Island. On the right is passed the *British* or *North Channel,* connecting on the east with Georgian Bay. Here are seen two small rocky islands belonging to the British Government, which command both channels of the river.

The *Nebish Rapids* are next passed by the ascending vessel, the stream here running about five knots per hour. The mainland of Canada is reached immediately above the rapids, being clothed with a dense growth of forest-trees of small size. To the north is a dreary wilderness, extending through to Hudson Bay, as yet almost wholly unexplored and unknown, except to the Indian or Canadian hunter.

LAKE GEORGE, twenty miles below the Saut, is another expansion of the river, being about five miles wide and eight miles long. Here a new channel has been formed, by dredging, which gives a greater depth of water than formerly.

The highlands to the north of Lake George present a wild and rugged appearance.

CHURCH'S LANDING, on Sugar Island, twelve miles below the Saut, is a steamboat landing; opposite it is SQUIRREL ISLAND, belonging to the Canadians. This is a convenient landing, where are situated a store and dwelling. The industrious occupants are noted for the making of *raspberry jam,* which is sold in large quantities, and shipped to Eastern and Southern markets.

Garden River Settlement is an Indian village ten miles below the Saut, on the Canadian shore. Here are a missionary church and several dwellings, surrounded by grounds poorly cultivated, fishing and hunting being the main employment of the Chippewa Indians who inhabit this section of country. Both sides of the river abound in wild berries of good flavor, which are gathered in large quantities by the Indians, during the summer months.

ST. MARY'S RIVER AND MACKINAC.

"The scenery of the St. Mary's River seems to grow more attractive every year. There is a delicious freshness in the countless evergreen islands that dot the river in every direction, from the Falls to Lake Huron, and I can imagine of no more tempting retreats from the dusty streets of towns, in summer, than these islands; I believe the time will soon come when neat summer cottages will be scattered along the steamboat route on these charming islands. A summer could be delightfully spent in exploring for new scenery and in fishing and sailing in these waters.

"And Mackinac, what an attractive little piece of *terra firma* is that island—half ancient, half modern! The view from the fort is one of the finest in the world. Perched on the brink of a precipice some two hundred feet above the bay—one takes in at a glance from its walls the harbor, with its numerous boats and the pretty village; and the whole rests on one's vision more like a picture than a reality. Everything on the island is a curiosity; the roads or streets that wind around the harbor or among the grove-like forests of the island are naturally pebbled and macadamized; the buildings are of every style, from an Indian lodge to a fine English house. The island is covered with charming natural scenery, from the pretty to the grand, and one may spend weeks constantly finding new objects of interest and new scenes of beauty. It is unnecessary to particularize—every visitor will find them, and enjoy the sight more than any description.

"The steamers all call there, on their way to and from Chicago, and hundreds of small sail vessels, in the fishing trade, have here their head-quarters. Drawn upon the pebbled beach or gliding about the little bay are bark canoes and the far-famed 'Mackinac boats,' without number. These last are the perfection of light sail-boats, and I have often been astonished at seeing them far out in the lake, beating up against winds that were next to gales. Yesterday the harbor was thronged with sail boats and vessels of every description, among the rest were the only two iron steamers that the United States have upon all the lakes, the 'Michigan' and the 'Surveyor,' formerly called the 'Abert,' employed in the Coast Survey.

"For a wonder, Lake Huron was calm and at rest for its entire length, and the steamer 'Northerner' made a beautiful and quick passage from Mackinac to this place. The weather continues warm and dry, and hundreds are regretting they have so early left the Saut and Mackinac, and we believe you will see crowds of visitors yet.   JAY."

St. Mary's River.

By a careful examination of the Government Charts of the Straits of Mackinac and River Ste. Marie, published in 1857, it appears that the *Point De Tour Light-House* is situated in 45° 57' N. Lat., being 36 miles to the eastward of Fort Mackinac. The width of the De Tour passage is about one mile, with a depth of water of 100 feet and upwards, although but 50 feet is found off the light, as you run into Lake Huron. *Drummond Island*, attached to the United States, lies on the east, while the main shore of Michigan lies to the west of the entrance. *Pipe Island*, 4 miles, is first passed on ascending the stream, and then *Lime Island*, 6 miles further. *St. Joseph's Island*, with its *old fort*, attached to Canada, lies 8 miles from the entrance. *Potagannissing Bay*, dotted with numerous small islands, mostly belonging to the United States, is seen lying to the eastward, communicating with the North Channel. *Mud Lake*, 6 miles further, is next entered, having an expanse of about 4 miles in width, when *Sailor's Encampment Island* is reached, being 20 miles from Lake Huron. The head of St. Joseph's and part of *Sugar Island* are reached 26 miles northward from the De Tour, where diverges the Canadian or North Channel, running into the Georgian Bay; this channel is followed by the Canadian steamers. The *Nebish Rapids* are next passed, and *Lake George* entered, 6 miles further, being 32 miles from Lake Huron. This lake or expansion of the river is 9 miles in length and 4 miles broad, affording 13 feet of water over the shoals and terminating at *Church's Landing*, lying opposite *Squirrel Island*, attached to Canada. *Garden River Settlement*, 3 miles, is an Indian town on the Canada side. *Little Lake George* is passed and *Point Aux Pins* reached, 3 miles further. From Little

Lake George to the *Saut Ste. Marie*, passing around the head of Sugar Island, is 8 miles further, being 55 miles from Lake Huron. The *Rapids*, or *Ship Canal*, extend for about one mile, overcoming a fall of 20 feet, when a beautiful stretch of the river is next passed and *Waiska Bay* entered, 6 miles above the rapids; making the St. Mary's River 62 miles in length. The channel forming the boundary line between Canada and the United States is followed by the ascending steamer from the lower end of St. Joseph's Island to Lake Superior, while a more direct passage is afforded for vessels of light draught through *Hay Lake*, lying west of Sugar Island and entering Mud Lake. Nothing can be more charming than a trip over these waters, when sailing to or from the Straits of Mackinac, thus having in view rich and varied lake and river scenery, once the exclusive and favored abode of the red man of the forest, now fast passing away before the march of civilization.

A NEW ROUTE.

The Detroit *Free Press* says: "Early last season Gen. Weitzel called the attention of the United States Government to the desirability of not only shortening the distance from the country below to Lake Superior, but urged that a route entirely within American waters might be utilized with comparatively little cost. The suggestion met with favor at once, and, through Gen. Humphrey, Gen. Weitzel was instructed to make the necessary surveys and report as soon as practicable.

"The point of departure from the old course is about the middle of Mud Lake, from which the line, as shown on the surveyors' map, passes through West Neebish Rapids to the Saut Canal, making a difference in favor of the new route of thirteen miles, affording a safe passage for vessels, in water having an average depth of sixteen feet, both night and day, whereas now portions of the old route can be traversed only in the day-time, as at East Neebish Rapids and the cut at Lake George. The only dredging necessary to open the new route to commerce will be about twelve feet for a distance of one and a half miles in West Neebish Rapids; at Hay Lake Flats, where four feet of dredging will be required over a space of three and a half miles, and a small amount of work at Sugar Island Rapids."

THE STRAITS OF MACKINAC.

THE OPENING FOR TWENTY-ONE YEARS.

The following Table gives the dates of the opening of the STRAITS for twenty-one years, including the present:—

1854.............. April 25	1860...............April 13	1865...............April 21	1870...............April 3
1855................. May 1	1861...............April 25	1866...............April 29	1871...............April 3
1856..............May 2	1862...............April 18	1867...............April 23	1872...............April 28
1857..............May 1	1863...............April 17	1868...............April 19	1873...............April 30
1858..............April 6	1864...............April 23	1869...............April 23	1874...............April 29
1859..............April 4			

ST. MARY'S SHIP CANAL

NUMBER OF VESSELS, TONNAGE AND PASSENGERS PASSING THROUGH THE CANAL.

Vessels, &c., Passing Through.	No. in 1870.	No. in 1871.	No. in 1872.	No. in 1873.
Sail Vessels	1,397	1,064	1,214	1,544
Steam Vessels	431	573	790	968
Total.............................	1,828	1,637	2,004	2,512
Tolls per ton........................	6 cents.	4½ cents.	4½ cents.	4½ & 3½ cts.
Total receipts for tolls.............	$41,896.00	$33,865.45	$41,232.44	$44,943.18
Total amount of tonnage.........	696,825	752,100	914,735	1,204,445
Increase of tonnage...............	....	....	162,635	289,710
Number of passengers............	17,158	15,859	25,230	30,966
Total receipts since opening the Canal.................................	....	....	$430,542.86	$275,541.04

PRINCIPAL HOTELS IN THE LAKE CITIES, &c.

	NAME.	PROPRIETORS.
BUFFALO, N. Y.	Mansion House,	L. L. Hodges.
	Tift House,	Tuthill Brothers.
NIAGARA FALLS, N. Y.	Cataract House,	Whitney, Jerrauld & Co.
	International Hotel,	Jas. T. Fulton.
(Canada Side)	Clifton House,	Colburn & Co.
NIAGARA, Ont.	Queen's Royal Hotel,	Henry Winnett, Manager.
ST. CATHERINES, Ont.	Stevenson House,	
HAMILTON, "	Royal Hotel,	
TORONTO, "	Queen's Hotel,	Thomas McGaw, Manager.
	Rossin House,	George P. Shears.
ERIE, Penna.	Ellsworth House,	F. H. Ellsworth.
	Reed House,	
CLEVELAND, Ohio,	Forest City House,	Terrill & Ingersoll.
	Kennard House,	David McClasky.
	Weddell House,	R. A. Gillette.
TOLEDO, Ohio.	Boody House,	Mr. DeGroff.
DETROIT, Mich.	Biddle House,	
	Michigan Exchange,	Edward Lyon.
	Russell House,	Witbeck & Chittenden.

TABLE OF DISTANCES,

From Toronto and Collingwood, to Saut Ste. Marie and Fort William, Canada.

PASSING THROUGH GEORGIAN BAY, THE NORTH CHANNEL AND LAKE SUPERIOR.

MILES.	STATIONS, &c.		MILES.
	Northern Railway of Canada.		
436	**TORONTO**............		0
422	Thornhill....................		14
418	Richmond Hill.............	4	18
406	Aurora......................12		30
402	Newmarket..................	4	34
398	Holland Landing.........	4	38
395	Bradford.....................	3	41
387	Gilford.....................	8	49
173	Allendale....................14		63
372	Barrie......................	1	64
356	Sunnidale 16		80
342	**Collingwood**.........14		94
	Steamboat Route.		
	(Georgian Bay.)		
312	Cape Rich....................30		124
296	Owen's Sound..............16		140
262	Cabot's Head...............34		174
242	Lonely Island..............20		194
226	Squaw Island..............16		210
216	Cape Smythe...............10		220
196	*She-ba-wa-nah-ning*........20		240
181	Man-i-tou-wah-ning......15		255
171	*Little Current*—Great Manitoulin Island. }..10		265
151	Clapperton Island.........20		285
121	Barrie Island...............30		315
86	Cockburn Island..........35		350

MILES.	STATIONS, &c.		MILES.
71	Drummond's Is., Mich...15		365
53	Bruce Mines, Can........18		383
45	St. Joseph Island.........	8	391
39	Campement D'Ours Is...	6	397
36	The Narrows...............	3	400
26	Sugar Island, Mich.......10		410
	(St. Mary's River.)		
24	Nebish Rapids.............	2	412
21	Lake George..............	3	415
14	Church's Landing......	7	422
10	Garden River Settle'nt..	4	426
0	**Saut Ste. Marie**...10		436
	(Ship Canal.)		
324	**Saut Ste. Marie**...		436
318	Point aux Pins, Can...	6	442
309	Gros Cap....................	9	451
299	Parisien Island............10		461
294	Goulois Bay and Point..	5	466
280	Sandy Islands..............14		480
275	Batchewanaung Bay.....	5	485
265	Mamainse Point...........10		495
190	Michipicoten Island......75		570
105	Slate Islands...............85		655
75	Ste. Ignace Island........30		685
60	Ent'ce to Neepigon Bay.15		700
15	Silver Islet........,......40		740
0	**Fort William,** Can.20		760

APPROACHES TO LAKE SUPERIOR via NIAGARA FALLS, TORONTO AND COLLINGWOOD, CANADA.

COLLINGWOOD ROUTE.

Of all the approaches to Lake Superior from the Atlantic Seaboard, or from Montreal and Quebec, nothing exceeds the *Collingwood Route* for grand and varied Lake, Island and River Scenery,—Niagara Falls and River,—Lake Ontario,—The Thousand Islands in the St. Lawrence River,—combined, with their shores, have no equal on the continent of America for sublimity and grandeur. Then the Georgian Bay with its innumerable islands, the North Channel and islands of great beauty, together with the lovely St. Mary's River and its Rapids repeats the inducements of pleasure travellers to select this favorite route in approaching or returning on their trips to the Upper Lakes.

The *Great Western Railway of Canada*, as well as the *Grand Trunk Railway* both connect with the above Line of Travel at Toronto; also, Steamers running on the St. Lawrence River and Lake Ontario. The *Northern Railroad of Canada*, 94 miles in length, runs from Toronto to Collingwood, connecting with a Line of Steamers for Saut Ste. Marie and Lake Superior.

Toronto, the Seat of Government for the Province of Ontario, with its beautiful Bay, fine Streets, Public Buildings, and good Hotels, with moderate charges, is an attractive place for Summer Tourists. The *Queen's Hotel* and the *Rossin House* are much frequented by pleasure travellers.

TORONTO TO COLLINGWOOD.

On leaving Toronto for Collingwood, via the *Northern Railway of Canada*, the route extends through an interesting section of country for most of the distance, 94 miles. Several villages and many highly cultivated farms are passed before arriving at the head of *Lake Simcoe*, a beautiful body of water, which is passed lying on the east.

NEW MARKET, is a flourishing incorporated town situated near the west side of Lake Simcoe, in the county of York, 34 miles from Toronto. It is one of the largest and most important stations on the line of the Northern Railway. Population about 3,000.

BELL EWART, 54 miles from Toronto, is a flourishing village on Lake Simcoe. It contains 3 churches, sawmills and other manufacturing establishments. Steamers ply from this place to Barrie, Beaverton, and Orillia, situated at the foot of the lake. A steamer also runs to Muskoka, and other landings northward, passing through an interesting section of country.

ALLENDALE, 63 miles, is a new settlement where commences a Branch Railroad running to Orillia, situated at the foot of Lake Simcoe.

BARRIE, 64 miles north of Toronto, is an incorporated town at the head of Kempenfelt Bay, on Lake Simcoe. It is the capital of the county of Simcoe, being surrounded by a fine section of country. It is distant from Collingwood 30 miles and from Penetanguishene 32 miles. Population about 3,500. A steamer runs from Barrie to the dif-

148

ferent landings on *Lake Simcoe*, which lies about 100 feet above Lake Huron, into which it empties its surplus waters through the Severn River.

Collingwood, 94 miles from Toronto, is an important town lying on the south shore of Georgian Bay. Here is carried on an extensive lumber business, and it offers great facilities to ship builders and other kinds of manufacturing; several of which are in operation. The completion of the *Northern Railway of Canada* made Collingwood the nearest route to Green Bay, Milwaukee and Chicago, as well as to Lake Superior. Steamers and sail vessels now run to all the above ports. A Line of Steamers of a large class now run from Collingwood to the Saut Ste. Marie and the ports on the North Shore of the above Inland Sea, connecting with a line of travel to FORT GARRY, Manitoba.

OWEN SOUND, capital of the county of Grey, is the name of a large and flourishing town situated on the south side of Georgian Bay, 40 miles southwest of Collingwood. Population about 3,500. Steamers run to and from Collingwood and other places. The *Toronto, Grey and Bruce Railroad* is constructed, running to Toronto, distant 122 miles.

GEORGIAN BAY, lying east of Lake Huron, is one of the purest and most interesting bodies of water of the Upper Lake System. Its headlands, harbors, and innumerable islands, forming groups, known as Limestone Islands, Indian Islands, and Parry's Island and Sound, altogether form labyrinths which it is impossible to describe. The islands on the northwest are called Lonely Island, Bustard Islands, Fox Islands, Squaw and Papoose Islands.

The NORTH CHANNEL, extending for upwards of 100 miles westward is another lovely sheet of water, embosoming a large number of virgin islands, covered with a thick foliage.

ROUTE FROM COLLINGWOOD, C. W., TO THE SAULT STE. MARIE.
THROUGH GEORGIAN BAY AND NORTH CHANNEL.

This is a new and highly interesting steamboat excursion, brought into notice by the completion of the *Northern Railway of Canada*, 94 miles in length, extending from Toronto to Collingwood, at the southern extremity of Georgian Bay.

NOTTAWASSAGA BAY, the southern termination of Georgian Bay, is a large expanse of water bounded by Cape Rich on the west, and Christian Island on the east, each being distant about 30 miles from Collingwood. At the south end of the bay lies a small group of islands called the *Hen and Chickens*.

CHRISTIAN ISLAND, lying about 25 miles from Penetanguishene, and 25 miles north-east of Cape Rich, is a large and fertile island, which was early settled by the Jesuits. There are several others passed north of Christian Island, of great beauty, while still farther northwest are encountered innumerable islands and islets, forming labyrinths, and secluded passages and coves as yet almost unknown to the white man, extending westward for upward of one hundred miles.

PENETANGUISHENE, Can., 50 miles north of Collingwood by steamboat route, situated on a lovely and secure bay, is an old and important settlement, comprising an Episcopal and Roman Catholic church, two hotels, several stores and storehouses, and has about 500 inhabitants. In the immediate vicinity are a naval and military depot and barracks, established by the British government. The natural beauties of the bay and harbor, combined with the picturesque scenery of the shores, make up a picture of rare beauty. Here may be seen the native Indian, the half-breed, and the Canadian *Voyageur*, with the full-blooded Englishman or Scotchman, forming one community. This place, being near the mouth of the River Severn, and contiguous to the numberless islands of Georgian Bay, is no doubt destined to become a favorite resort for the angler and sportsman, as well as for the invalid and seeker of pleasure.

On leaving *Collingwood* for Bruce Mines and the Saut Ste. Marie, the steamer usually runs direct across Georgian Bay to Lonely Island, passing Cabot's Head to the right, and the passage leading into the broad waters of Lake Huron, which is the route pursued by the steamers in the voyage to Mackinac, Green Bay, and Chicago. During the summer months the trip from Collingwood to Mackinac and Chicago affords a delightful excursion.

OWEN'S SOUND, or SYDENHAM, 50 miles west of Collingwood, although off the direct route to the Saut Ste. Marie, is well worthy of a passing notice. Here is a thriving settlement, surrounded by a fertile section of country, and containing about 2,500 inhabitants. A steamer runs daily from Collingwood to this place, which will, no doubt, soon be reached by ailroad.

LONELY ISLAND, situated about 100 miles west of Collingwood and 20 miles east of the Great Manitoulin Islands, is a large body of land mostly covered with a dense forest, and uninhabited, except by a few fishermen, who resort here at certain seasons of the year for the purpose of taking fish of different kinds. The steamer usually passes this island on its north side, steering for *Cape Smyth*, a bold promontory jutting out from the Great Manitoulin, and distant from Lonely Island about 25 miles.

SQUAW ISLAND and PAPOOSE ISLAND are seen on the northeast, while farther inland are the *Fox Islands*, being the commencement on the west of the innumerable islands which abound along the north shore of Georgian Bay.

LA CLOCHE MOUNTAINS, rising about 2,000 feet above the sea, are next seen in the distance, toward the north; these, combined with the wild scenery of the islands and headlands, form a grand panoramic view, enjoyed from the deck of the passing steamer.

SMYTH'S BAY is passed on the west, some eight or ten miles distant. At the head of this bay, on the great Manitoulin Island, are situated a village of Indians, and a Jesuit's mission, called We-qua-me-kong. These aborigines are noted for their industry, raising wheat, corn, oats, and potatoes in large quantities. This part of the island is very fertile, and the climate is healthy.

SHE-BA-WA-NAH-NING, signifying, in the Indian dialect, "*Here is a Channel,*" is a most charming spot, 40 miles distant from Lonely Island, hemmed in by mountains on the north and a high rocky island on the south. It is situated on the north side of a narrow channel, about half a mile in length, which has a great depth of water. Here are a convenient steamboat landing, a church, a store, and some ten or twelve dwellings, inhabited by Canadians and half-breeds. Indians assemble here often in considerable numbers, to sell their fish and furs, presenting with their canoes and dogs a very grotesque appearance. One resident at this landing usually attracts much attention—a noble dog, of the color of cream. No sooner does the steamer's bell ring, than this animal rushes to the wharf, sometimes assisting to secure the rope that is thrown ashore;

the next move he makes is to board the vessel, as though he were a custom-house officer; but on one occasion, in his eagerness to get into the kitchen, he fell overboard; nothing daunted, he swam to the shore, and then again boarding the vessel, succeeded in his desire to fill his stomach, showing the instinct which prompts many a biped office-seeker.

On leaving She-ba-wa-nah-ning and proceeding westward, a most beautiful bay is passed, studded with islands; and mountains upwards of 1,000 feet in height, presenting a rocky and sterile appearance, form an appropriate background to the view; thence are passed Badgley and Heywood Islands, the latter lying off Heywood Sound, situated on the north side of the Great Manitoulin.

MAN-I-TOU-WAH-NING, 25 miles north-west of She-ba-wa-nah-ning, is handsomely situated at the head of Heywood Sound. It is an Indian settlement, and also a government agency, being the place annually selected to distribute the Indian annuities.

LITTLE CURRENT, 25 miles west of She-ba-wa-nah-ning, is another interesting landing on the north shore of the Great Manitoulin, opposite La Cloche Island. Here the main channel is narrow, with a current usually running at the rate of five or six knots an hour, being much affected by the winds. The steamer stops at this landing for an hour or upward, receiving a supply of wood, it being furnished by an intelligent Indian or half-breed, who resides at this place with his family. Indians are often seen here in considerable numbers. They are reported to be indolent and harmless, too often neglecting the cultivation of the soil for the more uncertain pursuits of fishing and hunting, although a considerably large clearing is to be seen indifferently cultivated.

CLAPPERTON ISLAND and other islands of less magnitude are passed in the *North Channel*, which is a large body of water

about 120 miles long and 25 miles wide. On the north shore is situated a post of the Hudson Bay Company, which may be seen from the deck of the passing steamer.

COCKBURN ISLAND, 85 miles west of Little Current lies directly west of the Great Manitoulin, from which it is separated by a narrow channel. It is a large island, somewhat elevated, but uninhabited, except by Indians.

DRUMMOND ISLAND, 15 miles farther westward, belongs to the United States, being attached to the State of Michigan. This is another large body of land, being low, and as yet mostly uninhabited.

The next Island approached before landing at Bruce Mines is ST. JOSEPH ISLAND, being a large and fertile body of land, with some few settlers.

BRUCE MINES VILLAGE, Can., is situated on the north shore of Lake Huron, or the "North Channel," as it is here called, distant 290 miles from Collingwood, and 50 from the Saut Ste. Marie. Here are a Methodist chapel, a public-house, and a store and storehouse belonging to the Montreal Copper Mining Company, besides extensive buildings used for crushing ore and preparing it for the market; about 75 dwellings and 600 inhabitants. The copper ore, after being crushed by powerful machinery propelled by steam, is put into puddling troughs and washed by water, so as to obtain about 20 per cent. pure copper. In this state it is shipped to the United States and England, bringing about $80 per ton. It then has to go through an extensive smelting process, in order to obtain the pure metal. The mines are situated in the immediate vicinity of the village, there being ten openings or shafts from which the ore is obtained in its crude state. Horse-power is mostly used to elevate the ore; the whims are above ground, attached to which are ropes and buckets. This mine gives employment to about 300 workmen.

The capital stock of the company amounts to $600,000.

The *Wellington Mine*, about one mile distant, is also owned by the Montreal Mining Company, but is leased and worked by an English company. This mine, at the present time, is more productive than the Bruce Mines.

The Lake Superior *Journal* gives the following description of the Bruce Mine, from which is produced a copper ore differing from that which is yielded by other mines of that peninsula.

"Ten years ago this mine was opened, and large sums expended for machinery, which proved useless, but it is now under new management, and promises to yield profitably. Twelve shafts have been opened, one of which has been carried down some 330 feet. Some 200 or 300 men are employed, all from the European mines. Some of the ores are very beautiful to the eye, resembling fine gold. After being taken out of the shaft, they are taken upon a rail-track to the crushing-house, where they are passed between large iron rollers, and sifted till only a fine powder remains; from thence to the 'jigger-works,' where they are shaken in water till much of the earthy matter is washed away, after which it is piled in the yard ready for shipment, having more the appearance of mud than of copper. It is now mostly shipped to Swansea, in Wales, for smelting. Two years since,

1,500 tons were shipped to Baltimore and Buffalo to be smelted."

On resuming the voyage after leaving Bruce Mines, the steamer runs along St. Joseph Island through a beautiful sheet of water, in which are embosomed some few islands near the main shore.

CAMPEMENT D'OURS is an island passed on the left, lying contiguous to St. Joseph Island. Here are encountered several small rocky islands, forming an intricate channel called the "*Narrows*." On some of the islands in this group are found copper ore, and beautiful specimens of moss. The forest-trees, however, are of a dwarfish growth, owing, no doubt, to the scantiness of soil on these rocky islands.

About 10 miles west of the "Narrows," the main channel of the St. Mary's River is reached, forming the boundary between the United States and Canada. A rocky island lies on the Canadian side, which is reserved for government purposes, as it commands the main or ship channel.

SUGAR ISLAND is now reached, which belongs to the United States, and the steamers run a further distance of 25 miles, when the landing at the Saut Ste. Marie is reached, there being settlements on both sides of the river. The British boats usually land on the north side, while the American boats make a landing on the south side of the river, near the mouth of the ship canal.

INDIAN WIGWAMS, CANOES, &c. — Saut Ste. Marie.

LAKE SUPERIOR GUIDE.

Saut Ste. Marie, the capital of Chippewa county, Michigan, and a port of entry, is advantageously situated on St. Mary's River, or Strait, 355 miles N.N.W. of Detroit, being 50 miles above Lake Huron, and 15 miles from the foot of Lake Superior, in N. lat. 46° 30′; W. long. 84° 43′. The Rapids at this place, giving the name to the settlements on both sides of the river, have a descent of 20 feet within the distance of a mile, and form the natural limit of navigation. The *Ship Canal,* however, finished in 1855, on the American side, obviates this difficulty. Steamers and sail vessels of a large class now pass through the locks into Lake Superior, greatly facilitating trade and commerce.

The village on the American side is pleasantly situated at the foot of the Rapids, and contains a Presbyterian and a Roman Catholic church, 2 hotels, 8 or 10 stores and storehouses, and about 1200 inhabitants, having increased but slowly since its first settlement by the French in 1668. Many of the inhabitants, Indians and half-breeds in the vicinity, are engaged in the fur trade and fisheries; the latter being an important and profitable occupation, here being taken large quantities of white fish. Summer visitors annually flock to this place and the Lake Superior country for health and pleasure. The *Chippewa House,* a well-kept hotel, on the American side, situated near the Steamboat Landing, and one on the Canadian side of the river, both afford good accommodations.

The scene, as witnessed from the deck of the steamer on passing through the locks, is of the most interesting and exciting character. The Ship Canal — the River — the Islands — the two villages in sight on either side of the stream, and the Indians in their birch canoes, engaged in taking white fish below the Rapids, are all in view at the same time, presenting altogether a magnificent panorama.

Fort Brady, erected in 1823, is an old and important United States military post, contiguous to this frontier village. It commands the St. Mary's River at this point and approach to the Ship Canal.

Early in the present century the American Fur Company established a trading post at the Saut, which was kept up until the year 1848.

The mean annual temperature of Saut Ste. Marie is 40½° Fahr.; Spring, 37½°; Summer, 62°; Autumn, 43½°; Winter, 20°; it being situated near the northern limit of the temperate zone. Immediately to the north, in this latitude, the country is liable to killing frosts during the summer months, owing to the cold influence sweeping down from Hudson Bay, some 300 or 400 miles distant.

The *Marquette, Saut Ste. Marie,* and *Mackinac Railroad,* now being surveyed and constructed, will connect the Saut with both Mackinac and Marquette, affording an opportunity to travellers to reach this point at all seasons of the year.

A railroad is also chartered by the Canadian Government to construct a road eastward to Toronto, Montreal, etc.

153

Saut Ste. Marie, Algoma District, Canada, situated on the opposite side of the river, is a scattered settlement, where is located an old post of the Hudson Bay Company. Here is a steamboat landing, a public house and 3 or 4 stores, a stone court-house and jail, 3 churches, and 600 or 700 inhabitants. Indians of the Chippewa tribe reside in the vicinity in considerable numbers, they having the right to take fish in the waters contiguous to the Rapids. They also employ themselves in running the Rapids in their frail bark canoes, when desired by citizens or strangers — this being one of the most exhilarating enjoyments for those fond of aquatic sports.

The streams flowing into the St. Mary's River and Lake Superior, on the Canada side, are favorite resorts for anglers fond of pursuing the brook trout, which are here taken in large quantities during the summer months.

The country in the rear of this frontier settlement is settled for five or six miles; but a few miles farther to the north commences an endless wilderness, extending north to within the Arctic circle, being sparsely inhabited by Chippewas and roving tribes of Indians.

The primitive appearance of the towns on the American and Canadian banks of St. Mary's River, in connection with their surroundings, are of the most interesting character. The Ship Canal and locks are the only perceptible improvements made during the past century, while the mixed character of the population on the South Shore, consisting of Americans, French, half-breeds, and Indians, in connection with their English neighbors on the opposite side of the river, who are a more aristocratic class, remind one of something foreign to the general appearance of American villages. A steam ferry connects the two settlements.

Nothing but the projected railroads to connect with Montreal and Toronto, on the Canada side, and the railroad to be built from this point to Mackinac and Marquette, on the American side, will wake up these places from their "Rip Van Winkle" slumbers.

TROUT FISHING RESORTS.—In the vicinity of the Saut Ste. Marie are several streams where sportsmen resort for the purpose of taking speckled trout. The nearest points are the *Rapids* on both sides of St. Mary's River, and the small streams between the islands on the Canada side; on the American side, *Crystal Rapids*, two miles below the Saut. On St. Mary's River are several projecting points, from one to five miles below the Saut, where anglers resort. *Garden* and *Root Rivers*, on the Canada side of the river, below the falls, are fine trout streams.

On the North Shore, Canada side, are several fine trout fishing resorts, from fifteen to sixty miles from the Saut, where Indians or half-breeds with their canoes have to be employed, often camping out for several days. *Goulais Bay* and *Batcheewanaung Bay*, from twenty to forty miles, are the nearest points. On the latter bay enters Batcheewanaung River and Harmony River, both fine trout streams. Then farther northward, some thirty or forty miles, are the Montreal River and the Aguawa River, both celebrated trout streams, where are good boat harbors.

Fort Brady, Michigan, is situated on the southern bank of the Saut Ste. Marie, in 46° 30′ north latitude; altitude, 600 feet above the ocean. Mean annual temperature, 40° Fahrenheit.

"The military history of this post extends back to 1750, at which time the French claimed jurisdiction over all the territory north of the Ohio, and sought to establish posts at the more important places, for the purpose of controlling the lakes, and excluding the English as far as possible from obtaining a foothold on Lake Superior, as well as to establish a depot of supplies and afford protection to the traders.

"In 1820 the late Gen. Lewis Cass, then Indian agent for the North-west Indians, made a trip around the lakes, visiting the shores of Lake Superior, and afterwards of Lake Michigan, going as far as Chicago; and when he landed at this place, on his voyage up, the British flag was flying at the head of the rapids, near the termination of the canal. He proceeded in person to haul it down and raise his own in its place. There were nearly two thousand natives and French residents, whose sympathies were intensely Canadian, and consequently this act of his so enraged them that they were on the point of attacking him at once. Through the intervention of a few of the English half-breeds, the Indians were quieted, and the General allowed to go in peace. On his return, however, from his voyage around Lake Superior, Gen. Cass concluded a treaty with the Chippewas, on June 20, 1820, for the purchase of sixteen square miles of land, with a river front extending from a large rock near the national boundary, above the falls, to the Little Rapids, at the head of Sugar Island, *the Indians reserving the right to fish undisturbed.* This purchase constituted the original military district. Its boundary was about three and a half miles along the river, by four and one-third miles deep. This rock still remains

as one of the leading peculiarities on Ashman's Bay, being a large boulder lying in shoal water, about twenty rods from the ship channel, entering the upper end of the canal. In 1822 the Government of the United States determined upon its permanent occupancy, and accordingly Gen. Brady was directed to proceed, in the autumn of that year, to this place, with six companies of infantry, and erect a stockade and buildings.

"This post is considered one of the healthiest on the Upper Lakes, yet Dr. McDougall, in his report in 1837–38 to the Surgeon-General, establishes beyond controversy that want of attention to sanitary science cannot be allowed even in this northern region, except at the most imminent risks."

Fort Brady is at the present time (1873) garrisoned by two companies of the first United States infantry, under the command of Captain Kinzie Bates. Here are a park of artillery, officers' quarters, hospital and barracks, with grounds handsomely situated, overlooking the river and the opposite Canadian shore.

The Indians that have resided in this vicinity for the past one or two hundred years are the Chippewa tribe, a numerous body of peaceful Indians, whose habitations surround Lake Huron on the north, and Lake Superior on both shores, extending westward to the Upper Mississippi River.

St. Mary's Ship Canal. — The enlargement of the Ship Canal by the United States Government commenced in 1871, since when an appropriation of $1,000,000 has been granted. The capacity of the new lock will be 80 feet wide, 500 feet long, affording 18 feet of water, to be built of stone, in the most substantial manner. Lift of the lock 18 feet, overrunning the Rapids in St. Mary's River. The canal is one mile and a quarter in length, 80 feet wide at bottom, and 100 feet wide on water surface, accommodating vessels of the largest size navigating the Upper Lakes.

Improvements will have to be made in deepening St. Clair Flats and the St. Mary's River in order to accommodate vessels of a large tonnage. American and Canadian steamers and sail-vessels are almost constantly passing up and down through the locks of the canal during the season of navigation.

Location and General Description of the New Lock. — The axis of the Lock will be parallel to the axis of the present Locks, at a distance of 175 feet; the upper lock gates to be opposite the upper lock gates of the present Locks.

The chamber of the Lock will be 80 feet wide at the coping, and 450 feet in length from quoin to quoin. A set of guard gates will be placed 61¼ feet above the upper lock gates, and the walls continued 70 feet above the hollow quoins of the guard gates, including wing buttresses of 2 feet. At a distance of 61¼ feet below the lower lock gates, a set of guard gates will be placed, to open down stream; the wall on each side to extend 70 feet below the hollow quoin, for the guard gate, including a wing buttress of 2 feet.

The Lock will have a lift of 18 feet, and a depth of 16 feet of water on the mitre sills at the stage which has been assumed as that of ordinary low water.

Extremes of the water-level on Lake Superior, from June 30, 1872, to June 30, 1873, as measured above the Locks at the St. Mary's Ship Canal. — Highest water, November 27, 1872, 2½ feet above ordinary high water; lowest water, April 10, 1873, 2 feet below ordinary high water. Extreme variation 4½ feet, this being the most remarkable variation on record. The above phenomena was caused by a north-west wind blowing over Lake Superior, and the reverse by a south-east wind.

Opening and Closing of the Ship Canal.

The business of the canal began on the eleventh day of May (1872), and continued, without interruption, until the twenty-sixth day of November of the same year, — six and a half months.

In 1873, the first steamer passing through locks, upward bound, was the Keweenaw, May 11th, followed by other steamers and propellers, while the ice remained in the harbors of Marquette and Duluth for about one month later.

STATEMENT

Of Receipts for each year from the opening of the Canal, June 18, 1855, to the close of the season, November, 1873.

Receipts for 1855, 6 cts. per ton on steamers, etc........	$4,374 66
Receipts for 1856............	7,575 78
" " 1857............	9,406 74
" " 1858............	10,848 80
" " 1859............	16,941 84
" " 1860............	24,777 82
" " 1861............	16,672 16
" " 1862............	21,607 17
" " 1863............	30,574 44
" " 1864.........	34,287 31
Receipts for 1865, 4½ cts. per ton on sail vessel.....	22,339 64
Receipts for 1866............	23,069 54
" " 1867............	33,515 54
" " 1868............	25,977 14
" " 1869............	31,579 96
" " 1870............	41,896 43
Receipts for 1871, 4½ cts. per ton on steamers, etc.	33,865 45
Receipts for 1872............	41,232 44
Receipts for 1873, 3½ cts. per ton on steamers, etc.	44,943 18
Total..................	$475,486 04

Principal Places of Resort on Lake Superior and its Vicinity.

1. *Saut Ste. Marie*, with its Rapids, Ship Canal, and trout fishing resorts.
2. *Goulais Bay*, on the Canada side.
3. *Batcheewanaung Bay* and River.
4. *White Fish Point* and Fishing Grounds.
5. *Pictured Rocks* — Chapel, Cascade, Wreck Cliff, Grand Portail or Arched Rock, Zebra Cliff, Spirit Cave, Sail Rock, Miners' Castle, etc.
6. *Grand Island Harbor*, and Munising, with its romantic falls, etc.
7. *Marquette*, with its surroundings, the most fashionable resort on the Lake.
8. *Negaunee* and *Ishpeming*, together with Iron Mines — reached by railroad.
9. *Escanaba*, an iron port, favorably situated on Green Bay — reached by railroad.
10. *L'Anse*, a new and flourishing town on Keweenaw Bay — fine sailing and fishing.
11. *Houghton* and *Hancock* — copper mines and smelting works.
12. *Calumet* and *Hecla* Copper Mine, where is a large settlement.
13. *Copper Harbor* and Lake Fanny Hoe, near the end of Keweenaw Point.
14. *Eagle Harbor* and *Eagle River*, where are copper mines.

15. *Ontonagon* — copper and silver mines.
16. *Bayfield* and *Ashland*, fashionable resorts, where boating and fishing can be enjoyed.
17. *La Pointe* and the *Apostle Islands*.
18. *Duluth* and *Superior City*, together with the Dalles of the St. Louis River.

NORTH SHORE.

19. *Isle Royale*, with its copper mines and precious minerals.
20. *Pigeon River*, and romantic falls, situated on the boundary line.
21. *Fort William* and *Prince Arthur's Landing*, Canada—silver mines and amethysts.
22. *Silver Islet*, and Thunder Cape — silver region.
23. *Nepigon Bay* and River, with romantic scenery and famous trout fishing.
24. *Island of Michipicoten*, with its rich foliage, fishing, etc.
25. *Michipicoten Bay* and River, with grand scenery and good fishing.

All the above places of the South and North Shore are well worthy of a visit. They can be reached by the American or Canadian steamers. Other points of interest can be reached by canoes or sail-boats.

Islands in Lake Superior.

AMERICAN SIDE.

Grand Island and Light.	Manitou Island and Light.	Hermit Island.
Wood Island.	Isle Royale and Light.	Stockton Island.
Train Island.	*Apostle Islands.*	Oak Island.
Middle Island.	Michigan Island & Light.	Manitou Island.
Granite Island and Light.	Outer Island and Light.	Ironwood Island.
Huron Islands and Light.	Madeline Island.	Raspberry Island & Light,
Traverse Island.	Basswood Island.	and twelve others.

CANADIAN SIDE.

Parisien Island.	Michipicoten Island.	Welcome Islands.
Maple Island.	Slate Islands.	Islands in Thunder Bay.
Sandy Islands.	Pic Island.	Pie Island.
Montreal Island.	Simpson's Island.	Thompson's Island.
Lizard Islands.	Ste. Ignace Island.	Spar Island.
Leach Island.	Silver Islet.	Jarvis' Island.
Caribou Island.	Porphyry Islands.	Victoria Island.

STEAMBOAT EXCURSION — South Shore, Lake Superior.

SAUT STE. MARIE, situated on St. Mary's River, or Strait, 55 miles above Point de Tour, on Lake Huron, and 15 miles below Tonquamenon, or White Fish Bay, is a place of great interest to travellers, and the place of embarkation for fishing and pleasure parties, during the Summer months.

From this point is afforded a grand view of the Rapids and Islands lying in the river, while the scene is usually enlivened by seeing Indians taking white fish by means of scoop-nets.

On leaving the Ship Canal, on the upward trip, a beautiful stretch of the river is passed before arriving at POINT AUX PINS, 7 miles, situated on the Canada side. Here is a good steamboat landing and a desirable place for fitting out fishing parties. *Waiska Bay* is next entered, being an expansion of the river of about 5 miles in length — *Round Island Light* is seen on the south.

Iroquois Point and *Light*, on the American side, and *Gros Cap*, on the Canada side, are next passed, 15 miles from the Saut. The latter is a bold promontory, rising 500 or 600 feet above the water.

Tonquamenon, or *White Fish Bay,* is now entered, presenting a wide expanse of water, being about 25 miles long and as many broad, with a depth of 300 or 400 feet. Here a scene of wonder is presented to the view from the deck of the steamer — *Parisien Island,* attached to Canada, is passed on the right, while the highlands toward the north rise to the height of 800 or 1,000 feet.

Goulais Bay and *Batcheewanaung Bay,* being famous resorts for trout fishing, are seen toward the north-east, and in the far distance can be discerned *Mamainse Point,* where are found copper ore and other metals, being the commencement of the mineral region.

WHITE FISH POINT AND LIGHT, 40 miles from the Saut, is another object of great interest to the mariner.— LAKE SUPERIOR, stretching about 460 miles in a north-west direction, with an average width of about 100 miles, here presents a grand appearance from the deck of the passing steamer. It lies 600 feet above the sea, its greatest depth being 900 feet, extending 300 feet below the level of the ocean; estimated area, 32,000 square miles. Nearly two hundred creeks and rivers are said to flow into the lake, a few of which are navigable for steamers from 2 to 20 miles. Its principal affluents are the St. Louis River, Pigeon River, Kaministiquia River and Nepigon River.

"Father of Lakes! thy waters bend
  Beyond the eagle's utmost view,
When, throned in heaven, he sees thee send
  Back to the sky its world of blue.

"Boundless and deep, the forests weave
  Their twilight shade thy borders o'er,
And threatening cliffs, like giants, heave
  Their rugged forms along thy shore."

On passing White Fish Point the American steamers usually run near the South Shore of the Lake, having the land continually in sight along the *Upper Peninsula of Michigan,* extending westward to near the Apostle Islands, a distance of about 400 miles, passing in their course around Keweenaw Point.

Running along the coast from White Fish Point, westward, for about 50 miles, the shore presents high sandy bluffs, with no harbor or place of refuge for the mariner, although the mouths of Two Hearted and Sucker Rivers are passed.

GRAND MARAIS HARBOR, 45 miles west of White Fish Point, with 9 fathoms within the bar, would be rendered a secure and commodious harbor by the construction of a canal, or dredging for a short distance inland. "The harbor

158

is about 2 miles long, and is from 500 to 1,000 yards in width. The western or largest portion is a mile and one-half in length, and lies parallel with the shore of the Lake, with only a narrow strip of sand beach between. In many cases this is but a few rods wide, and at no point is it more than ten feet higher than the water. The eastern or further end curves inland, and is a full half mile in width. The entrance to the harbor is about five-eighths of a mile in width, and is over a bar where there is from four to eight feet of water at all times. This bar is very narrow, is composed of about 6 feet of sand resting on a clay and gravel bottom, and, to all appearance, never changes in shape under any circumstances. From the formation of the coast but very little change is likely to be made by the action of the waters of the lake, even were cribs built out, and we should not anticipate any trouble from the filling in of the channel. This will not certainly occur from any sedimentary deposit, by reason of a current out of the harbor, there being no perceptible one, only two or three small streams emptying into the harbor, and the whole of them put together not carrying water enough to drive a single saw. The harbor, which might perhaps more appropriately be called a small lake, *will safely hold our entire lake marine.* The water inside deepens very rapidly, and once over the bar, vessels will lie in ten fathoms at one-half their length from the shore. So bold, indeed, are the shores, that a steamer could lie close enough for her gang-planks to reach from her decks to the land."

In all the navigation on Lake Superior, there is none more dreaded by the mariner than that from White Fish Point to Grand Island, and this is especially the case late in the season, when the fall storms make navigation the most hazardous, and the heaviest and most valuable freights are on transit. This coast is ex-

posed to every wind that blows from the Lake, the sweep being of its full width and length, and there is not at present a single place where a landing can be made, or a lake steamer or vessel run for safety, in the whole distance of 80 odd miles. The entire coast is made up either of the towering cliffs at the western end, the bleak hills of sand at the Sauble Banks, coming down to the water's edge and offering no landing, except for a small boat in pleasant weather, or the dull low beach beyond — stretching away for nearly 50 miles farther, with one single insignificant creek — the Two Heart River — that a Mackinac boat can possibly enter if in smooth water, excepting always the harbor of Grand Marais — to which there is no entrance for large vessels. This beach is strewed with wrecks from the Pictured Rocks to White Fish Point.

Point Sable, 10 miles farther, is elevated 300 feet above the Lake, but affords no harbor in its vicinity, although two small streams enter from the south. Off this point soundings have been made, showing a depth of 120 fathoms, or 720 feet.

The PICTURED ROCKS, 18 miles west of Point Sable, constitute one of the greatest wonders of Lake Superior, extending along the coast from the Chapel some 8 or 10 miles to Miner's Castle. The principal objects of attraction are the *Arched Rock,* or *Grand Portail,* and *Sail Rock,* which can be seen distinctly from the passing steamer. The former is a bold promontory where is an excavation, worked by the waves, extending about 200 feet under a bluff, which juts out into the waters of the Lake. The Pictured Rocks can be seen to the best advantage, from a distance, when there is a favorable sunlight effect, or by a close view, which is obtained by coasting along the precipitous rocks in a small boat, during calm weather. For upwards of 100 miles the coast on the South Shore presents an almost unbroken wilderness. The soil on the summit of the cliffs is tolerably good, as indicated by maple trees.

Miner's River, at the mouth of which stands a bold promontory called Miner's Castle, is a beautiful clear trout stream, falling into the lake over rapids.

Sand Point, 10 miles east of the Pictured Rocks, is at the eastern side of *Grand Island Harbor*, one of the most secure bodies of water to be found on the South Shore. Here Lake Superior has its greatest width, it being about 150 miles to Nepigon Bay.

MUNISING, 2 miles farther, being 90 miles west of White Fish Point, is the first landing or settlement lying on the South Shore. For the whole of this distance the mariner is threatened with shipwreck in stormy weather, which usually occurs during the spring and autumn months. At Munising is a small settlement, and a good hotel for the accommodation of visitors desirous of fishing and visiting the Pictured Rocks.

The *Schoolcraft Furnace* is located a short distance west of Munising, where is a small stream entering the bay, on which is a fine fall of water.

ONOTA, 6 miles farther, is a new settlement, where is located Bay Furnace. This is the county-seat of Schoolcraft county.

GRAND ISLAND, 8 miles long and 4 miles wide, surrounded by bold shores, being elevated 300 or 400 feet, is a fine piece of land, being for the most part heavily wooded with hard and soft wood timber. Its shores are famous for its fisheries and romantic scenery. Here is a small settlement on the south end of the island; on the north end stands a lighthouse, erected on a high point of land, being distant 38 miles east of the Marquette Light. "The cliffs on the north bank are broken by the waves into picturesque caverns, pillars, and arches of immense dimensions." — *Bayfield*.

Wood Island is a small body of land lying a short distance west of Grand Island.

Train Point, 6 miles west of Grand Island, is a most romantic headland. *Train Island*, 4 miles farther, is another picturesque body of land.

Laughing Fish Point, 16 miles west of Grand Island, is a rocky projection, where may be witnessed fine scenery.

Sable River, Fish River, Chocolate River, and Carp River are all small streams flowing into the Lake between Laughing Fish Point and Marquette.

HARVEY, situated at the mouth of the Chocolay, is a small village, 3 miles east of Marquette, where is located a blast furnace.

On the completion of the *Marquette, Saut Ste. Marie and Mackinac Railroad*, now being constructed, this whole section of country along the South Shore will be accessible to tourists. Numerous small streams will be crossed, abounding in brook trout.

Steamboat Excursion from Marquette to Grand Island,

A steamer runs from Marquette to Grand Island, Munising, etc., passing through Grand Island Harbor, affording a delightful steamboat excursion. Several islands are passed, and the steamer runs to within a few miles of the far-famed *Pictured Rocks*. A good hotel is located at Munising, on the main land, opposite Grand Island, where is convenient steamboat landing.

Marquette, the chief city of the Upper Peninsula, the county-seat of Marquette Co., and a port of entry, is advantageously situated on the south shore of Lake Superior, in N. latitude 46° 32′, W. longitude 87° 33′, having a mean annual temperature of 40° Fahr. The harbor, formed by the Bay of Marquette, is safe and commodious, being protected from all but north-east winds: when blowing in that direction, vessels are obliged to anchor off the shore for safety. The United States Government have erected substantial piers, or breakwaters, for the further protection of the numerous steamers and sailing vessels which frequent the harbor, taking, annually, an immense amount of iron ore and pig metal to the eastern ports on Lake Erie. A Lighthouse stands on a point of land immediately north of the anchorage, as a guide for mariners.

The settlement of Marquette was commenced in July, 1849; in 1855, on the completion of the Ship Canal at the Saut Ste. Marie, commenced the shipment of iron ore to the Eastern markets. In 1859 it was incorporated as a village, and as a city February, 1871, being now governed by a Mayor, Recorder, and Common Council. It contains a court-house and jail; a public hall; 1 Episcopal, 1 Presbyterian, 1 Baptist, 1 Methodist, and 1 Roman Catholic church; a union high school and 3 ward schools; several hotels and taverns—the *North-western Hotel*, the *Coles House* and *Tremont House*—the former favorably situated, overlooking the lake. There are also several well-kept private boarding-houses, 3 banks, 30 or 40 stores and storehouses, 1 printing-office, besides a large foundry, a rolling-mill and blast furnace, and machine shops and factories of different kinds. Population in 1860, 1665; in 1870, 4000; now, 6000. Gasworks and water-works are in operation —the latter affording an abundant supply of pure water taken from Lake Superior.

11

There are four extensive piers, arranged for the transshipment of iron ore and pig metal, on which the railroad trains deliver and receive a great amount of freight. Here is the terminus of the *Marquette, Houghton and Ontonagon Railroad*, passing along the south shore of Lake Michigamme, to be extended westward to L'Anse and Ontonagon, and connect with the other railroads running south and west. The *Peninsula Railroad of Michigan* also unites with the above road, extending to Escanaba, on Green Bay, 65 miles.

This flourishing Lake City is closely identified with the extensive Iron Mines in the vicinity, being from 12 to 30 miles distant, situated on an elevated ridge, some 700 or 800 feet above the waters of Lake Superior, being known as the *Iron Mountain*. There are now about fifty mines extensively and profitably worked, being owned by separate companies—the Jackson Iron Company, the Cleveland Iron Company, and the Lake Superior Company having separate docks for the shipment of iron ore.

The drives from Marquette to Harvey, 4 miles, running along the beach; to Mt. Menard, 2½ miles, and to Collinsville and Forrestville, are all worthy of attention. The boating and sailing in the Bay of Marquette, and to the islands and mouths of several creeks or rivers, where good trout fishing is to be found, is a favorite source of amusement.

Negaunee, Marquette County, Michigan, 12 miles west of Marquette and 62 miles north of Escanaba, is situated on the line of the *Marquette, Houghton and Ontonagon Railroad*, at its junction with the *Peninsula Division* of the Chicago and North-western Railroad. It is in the immediate vicinity of the *Iron Mountain*, and is a flourishing village of about 3500 inhabitants. Here are situated the *Jackson Mine*, the *McComber Mine*, the *Grand Central*, and the *Negaunee Mine;*

also the *Pioneer Furnaces*, worked by the Iron Cliffs Company, altogether giving profitable employment to several hundred workmen.

The village contains a town hall, 2 railroad depots, 3 churches, 2 hotels—the *Ogden House* and *Jackson House*,—4 banks, 30 or 40 stores, and 1 steam sawmill. Thousands of strangers annually visit these celebrated mines within a range of 25 miles, now producing altogether upwards of 1,000,000 tons of ore yearly, most of which is shipped from Marquette, Escanaba, and L'Anse to Eastern markets.

There are several Iron Furnace Companies in the vicinity of the Mines, which produce annually a large amount of pig metal of a superior quality. The amount produced in 1873 was 71,507 tons. This industry must rapidly increase with the product of the mines.

Ishpeming, situated on the line of the Marquette, Houghton and Ontonagon Railroad. Three miles west of Negaunee is another flourishing mining town, where is located the Cleveland Mine, the Lake Superior Mine, Barnum Mine, New York Mine, New England Mine, Williams Mine, Winthrop Mine, and the Pittsburg and Lake Angeline Mine. The village contains a town hall, 3 churches, 2 banks, a hotel, and 20 or 30 stores, besides a foundry and machine shop and several fine residences. Population about 5000, most of whom are employed in the different Iron Ore Mines.

Beyond Ishpeming, on the line of the railroad, are situated *Winthrop*, 3 miles, *Greenwood*, 3 miles, *Clarksburg*, 4 miles, *Humboldt*, 1 mile, and *Champion*, 4 miles, before arriving at *Lake Michigamme*, situated 34 miles west of Marquette.

Champion Furnace and Mine, 32 miles from Marquette. Here is a mining settlement containing about 2000 inhabitants. The mine is located about one mile south-west of the furnace, where is a large deposit of red and black oxides of iron ore; a part of the ore is taken to the furnace and made into pig iron, the balance being sent to Eastern markets via Marquette.

———

Michigamme, Marquette County, Michigan, is a new mining town handsomely situated at the west end of Lake Michigamme, 38 miles from Marquette and 25 miles from L'Anse, both shipping ports, and in the immediate vicinity of the Michigamme iron range, where several mines are already opened. In June, 1873, this town was destroyed by fire, the inhabitants having to flee for their lives, the surrounding woods being also on fire. The new town was immediately rebuilt, and upwards of one hundred stores and dwellings erected within a few months, now presenting a lively appearance. Here is a large steam saw-mill and an iron furnace being erected, contiguous to the Michigamme iron mine.

Lake Michigamme * is a most beautiful sheet of water, studded with several wooded islands, while the shore is very irregular, presenting many points and indentations of lively appearance. It abounds in fish of various kinds, affording fine sport for the angler. It is fed by Michigamme River, the outlet flowing into the Menominee, which enters Green Bay. In this vicinity deer and other game abounds. West of the lake is an unexplored wilderness, heavily timbered, and where iron is supposed to abound.

———

* Lake Michigamme, called by the Indians "*Ma-she-ga'me*" (large lake). The Ojibwas classify lakes into three kinds: *Sa-ga-e'ga*, small lake; *Ma-she-ga'me*, large lake; and *Git-che-ga'-me*, great lake. This last is applied to the "Great Lakes" indiscriminately, and to the ocean.

Marquette to L'Anse,

Via *Marquette, Houghton, and Ontonagon Railroad.*

Miles.	Stations.		Miles.
63	MARQUETTE		0
60	Bancroft		3
57	Bruce	3	6
56	Morgan	1	7
55	Eagle Mills	1	8
52	Carp.	3	11
51	NEGAUNEE	1	12
48	ISHPEMING	3	15
44	Saginaw	4	19
42	Greenwood	2	21
38	Clarksburgh	4	25
37	Humboldt	1	26
32	Champion.	5	31
26	MICHIGAMME	6	37
24	Spurr Mine.	2	39
16	Sturgeon	8	47
10	Summit	6	53
7	Palmer	3	56
0	L'ANSE	7	63

The *Marquette, Houghton and Ontonagon Railroad,* completed 63 miles, commences at Marquette and runs westward by an ascending grade to Negaunee, 12 miles, where it connects with the *Peninsula Divison* of the Chicago and Northwestern Railroad; both roads here first strike the iron range of Lake Superior. Ishpeming, 3 miles farther, is the centre, at present, of the iron mining interest, from whence immense quantities of iron are sent to Eastern markets. At Humboldt, 26 miles from Marquette, runs a branch railroad, 9 miles in length, to the Republican and Kloman iron mines. Champion, 31 miles, and Michigamme, 37 miles from Marquette, are important stations; the latter situated on the shore of Lake Michigamme. The road continues onward by an ascending grade to the Summit, 1167 feet above Lake Superior, then descends more than one hundred feet to the mile to L'Anse, situated at the head of Keweenaw Bay.

Escanaba, the county-seat of Delta County, Michigan, is an important lake port, favorably situated for the shipment of iron ore, on Little Bay de Noc, the north-western arm of Green Bay, in lat. 45° 36' N., long. 87° 06' W., having a mean annual temperature of 41° Fahr. It is the terminus of the *Peninsula Division* of the Chicago and North-western Railroad, being 358 miles north of Chicago, and 74 miles south of Marquette, by railroad. It was first settled as a village in 1863, and now numbers about 2000 inhabitants, and is fast increasing in wealth and population. Here are a bank, 3 churches, 15 stores, 4 hotels, and a number of fine residences. It is destined to become a favorite resort during warm weather. The *Tilden House* is handsomely situated, facing the bay, with pleasure-grounds attached. Escanaba is favorably situated for manufacturing purposes, here being already in operation an extensive furnace, erected in 1872 at a cost of $225,000, machine shops, etc. The Ore Dock at this place is very large and commodious, from whence is annually shipped several hundred thousand tons of iron ore to Eastern and Southern markets. Steamers and sail-vessels run between this port and Chicago; also, to ports on the Lower Lakes during the season of navigation.

At the entrance to Little Bay de Noc stands a light-house, where is a large and well-protected harbor. Here usually may be seen a large number of vessels taking in loads of iron ore and pig metal.

At *Day's River Station,* 13 miles north of Escanaba, the railroad track crosses a fine trout stream. Here farming is successfully prosecuted, the soil producing fine crops of hay, oats, rye, wheat, potatoes, and other garden vegetables, the first frost here being in the latter part of September, and the last in the early part of May. Strawberries, raspberries, whortleberries, and cranberries here grow to perfection.

Pleasure Excursion — Marquette to Houghton.

There are but few if any more pleasant or interesting trips than from Marquette to Houghton and return by way of lake and rail. A steamer of a small class leaves Marquette every evening at 6.30, and affords passengers a delightful night voyage to Houghton. Her accommodations are ample for comfort, and her officers very efficient in the discharge of their duties. The line steamers of a larger class also run to and from Houghton and Marquette. A day at Houghton and Hancock, with quarters at the Douglass House, can be spent in pleasure or business, as circumstances may require. The copper mines in the vicinity of the towns are well worthy of a visit.

On the return, the steamer Ivanhoe can be taken at 8 o'clock in the morning from Houghton, making the trip — a most delightful one in all respects in favorable weather — to the bright and promising town of L'Anse, which has fair prospects of becoming an iron metropolis of considerable importance. Here we are transferred to the Marquette, Houghton and Ontonagon Railroad, and whirled across the greatest iron region in the world, passing Lake Michigamme and numerous iron mines. Resident citizens should not neglect to make this trip as often as they need recreation, and the tourist will miss the most interesting portion of the Lake Superior journey if he does not include this in his programme.

HOTEL ACCOMMODATION.

By an increase of hotel accommodation, affording good fare at reasonable rates, this whole section of country would be made a great summer resort. Here pure air and water, with an invigorating climate, can be enjoyed by the invalid and seeker of pleasure.

Stannard's Rock.

This dangerous rock, or reef, lies east of the track between Marquette and Keweenaw Point, distant 30 miles east-south-east, and in a calm time was invisible. Recently the United States Government have located thereon a beacon 33 feet above water. It has a base of cut stone 9 feet in diameter and 8 feet high, surmounted by a wrought-iron shaft, with a ball and cage on the top painted red. To the northward and westward of the beacon a rocky reef extends 30 by 320 yards — least water, 2 feet. One thousand and four hundred feet west of the beacon are two detached rocks with 8 feet of water. The soundings in proximity to the beacon and reef are from $3\frac{1}{2}$ to 10 fathoms. Bottom, rock and gravel. Vessels should not approach the beacon to the northward nearer than three-fourths of a mile.

Marquette to Portage Entry.

On leaving Marquette for Portage Entry or Keweenaw Point, both distant 70 miles; the steamer runs north to Granite Island, 12 miles; from thence N. by W. 58 miles to Keweenaw Point. If steering for Portage Entry, N.W. by W., 13 miles from Granite Island to Big Bay Point. Big Bay, Salmon Trout River, and the Huron Mountains lie to the west; the headlands and the mountain scenery here present a fine appearance. *Huron Islands* and Light are 22 miles farther, running N.W. by W. This is a rocky group of islands and dangerous to navigation. Huron Bay and Point Abbaye are passed on the south-west; the steamer running 23 miles west to Portage Entry; to Houghton, 14 miles farther.

On the south lies *Keweenaw Bay*, a fine expanse of water, extending 20 miles to its head, where is a new and thriving settlement. Here is also a Methodist and Roman Catholic Mission on the opposite sides of the bay.

Huron Bay, lying east of Keweenaw Bay, and south-west of Huron Islands, is land-locked, affording a most secure harbor, being 14 miles long and a good depth of water. In this bay is situated the new town of *Fairhaven*, being the outport for the Huron Bay Slate Quarries, located 4½ miles from the bay, in Township 51, Range 31. A tram railway is being constructed to extend from the mines to the steamboat landing. The soil on the shore of Huron Bay is a sandy loam, very deep, warm, and rich, producing good crops.

Arvon, Houghton County, Michigan, is a new location, where is found slate of a good quality, and iron ore. It is situated on Dashing River, which empties into Huron Bay, Lake Superior. In this vicinity is a slate belt hundreds of feet in width, and is considered inexhaustible. This slate and iron will be shipped from Huron Bay, the present season, to Eastern and Southern markets.

L'Anse, Houghton County, Mich., is a new town favorably situated at the head of Keweenaw Bay, where is a long pier and steamboat landing, also a commodious ore dock for the shipment of iron ore. It is surrounded by picturesque scenery, overlooking the Bay of L'Anse and adjoining shores. The harbor is one of the finest on the whole chain of lakes, being perfectly protected and secure in all weather. The water power of Fall River, here emptying into the bay, is ample to drive a large amount of machinery, being suitable for blast furnaces and mills of every kind. Large deposits of iron ore are found within eight or ten miles, while a few miles southward is located the Spurr Mountain and Michigamme Iron Mine, producing magnetic ore of a high standard. At L'Anse are two churches, a public school-house, a bank, two hotels, and one of a large class in the course of construction, a blast furnace, a public warehouse, fifteen stores, a brewery, a saw-mill, and machine-shops.

The *Marquette, Houghton and Ontonagon Railroad*, 63 miles in length, terminates at L'Anse, forming, in part, a through line of travel to Green Bay and Chicago. Steamers of a large class run to this place from ports below; and a steamer runs daily to and from Houghton, 30 miles, forming, with the railroad, a through line of travel to Marquette.

A *Methodist* and *Roman Catholic Mission House* are both situated about four miles north of L'Anse, on opposite sides of the bay, where are Indian settlements numbering several hundred souls.

On Fall River, and at the head of L'Anse Bay, is good trout fishing; the latter can be approached by row-boats, starting from the landing at L'Anse.

Portage River and *Lake* are navigable for steamers of a large class, by means of an artificial channel running through to the lake for about 7 miles. The lake is next entered, and soon Houghton and Hancock come into view, being distant 14 miles from Lake Superior, by the circuitous channel. Here are extensive copper mines, which are profitably worked by a number of mining companies.

Houghton, Michigan, the county-seat of Houghton County, and a port of entry, is situated on the south side of Portage Lake, 14 miles above Portage Entry, and 10 miles from Lake Superior, lying on the north-west. By means of a river improvement and ship canal, steamers can run through Portage Lake into Lake Superior on both sides of Keweenaw Point — thus forming one of the most capacious and secure harbors of the Upper Lakes, it being land-locked and protected by high hills on both sides. The settlement of Houghton was commenced in 1854, and incorporated as a village in 1861. It now contains a court-house and jail; 1 Episcopal, 1 Methodist, and 1 Roman Catholic church; 1 bank;

4 public houses, the *Douglas House* and the *Butterfield House* being the most frequented by visitors; 12 or 15 stores and several extensive warehouses; Houghton Copper Works or Rolling Mill, capital $250,000; 2 large stamp-mills, using steam power; 2 steam saw-mills; 2 breweries, and several other manufacturing establishments. The population of the town is estimated at 2000. This flourishing mining town, lying on a side-hill rising 300 or 400 feet, is identified with the copper mines in its immediate vicinity. There are several mines worked to a large extent, besides others of less note, which will, no doubt, soon be rendered productive. The mineral range of Keweenaw Point, some 4 to 6 miles in width, extends through all this section of country, being as yet only partially explored. The Isle Royal, Huron, Portage, Atlantic, Sheldon, and Columbian are the principal mines worked on the south side of the lake.

PORTAGE LAKE is an irregular body of water, about twenty miles in length, extending nearly across Keweenaw Point to within two miles of Lake Superior. Steamers and sail-vessels drawing 12 feet can pass through Portage Entry, and navigate the lake with safety. This body of water was an old and favorite thoroughfare for the Indians, and the Jesuit Fathers who first discovered and explored this section of the country. A canal of two miles in length would render this portage route navigable for steamers and sail-vessels navigating Lake Superior, thereby reducing the distance over 100 miles. During the winter months the atmosphere is very clear and transparent in the vicinity of Houghton, and all through Keweenaw Point; objects can be seen at a great distance on a clear day, while sounds are conveyed distinctly through the atmosphere, presenting a phenomenon peculiar to all northern latitudes. This is the season of health and pleasure to the permanent residents.

Hancock, Houghton County, Michigan, is situated on the north side of Portage Lake, opposite to the village of Houghton, with which it is connected by a steam ferry. The town was first laid out in 1858, and now contains about 2000 inhabitants, including the mining population on the north side of the lake, its sudden rise and prosperity being identified with the rich deposit of native copper, in which this section of country abounds. The site of the village is on a side-hill rising from the lake level to a height of about 500 feet, where the opening to the mines is situated. Here is 1 Congregational, 1 Methodist, and 1 Roman Catholic church; 2 banks; 2 public houses; the *Sumner Mine* and stamping-mill; a number of stores and warehouses; 1 steam saw-mill, 1 barrel factory, 1 foundry and machine-shop, and other manufacturing establishments. In the vicinity are 4 extensive steam stamping-mills, worked by the different mining companies—the Quincy, Pewabic and Franklin.

The *Portage Lake Smelting Works* is an incorporated company, turning out annually a large amount of pure merchantable copper. The business of the company consists of fusing and converting the mineral into refined metal or ingots, for manufacturing purposes. *Mineral Range Railroad*, 11 miles in length, is finished to Calumet.

The *Portage Lake and River Improvement Company* was chartered in 1861, for the improvement of Portage River, entering Keweenaw Bay, 14 miles below Houghton. An entrance was cut from the bay into the river, 14,000 feet long, 100 feet wide, and 12 feet deep, and a pier built out into the bay on the east side of the channel entrance. The channel of the river was also dredged so as to give a depth of 11 feet into Portage Lake, thus enabling the largest class of lake steamers to land at the wharves in Houghton and Hancock.

Portage and Lake Superior Ship Canal.
— This important work was commenced in 1868 and finished in 1873, at a cost of about $2,500,000. Its length is 2¼ miles, with piers 600 feet in length, extending out into Lake Superior on the north, affording a safe entrance for downward bound vessels. The canal is 100 feet wide, 15 feet deep, with banks rising from 20 to 35 feet above the water. At its southern entrance into Portage Lake, 8 miles above Houghton, it runs through a low marshy piece of ground, then enters the lake about half a mile wide. Below Houghton it connects with the Portage Lake and River Improvement, 14 miles in length, making the distance across from lake to lake, 24 miles.

Steamboat Excursions. — By means of the *Ship Canal* connecting Portage Lake with Lake Superior, 10 miles north-west of Houghton, and the *Portage Lake and River Improvement*, running 14 miles east of Houghton, is afforded most delightful steamboat excursions during the summer months.

On the upward trip the steamers run through the Lake, about half a mile in width, and then enter the Ship Canal, 2¼ miles in length, passing on to the broad waters of Lake Superior, where is presented a wide expanse of pure waters and billowy waves.

On the downward trip, bound for L'Anse or Marquette, the steamer runs east for a few miles, and then enters an expanse known as Dollar Bay — then enters the wide waters of Portage Lake, by some called Boot-Jack Lake, from its peculiar shape, resembling a human leg and foot, where enters the outlet of Torch Lake, a romantic sheet of water. Passing downward through Portage River, a winding stream, you soon pass *Edgerton's Landing*, and enter Keweenaw Bay or Lake Superior. Here is a long pier, a beacon light, and light-house. Proceeding southward, a beautiful expanse of water is passed over before arriving at L'Anse, a new and flourishing town, situated at the head of the bay, where is a well-kept Hotel. Here is found good fishing and boating, affording fine sport for anglers and pleasure seekers.

Calumet, Houghton co., is an important copper mining settlement, situated 12 miles north-east of Hancock, containing a population of about 3,000 inhabitants. Here is located the Calumet and Hecla Mine, one of the most productive copper mines in the world. In the vicinity are situated the Allouez, Kearsarge, and Schoolcraft Mines.

The Houghton *Mining Gazette* says: "The history of copper mining in the new and old world fails to record anything approaching a similar richness to that of the *Calumet and Hecla Mine* of Houghton county, Lake Superior, and, in its progressive developments under ground, one is confined to the ejaculation, 'wonderful!' The total mineral product of the Calumet and Hecla Mine for the year 1873 foots up the enormous figure of 11,551 tons." Other Copper Mines of nearly equal value are located on the Mineral Range, extending from Portage Lake to Copper Harbor.

The *Mineral Range Railroad* runs through Calumet from Hancock, toward the extreme end of Keweenaw Point. A railroad also runs to the company's Stamp Mills, located on Torch Lake, from whence large amounts of copper ore are annually shipped.

On resuming the outward trip for Keweenaw Point and Copper Harbor, the former 50 miles distant, the steamer runs north-east 50 miles to Manitou Island Light. This stretch presents a fine view of Mount Bohemia and Mount Houghton, as well as the head-lands along the coast.

BETE GRISE BAY and LAC LA BELLS are beautiful bodies of water, the latter being connected with the bay by means of a canal. In the vicinity of the lake are veins of copper ore of a rich quality. This place should be improved, and made a resort for seekers of health and pleasure. It is in the immediate vicinity of Mount Houghton, and of several fine trout streams.

KEWEENAW POINT is a bold projection jutting out into Lake Superior, and one of the most remarkable features of this Inland Sea. It may be said to extend 60 miles from S.W. to N.E., with an average width of 15 miles. For this distance, and beyond, the region is celebrated for rich copper mines, producing yearly large quantities of copper, which is shipped to eastern markets.

MANITOU ISLAND, situated in 47° 25′ N. lat., is a desolate island, on which is located a light-house to guide the mariner approaching Keeweenaw Point. Gull Rock is an Islet lying between Manitou Island and Keweenaw Point, on which is situated a light-house; the steamers usually running between this light and the main land.

Copper Harbor, Mich., is situated near the extreme end of Keweenaw Point, in N. lat. 47° 30′, W. long. 80°; the harbor, although somewhat difficult to enter, is one of the best on Lake Superior, being distant 250 miles from Saut Ste. Marie, and about the same distance from Duluth, lying on the direct route from Marquette to Isle Royale and Silver Islet. The village contains about 200 inhabitants, a church, a public house, and two or three stores, and the Clarke Copper Mine. This harbor is destined to rise in importance as the commerce of the Lakes increase.

Fort Wilkins, formerly a United States Military Post, is situated a short distance from the steamboat landing, where is a beautiful Lake called *Fanny Hoe.* In the vicinity are copper mines, which have been extensively worked, and are

well worthy of a visit. Steamers often run direct from this port to Silver Islet, on the North shore, 75 miles distant.

The course pursued in running from Copper Harbor to Ontonagon, 85 miles, is about S.W. by W., passing Agate Harbor, Eagle Harbor, Eagle River, and the north entrance to Portage Lake, where a Ship Canal is constructed. This part of the trip around the South Shore is uninteresting, so far as scenery is concerned. The ground rises to a moderate height, presenting no object of interest for a number of miles.

Eagle Harbor, 16 miles west of Copper Harbor, is a secure steamboat landing, with a light-house at its entrance. In the village are 2 churches, a hotel, several stores and store-houses, and about 500 inhabitants. The town was first settled in 1845. This is the shipping-port for several copper mines in the immediate vicinity; the Amygdaloid, Central, Copper Falls, Delaware, Pennsylvania, and Petheric mines.

Eagle River, 10 miles farther west, is the county town for Keweenaw county, situated at the mouth of a stream of the same name, where is a small harbor. Here is a court-house and jail, a church, a hotel, 4 stores, and several store-houses. It is the outlet for several rich copper mines, producing mass and stamped copper. The celebrated Cliff mine, the Phœnix, the North American, the St. Clair, and the Eagle River mine, are the principal mines in operation. The *Mineral Range Railroad,* running from Hancock, when completed, will extend to this place. The mouth of the *Lake Superior Ship Canal* is passed about 20 miles south-west of Eagle River, where is a small settlement.

On the north side of Keweenaw Point, in the distance, are seen the high lands which form the rich copper range of this region, extending S.W. for about 80 miles.

Ontonagon, Ontonagon co., Mich., 336 miles from the Saut Ste. Marie, is situated at the mouth of the river of the same name. The river is about 200 feet wide at its mouth, with a sufficient depth of water over the bar for steamers. The village contains 3 churches, a good hotel, smelting works, 2 steam tanneries, 1 grist-mill, 2 steam saw-mills, and 10 or 12 stores and store-houses, and 800 inhabitants.

In this vicinity are located the Minnesota, the National, the Rockland, and several other productive copper mines. The ore is found from 12 to 15 miles from the landing, being imbedded in a range of high hills traversing Keweenaw Point from N.E. to S.W. for about 100 miles. Silver is here found in small quantities, intermixed with the copper ore, which abounds in great masses.

The *Silver Mines,* situated on Iron River, 12 or 15 miles west of Ontonagon, are attracting great attention.

A good plank road runs from Ontonagon to near the Adventure Mine, and other mines, some 12 or 14 miles distant, where commences the copper range of hills.

The *Marquette, Houghton and Ontonagon Railroad,* and the *Oshkosh and Ontonagon Railroad,* are both in progress of construction, and when completed will afford great facilities to this section of country. The distance from Ontonagon to Isle Royale, attached to Michigan, is 60 miles, and to the mouth of Pigeon River, Minn., is about 70 miles.

Rockland, Ontonagon county, situated 13 miles south-east of Ontonagon, is a flourishing mining village. In the vicinity are located the Minnesota, National, Rockland, and Superior Copper Mines. The Minnesota Mine was formerly the most productive mine on Lake Superior, producing large quantities of mass copper, but is now eclipsed by the Calumet and Hecla.

The PORCUPINE MOUNTAIN, lying 20 miles west of Ontonagon, is a bold headland that can distinctly be seen at a great distance, rising some 1,300 feet above the lake surface.

Michigan Island and Light, lying 60 miles west of Ontonagon, is the next object of interest. This is the easternmost of the group of islands known as the APOSTLE ISLANDS; they consist of some 20 islands of different sizes, most of which are uninhabited, being for the most part clothed with forest trees of a small growth. *Madeline Island,* the largest of the group, is in part cultivated. These islands are all attached to Ashland county, Wisconsin.

The names of the Islands, as designated on the Government chart, are as follows: Michigan (Light), Madeline, Basswood, Hermit, Stockton, Oak, Manitou, Outer (Light), Cat, Ironwood, South Twin, North Twin, Otter, Rocky, Bear, Devil, Raspberry (Light), York, Sand, Eagle, and Steamboat. Good anchorage and protection from all winds can be found anywhere within the Apostle Islands, with a depth of water from six to twenty fathoms.

Chaquamegon Bay and Point, situated south of Madeline Island, form a capacious harbor, at the head of which is situated the town of Ashland.

La Pointe, 78 miles west of Ontonagon, situated on the south end of Madeline Island, the largest of the *Apostle Islands,* is one of the oldest settlements on Lake Superior; it was first peopled by the French Jesuits and traders in 1680, being 420 miles west of the Saut Ste. Marie, which was settled about the same time. The mainland and islands in this vicinity have been for many ages the favorite abode of the American Indian, now lingering and fading away as the country is being opened and settled by the white race.

The village now contains 200 inhabitants, most of whom are half-breeds and French. Here is an old Roman Catholic church, and a Methodist church; 2 hotels, 2 stores, and several coopering establishments for the making of fish barrels. The harbor and steamboat landing are on the south end of the island, where may usually be seen fishing-boats and other craft navigating this part of Lake Superior.

Wheat, rye, barley, oats, peas, potatoes, and other vegetables are raised on the island. Apples, cherries, gooseberries and currants are raised in the gardens at La Pointe. The wild fruits are plums, cranberries, strawberries, red raspberries, and whortleberries. The principal forest-trees on the island are maple, pine, hemlock, birch, poplar and cedar.

———

Bayfield, capital of Bayfield Co., Wis., is favorably situated on the southern shore of Lake Superior, 80 miles east of its western terminus, and 3 miles west of La Pointe, being 80 miles west of Ontonagon. The harbor is secure and capacious, being protected by the Apostle Islands, lying to the north-east. The town plot rises from 60 to 80 feet above the waters of the lake, affording a splendid view of the bay, the adjacent islands and headlands. Its commercial advantages are surpassed by no other point on Lake Superior, being on the direct route to St. Paul, Minn., and the Upper Mississippi. Here are situated a Presbyterian, a Methodist, and a Roman Catholic church, 2 hotels, 4 stores, 2 warehouses, 1 steam saw-mill, and several mechanics' shops. Population in 1870, 400.

LA POINTE BAY, on the west side of which is situated the port of Bayfield, is a large and safe body of water, being protected from winds blowing from every point of the compass. The shores of the islands and mainland are bold, while the harbor affords good anchorage for the whole fleet of the lakes.

The Indian Agency for the Chippewa tribe of Indians residing on the borders of Lake Superior, have their headquarters at Bayfield.

———

Ashland, Ashland Co., Wisconsin, is most favorably situated at the head of Cha-qua-me-gon Bay, 18 miles south of Bayfield and 70 miles east of Superior City, by proposed railroad route. The town is located on a level table-land, elevated about thirty feet above the waters of the bay, having a depth of 12 or 15 feet. It is regularly laid out on streets running parallel to the water's edge for a distance of two miles, and presents a fine appearance. Here are erected three docks or steamboat landings and a long railroad dock, 4 warehouses, 16 stores, 2 churches, 6 hotels or taverns, 2 steam saw-mills, 1 sash, door and blind factory, and 1 printing-office and weekly paper, the *Ashland Press.* The village contains about 1,000 inhabitants. A small steamer runs daily between Ashland, La Pointe and Bayfield, while larger steamers stop on their upward and downward passage from Duluth, Minn. The harbor is one of the largest and safest on Lake Superior.

The *Wisconsin Central Railroad* terminates at Ashland, extending southward to Menasha, Wis., 240 miles, there connecting with railroads running north, south and east, forming in part a through line of travel to Milwaukee, Chicago, etc. Here are the headquarters of the Northern Division of the above railroad, where will be erected machine-shops, etc.

Houghton Point, 9 miles south of Bayfield, and the same distance north of Ashland, is a delightful location, where is being formed a settlement for permanent residence. Here it is intended by the proprietor, F. Prentice, Esq., to erect a Summer Hotel for the accommodation of visitors resorting to Lake Superior for health and pleasure. A steamer stops daily at the landing on the route between Bayfield and Ashland.

The *Penoka Iron Range*, lying 25 miles south of Ashland, is an immense deposit of magnetic ore of a rich quality, extending 30 or 35 miles east and west, elevated from 800 to 1,200 feet above Lake Superior. The line of the railroad runs through this range, and the iron ore will be shipped from Ashland. The Iron Mining Companies already formed are the Ashland Iron Company, Magnetic Iron Company, La Pointe Iron Company, and the Wisconsin Iron Mining and Smelting Company.

The three northern counties of Wisconsin bordering on Lake Superior are *Ashland*, including the Apostle Islands, *Bayfield* and *Douglas*. These counties are but sparsely populated, but will, no doubt, rapidly increase when the railroads now in progress of construction shall be completed from Duluth and Superior City to Montreal River, and thence to Marquette, Mich., also, the proposed railroads running south to different points from Ashland.

Numerous small streams flow into Lake Superior, rising in the northern portion of the above counties, while the southern portion is drained by the St. Croix River and Chippewa River, flowing south into the Mississippi River. On the borders of the latter streams there is an immense amount of pine timber and much good farming land, producing wheat, oats, potatoes and other vegetables. The lumbering business and fishing are at the present time the main support of the inhabitants.

The principal streams flowing into Lake Superior from Wisconsin are Montreal River, Bad River, White River, Fish River, Pike's River, Sand River, Siskowit River, Brule or Burnt Wood River, and Nemadji River.

Odanah, Ashland Co., is an Indian village situated on Bad River, 4 miles from the lake. At this place is established the *Odanah Mission*, where are two churches, a Presbyterian and Roman Catholic, a school-house and a number of dwellings. Here are good farming lands, being cultivated by the Indians under the direction of the government farmer.

Maskeg or *Bad River* rises in the southern part of Ashland County, and flows northerly through the Penoka Iron Range into Lake Superior, affording good water contiguous to the mines. The valley contains much good farming land, and abounds in timber of different kinds.

Montreal River, forming in part the boundary between the States of Wisconsin and Michigan, enters Lake Superior some 10 or 12 miles east of Bad River.

THE APOSTLE ISLANDS.—The following description of these romantic islands is copied from *Owen's Geological Survey of Wisconsin, etc.*

"When the waters of Lake Superior assumed their present level, these islands were doubtless a part of the promontory, which I have described as occupying the space between Chaquamegon Bay and Brule River. They are composed of drift-hills and red clay, resting on sandstone which is occasionally visible. In the lapse of ages, the winds, waves, and currents of the lakes cut away channels in these soft materials, and finally separated the lowest parts of the promontory into islands, and island-rocks, now twenty-three in number, which are true outliers of the drift and sandstone.

"At a distance they appear like mainland, with deep bays and points, gradually becoming more elevated to the westward. '*Ile au Chêne*,' or *Oak Island*, which is next the Detour (or mainland), is a pile of detached drift, 250 or 300 feet high, and is the highest of the group.

Madeline, ' Wau-ga-ba-me' Island, is the largest (on which lies La Pointe), being 13 miles long, from north-east to south-west, and has an average of 3 miles in breadth. " Muk-quaw " or *Bear Island,* and " Eshquagendeg " or *Outer Island,* are about equal in size, being six miles long and two and a half wide.

" They embrace in all an area of about 400 square miles, of which one-half is water. The soil in some places is good, but the major part would be difficult to clear and cultivate. The causes to which I have referred, as giving rise to thickets of evergreens along the coast of the lake, operate here on all sides, and have covered almost the whole surface with cedar, birch, aspen, hemlock, and pine. There are, however, patches of sugar-tree land, and natural meadows.

" The waters around the islands afford excellent white fish, trout, and siskowit, which do not appear to diminish, after many years of extensive fishing for the lower lake markets. For trout and siskowit, which are caught with a line in deep water, the best ground of the neighborhood is off Bark Point or ' Point Ecorce' of the French. Speckled or brook trout are also taken in all the small streams.

" That portion of the soil of the islands fit for cultivation, produces potatoes and all manner of garden vegetables and roots in great luxuriance. In the flat, wet parts, both the soil and climate are favorable to grass, and the crop is certain and stout. Wheat, oats, and barley do well on good soil when well cultivated.

" In regard to health, no portion of the continent surpasses the Apostle Islands. In the summer months they present to the residents of the South the most cool and delightful resort that can be imagined, and for invalids, especially such as are affected in the lungs or liver, the uniform bracing atmosphere of Lake Superior produces surprising and beneficial effects."

On leaving Bayfield for Duluth, the steamer runs northward to Oak and Raspberry Islands, then westward, passing within sight of both shores of the lake, which here contract to 20 or 30 miles in width, narrowing as you approach the mouth of St. Louis River, where stands a light-house and Superior City — Duluth being situated at the head of Minnesota Point, some six miles farther. Here a grand view is presented, taking in both the Wisconsin and Minnesota shores — the former is low and wooded, while the latter rises by a steep ascent some 300 or 400 feet above the water's edge for a considerable distance.

The North Shore of Lake Superior, attached to the State of Minnesota, extends N.E. by E. about 150 miles, from Duluth to Pigeon River and Isle Royale. This coast is highly interesting, with occasional good harbors,—Agate Bay, Burlington Bay, Encampment Island, Beaver Bay, and Good Harbor Bay affording the best landings. The land in many places is elevated and clothed for the most part with evergreens, such as pine, spruce, hemlock, etc.

The **City of Superior,** the county-seat of Douglas County. It is in N. lat. 46° 38′ ; W. long. 91° 03′ ; mean annual temperature 40° Fahr. It is advantageously situated on the south side of the Bay of Superior, formed by the Minnesota and Wisconsin Points; between these points is the entry to the harbor through which is discharged the waters of the St. Louis and the Nemadji Rivers; the former being one of the largest tributaries of Lake Superior. The United States Government recently made an appropriation of $189,000, which, in the opinion of competent engineers, is sufficient to make the harbor of Superior what it was by nature designed to be, — one of the finest harbors on the continent; it is seven miles in length and three-fourths of a mile in width, affording a sufficient depth of water for the largest class of Lake Steamers.

Superior was surveyed and laid out in June, 1854, rapidly increasing in population until the year 1857; since that period it has slowly increased in numbers, there now being about 1,200 inhabitants. Here is a court-house and jail, four churches, one hotel (the *Avery House*), ten stores and store-houses, three steam saw-mills, one steam planing-mill and sash factory, one printing-office, one tannery, and several mechanic shops. The fisheries off the Minnesota and Wisconsin Points are extensively and profitably carried on, where are annually taken large quantities of white-fish, mackinac trout, and siskowit. This town, favored by natural advantages, is destined soon to have railroad facilities. A road is already projected to pass eastward from Duluth, across Rice's Point, toward the Straits of Mackinac, and another to approach Superior from the south-east.

Minnesota Point is a strip of land jutting out into the Lake some seven or eight miles, on the end of which is a light-house, to guide the mariner into the Bay of Superior.

Duluth, St. Louis Co., Minn., a city and port of entry, is most advantageously situated at the head of Lake Superior, near the mouth of the St. Louis River, in N. lat. 46° 38′, W. long. 92° 10′, being elevated 600 feet above the Ocean, the ground rising in the rear to the height of 300 feet. Mean annual temperature, 40° Fahr. It is distant from Buffalo, *via* the Great Lakes, 1,200 miles, and from Quebec, *via* the St. Lawrence River, 1,750 miles; from St. Paul, by railroad, 155 miles, and from Chicago, 565 miles; from the Red River of the North, 252 miles; from the Upper Missouri River, 450 miles, and from Puget Sound, by the proposed line of the *Northern Pacific Railroad*, 1,750 miles; making the dis-

tance across the Continent, by Railroad and Steamer, with one transshipment, 3,500 miles. This distance can be shortened 500 miles by the proposed International Route through Canada.

This City, in four years, has increased from a small village to a flourishing mart of commerce. It is now an incorporated City with an active population of about 4,000 inhabitants, being governed by a mayor and board of aldermen. The streets are wide and regular, crossing each other at right angles. Within the past three years nearly fifteen miles of streets have been opened, affording ready access to all parts of the town. A breakwater is completed nearly 500 feet from the shore, behind which are located the docks of the Lake Superior and Mississippi Railroad and the grain elevator. Alongside the docks vessels can lay in perfect security during the severest gales. The Citizens' dock, 40 feet in width, and extending 600 feet into the Lake, is completed, affording ample accommodation for steamers and other lake craft. The Ship Canal through Minnesota Point, which affords access to the inner harbor in the Bay of Duluth, has been completed, disconnecting the point with the mainland. This canal will be 1,200 feet in length and 250 feet wide, and afford 14 feet of water. The Union Improvement and Elevator Company have finished an elevator with a storage capacity of 500,000 bushels of grain, and capable of unloading twenty cars per hour, and of handling 10,000,000 bushels during the season of navigation. There is also erected three extensive warehouses, being about 600 feet long by 60 feet wide. Granite is found in large quantities in and about Duluth. The machinery for dressing this stone is now on the ground, and the buildings necessary for its working have been erected. This trade will, no doubt, become large and important, as the stone is well adapted for building purposes.

Here is a United States land-office, a court-house and jail, 6 churches of different denominations, 4 public schools, a public library and reading-room, and a Young Men's Christian Association; 3 banks, 3 printing-offices, a foundry, machine shops, 3 steam saw-mills, 2 planing-mills, and several other manufacturing establishments; numerous stores for wholesale and retail purposes; 3 hotels, the *Clark House* and *Bay View House* being situated near the Railroad and Steamboat landings. The buildings of all kinds now number about 700, some of them being first class brick structures.

An extensive and magnificent view is afforded from the top of the *Granite Quarry*, within the limits of Duluth, elevated about 100 feet; it is of the most grand and interesting character imaginable when taking in all surroundings. Looking eastward on the right hand lies St. Louis Bay and Rice's Point; in front is seen the beautiful Bay of Superior, Minnesota Point, and the City of Superior in the distance; while on the left hand is seen the City of Duluth, the inner and outer harbor, and the broad waters of Lake Superior extending eastward as far as the eye can reach, here and there being visible lake craft of different dimensions, while the moving train of cars at your feet present altogether a sight unequalled in any other part of the country. The future of this grand prospect it is hard to conceive, when here will be centred two rival cities numbering their tens of thousands of inhabitants, and the surrounding waters alive with commerce.

HARBOR IMPROVEMENTS.

A breakwater has been built by the Lake Superior and Mississippi Railroad, at a cost of about $85,000, and the necessarily accompanying docks, about $60,000 more; a Citizens' Dock, in the nature of an outside harborage, costing nearly $50,000; the Ship Canal through Minnesota Point, constituting a safe and reliable entrance for the Inner or Bay Harbor, costing the city over $100,000.

The *Dyke*, extending from Rice's Point to Minnesota Point, was finished in March, 1872; it gives to Duluth an inner harbor in the bay of two square miles, the expense of which was $40,000.

RAILROADS.

Here commences the *Lake Superior and Mississippi Railroad* and the *Northern Pacific Railroad*, the one running to St. Paul, 155 miles, the other being completed to Bismarck, on the Missouri River, 450 miles, connecting with the Upper Mississippi River and the Red River of the North.

STEAMBOAT LINES.

Lines of steamers run from Buffalo, on Lake Erie, and Chicago, on Lake Michigan, about 1,200 miles, to Duluth. Steamers also run from Collingwood and Sarnia, Canada, to Duluth, altogether carrying large numbers of passengers and an immense amount of freight to and from the lower lake ports. Steamers also run from Duluth to Ontonagon, Houghton and Marquette, landing at the intermediate ports on the South Shore.

Mariner's Guide to the Upper Lakes.

LIGHT-HOUSES ON LAKE HURON.

1. *Fort Gratiot,* 43° N. lat., situated at the head of St. Clair River, Mich.; elevated 82 feet. Marks entrance from Huron into St. Clair River, Goderich, Canada, N.E. by N. 63 miles.

2. *Point aux Barques,* 44° 1′ N. lat., on eastern side of entrance into Saginaw Bay; elevated 88 feet. Thunder Bay Island light, N. by W. ¼ N., 74 miles.

3. *Charity Island,* at the mouth of Saginaw Bay; elevated 45 feet. Saginaw Bay light-house S.W. ¾ S., 35 miles.

4. *Tawas,* on Ottawa Point, north-west shore of Saginaw Bay, Mich.; elevated 54 feet. Charity Island light bears S. 15½ miles.

5. *Sturgeon Point,* on the western shore of Lake Huron, 24 miles south of Thunder Bay Island; elevated 69 feet. Point aux Barques light, S.S.E. ¼ E., 54 miles.

6. *Thunder Bay Island,* 45° 2′ N. lat., north side of entrance to Thunder Bay, Mich.; elevated 59 feet. Sturgeon Point light, S. by W. 23 miles. Great Duck Island, Canada, N. by E. ½ E., 43 miles.

7. *Presque Isle,* about 25 miles northwest of Thunder Bay light; elevated 123 feet. A coast light on the northern point of the Peninsula. The Detour light bearing N.N.W. ¼ W., 48 miles.

8. *Spectacle Reef,* situated to the eastward of the entrance to the Straits of Mackinac, in Lake Huron; elevated 32 feet. Bois Blanc light-house W. by N., 14 miles; Mackinac Island (Fort Holmes), W. by N. ⅜ N., 24 miles; Detour light-house N.E. ½ N., 16½ miles.

9. *Cheboygan,* on the mainland opposite Bois Blanc Island; elevated 37 feet. Marks entrance into the south channel of the Straits of Mackinac. Old Fort Mackinac on the main, N.W. by W. ¼ W., 17 miles.

10. *Bois Blanc,* on the north side of Bois Blanc Island, in the Straits of Mack-

inac; elevated 53 feet. The Detour light bearing E.N.E., 26 miles. Fort Mackinac W.N.W. ½ W., 9½ miles.

11. A Light-house is to be erected at Mackinac Island, and one is erected near Old Mackinac, in the Straits of Mackinac, at the entrance to Lake Michigan.

12. *Detour,* 45° 57′ N. lat., situated at the mouth of the River St. Mary, Lake Huron, Mich.; elevated 75 feet. Marks entrance from Lake Huron into St. Mary's River. Island of Mackinac 30 miles west. Saut Ste. Marie, 55 miles N.N.W. Presque Isle light S.S.E. ½ E., 47½ miles.

LIGHT-HOUSES ON LAKE SUPERIOR.

1. *Round Island,* 46° 26′ N. lat., situated in White Fish Bay, near entrance to St. Mary's River; elevated 50 feet. Light on square stone tower, rising from the keeper's dwelling. A guide to and from the entrance to Lake Superior.

2. *Point Iroquois,* on south shore of White Fish Bay; elevated 72 feet. The light is on a red brick tower, connected with the keeper's dwelling. Parisien Island, N. by W. ½ W., 12 miles. White Fish Point, N.W. ¾ N., 25 miles.

3. *White Fish Point,* 46° 46′ N. lat., on White Fish Point, Mich.; elevated 76 feet. Iron pile structure. A guide to and from White Fish Bay, at the S.E. extreme of Lake Superior. The southwest point of the shoal off Caribou Island, Canada, bears N.W. ½ W., 55 miles; Manitou Island, W.N.W. ¼ W., 131 miles; Montreal Island, Can., N. by E., 37 miles.

4. *Big Sable,* at Big Sable Point, on the south shore of Lake Superior, between White Fish Point and Grand Island, distant 19 miles from Grand Portal (Pictured Rocks), 27 miles to Grand Island light, and 60 miles from Marquette.

5. *Grand Island,* on north point of

Grand, on a high bluff; elevated 205 feet. Stannard Rock Beacon, N.W. by N. ½ N., 50 miles. Manitou Island light, N.W. ⅞ N., 73 miles. Marquette light, W. ¼ S., 33 miles.

6. *Grand Island Harbor*, — a guide through the eastern passage to the anchorage at Grand Island Harbor and the wharf at Munising; elevated 34 feet.

7. *Grand Island Harbor*, (front light,) on the mainland at west entrance to Grand Island Harbor; elevated 23 feet. Front light in wooden tower. Rear light on keeper's frame dwelling. These lights are a guide for entering the west channel. With the north point of Wood Island, bearing W. ⅞ of a mile, and the lights in range, steer S. by E. ½ E., 5 miles, toward the range lights.

8. *Marquette*, 46° 32′ N. lat., situated on the north point of Marquette Harbor; elevated 77 feet. Stannard's Rock Beacon, N. ⅞ E., 43½ miles; Grand Island Light, E. ¼ N., 34 miles.

9. *Granite Island*, on Granite Island, 12 miles N.W. of Marquette; elevated 93 feet. Keweenaw Point, N. by W. ½ W., 48 miles; Stannard's Rock, N. by E. ½ E., 32 miles.

10. *Huron Island*, on the West Huron Island, off Huron Bay; elevated 197 feet. Manitou Island light, N. by N.E. ⅞ E., 37 miles; Traverse Island, N.W. by W., 14½ miles. Portage Entry light, W. 20 miles.

11. *Portage River*, near mouth of Portage Entry, western shore of Keweenaw Bay; elevated 68 feet. Manitou Island light, N.E ½ E., 50 miles. To Houghton and Hancock, *via* Portage Lake, 14 miles; to head of Ship Canal, Lake Superior, 23 miles.

12. *Portage Range*, (front light,) on prolongation of cut from Keweenaw Bay; elevated 21 feet. Guide into Portage River. Front light on wooden tower. Rear light on keeper's dwelling.

13. *Manitou*, on the east point of Manitou Island. Iron pile structure, connected with keeper's house; elevated 81 feet. Stannard's Rock, S.E. 23½ miles; the east point of Isle Royale, N.W.¾ N., 66 miles.

14. *Gull Rock*, on a rocky islet between Manitou Island and Keweenaw Point; elevated 50 feet. Tower and keeper's dwelling connected, built of Milwaukee brick. A guide through the passage between Gull Rock and Keweenaw Point.

15. *Copper Harbor*, 47° 28′ N. lat., situated on the eastern point to the entrance to Copper Harbor; elevated 65 feet. Tower and keeper's dwelling connected. S.W. point of Isle Royale, W.N.W., 68 miles; Passage Island, (route to Silver Islet,) N.N.W. ¼ W., 58 miles; Saut Ste. Marie, E.S.E., 200 miles.

16. *Copper Harbor Range*, (front light,) on Fort Wilkins reserve; elevated 22 feet. To guide vessels into Copper Harbor.

17. *Eagle Harbor*, on the west point of the entrance to Eagle Harbor. Light on keeper's dwelling; elevated 47 feet. The north point of the outer or easternmost of the Apostle Group, W. by S. ⅛ S., 105 miles.

18. *Eagle River*, near the entrance to Eagle River, south shore of Lake Superior. Light on keeper's dwelling; elevated 61 feet. Michigan Island light, W.S.W. ⅜ W., 100 miles.

19. A Light-house is to be erected at the entrance to the Portage Ship Canal, on the south shore of Lake Superior.

20. *Ontonagon*, 46° 52′ N. lat., at mouth of Ontonagon River; elevated 47 feet. House and tower of Milwaukee brick. The S.W. point of Isle Royale, N. ⅛ E., 67 miles.

21. *Michigan Island*, on the southern point of Michigan Island, the most eastern of the Apostle Group, Wisconsin; elevated 129 feet. A guide through the passage between Madeline and Michigan Islands, to Bayfield and La Pointe. Ontonagon, E., 56 miles.

22. A Light-house is erected on the

Outer Island of the Apostle Group of Islands, to guide mariners to the head of Lake Superior, distant 80 miles S.W. of Isle Royale, and about 90 miles E. of Duluth.

23. *La Pointe*, on Chaquamegon Point, the south-eastern entrance to La Pointe Harbor, and to the northward of Ashland and southward of Bayfield; elevated 42 feet. Light on keeper's dwelling, painted white.

24. *Raspberry Island*, on the southwest point of the island, one of the Apostle Group; elevated 77 feet. A guide between the Main and Raspberry Island, and to Bayfield.

25. *Minnesota Point*, 46° 46′ N. lat., situated at the head of Lake Superior, mouth of St. Louis River; elevated 50 feet. A guide into Superior Bay and Duluth. The S.W. point of Isle Royale, N.E. by E. ⅞ E., 150 miles; Silver Islet, Canada, N.E. ¾ E., 180 miles.

26. *Duluth*, on the end of the south pier of Duluth, Minn.; a guide into the canal or inner Bay.

Table of Distances,

AT WHICH OBJECTS CAN BE SEEN ON THE LAKES AT DIFFERENT ELEVATIONS.

Height in feet.	Distances in miles.	Height in feet.	Distances in miles.
10	4·184	100	13·221
20	5·916	150	16·208
30	7·245	200	18·702
40	8·366	300	22·917
50	9·354	400	26·450
60	10·246	500	29·583
70	11·067	600	32·406
80	11·832	800	37·418
90	12·549	1000	41·883

Statute miles.

EXAMPLE. — Detour Light, 75 feet high, visible. .... 11·456

Add for height of observer's eye, on deck 10 feet.............. 4·184

Distance of Light............... 15·406

Duluth to Isle Royale and Pigeon River.

The trip along the North Shore of Minnesota to the mouth of Pigeon River and Isle Royale, which lies opposite, is one of great interest. The harbors, headlands, islands, and objects of interest are numerous. Then the inside passage along the Canada shore, passing several islands, is still more interesting, until you enter Thunder Bay, when the scene is by far more grand and imposing — having in full view Pie Island, M'Kay's Mountain, and Thunder Cape, the latter rising 1350 feet in height, with almost perpendicular walls.

GRANITE POINT and KNIFE RIVER are passed 18 miles N.E. of Duluth, near where is the settlement of *Buchanan* and Isle aux Roches.

AGATE BAY, 25 miles, and *Burlington Bay*, 2 miles farther, are much frequented

by tourists. Starting from Duluth or Superior City, a small steamer runs frequently to these harbors during the summer months, carrying parties of pleasure.

ENCAMPMENT ISLAND and RIVER, 35 miles from Duluth, is another interesting locality. Here the shores are remarkably bold and precipitous, rising from 800 to 1000 feet above the water.

BEAVER BAY and SETTLEMENT, 18 miles farther, is a small village inhabited mostly by Germans. Here the shore presents a rugged appearance, with high lands rising in the background.

The *Palisades*, 6 miles farther, presents a grand appearance from the water's edge. The rock rises to the height of over 300 feet, presenting perpendicular columns. *Baptism River*, 2 miles, is a

12

mountain stream, rising in an eminence called Saw-Teeth Mountains.

The numerous rivers and creeks falling into Lake Superior along the North Shore, from Duluth to Fort William, Canada, for a distance of about 180 miles, are generally rapid streams of moderate length. They are mostly pure water, abounding in speckled trout and other kinds of fish. In the rear it is almost an unbroken wilderness, although supposed to be rich in minerals of different kinds,—copper, iron, silver, and gold,—the latter being found in the vicinity of Vermillion Lake, lying in a north-easterly direction some fifty or sixty miles.

Petit Marais and the mouth of *Manitou River* are next passed. *Two Islands River* is named from two beautiful islands which lie off its mouth.

Temperance River, another romantic stream, rising in Carlton's Peak, is thus named owing to its having no *bar* at its mouth. This locality should become a favorite resort for invalids and others seeking health and pleasure. Here are a succession of waterfalls of great beauty, some only 200 or 300 feet from the Lake shore.

Good Harbor Bay, 100 miles N.E. of Duluth, is a safe harbor, as its name implies. *Terrace Point, Rock Island*, and *Grand Marais* are all in sight, while small streams flow into the lake, alive with speckled trout.

Grand Marais Harbor, Minn.—This harbor is located on the north shore of Lake Superior, about 115 miles east of Duluth. It is the only large and secure harbor between Duluth and Grand Portage, a distance of about 150 miles. The harbor is of elliptical shape, and is 2500 feet long by 1400 feet wide. The mouth of the harbor is 1000 feet wide; the depth, at the entrance, 24 feet, with a rock bottom. After entering the harbor the depth increases, and the bottom is composed of gravel and boulders.

Grand Portage Bay and *Island*, 40 miles farther, is a place of considerable resort. Here is an Indian village and Reservation, where is a Roman Catholic Mission, a block-house, and some 12 or 15 dwellings. From this point commences a portage route to Pigeon River, Rainy Lake, and Lake of the Woods.

Waus-wau-Goning Bay is a beautiful expanse of water, encircled in part by highlands, rising to the height of 1000 feet, and several beautiful wooded islands. From this bay to the south end of Isle Royale, the distance is 20 miles.

Pigeon Bay and River, forming the north-west boundary between the United States and Canada, is well worthy of a visit. The bay contains several beautiful islands and islets, where is excellent fishing. *Pigeon River Falls*, situated about half a mile above the landing, are exceedingly grand and imposing. Here the water falls about 90 feet, almost perpendicular, into a chasm, then rushes through a gorge into the bay.

Parkersville is a small settlement situated at the mouth of the river, where also is to be seen Indian huts and wigwams constructed of birch bark. This place, no doubt, is soon destined to become a place of resort during the summer months; it is situated about 150 miles N.E. of Duluth.

Isle Royale, Houghton Co., Mich., being about 45 miles in length from N.E. to S.W., and from 6 to 12 miles in width, is a rich and important island, abounding in copper ore and other minerals, and also precious stones. The principal harbor and only settlement is on *Siskowit Bay*, being on the east shore of the island, 50 miles distant from Eagle Harbor, on the main shore of Michigan.

The other harbors are Washington Harbor on the south-west, Todd's Harbor on the west, and Rock Harbor and Chippewa Harbor on the north-east part of the island. In some places on the west

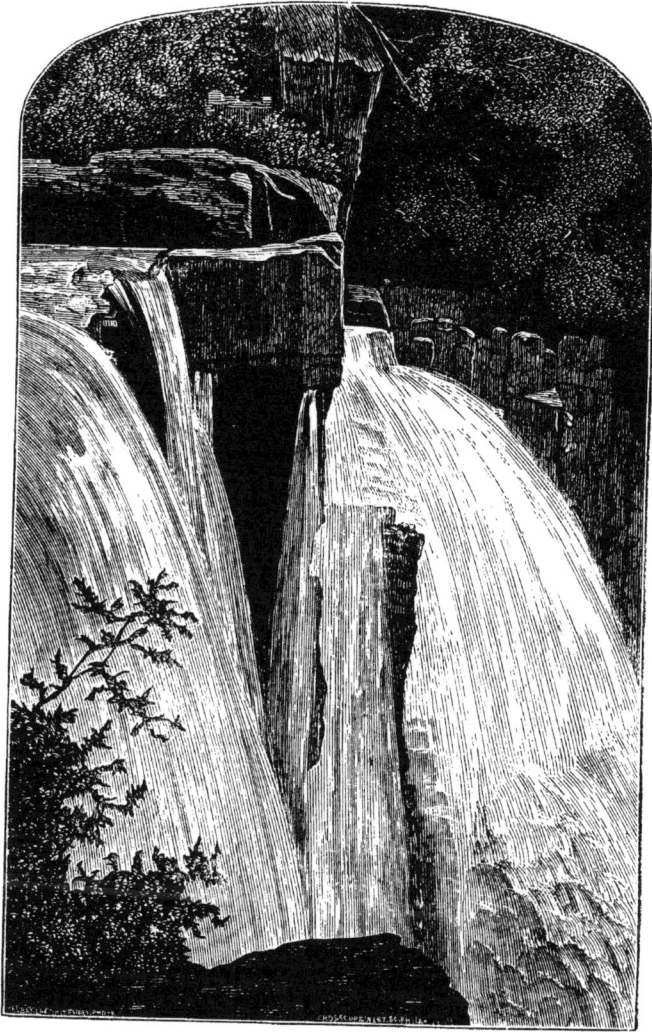

GREAT FALLS — Pigeon River.

are perpendicular cliffs of green stone, very bold, rising from the water's edge, while on the eastern shore conglomerate rock or coarse sandstone abounds, with occasional stony beach. On this coast are many islets and rocks of sandstone, rendering navigation somewhat dangerous. Good fishing-grounds abound all around this island, which will, no doubt, before many years, become a favorite summer resort for the invalid and the sportsman, as well as for the scientific tourist.

SISKOWIT LAKE is a considerable body of water lying near the centre of the island, which apparently has no outlet. Other small lakes and picturesque inlets and bays abound in all parts of the island. Hills, rising from 300 to 400 feet above the waters of the lake, exist in many localities throughout the island, which is indented by bays and inlets.

Trip along the South Shore of Lake Superior.

On leaving DULUTH by the American steamers bound for Marquette and the Saut Ste. Marie, the boat pursues an easterly course until the Apostle Islands are reached, about 60 miles distant — the Minnesota and Wisconsin shores being both in sight. The first island passed, on the right, is called Steamboat Island; then comes in succession Sandy Island, York Island, and Raspberry Island; on the latter there is a light-house. Oak Island is passed on the left, it being a large, elevated, and wooded body of land; then comes Hermit's Island and Bass Island; on the latter there is a valuable stone quarry of red sandstone. Here the view is most beautiful in fine weather, there being in full view a number of islands, altogether numbering twenty, of different sizes, of which Madeline Island is the largest, where is located, on its south end, *La Pointe*, an old settlement.

BAYFIELD, Wis., 80 miles east of Duluth, by water, is the first regular landing. The hotel accommodations are good, and no place on the shores of Lake Superior affords greater inducements for the seeker of health and pleasure. Boating, sailing, and fishing can be enjoyed during the summer months.

ASHLAND, situated 18 miles south of Bayfield, with which it is connected by a steam-ferry, as well as La Pointe, and a new landing at Houghton Point, all add to the attractions which *Cha-quam-e-gon Bay* and its vicinity affords to tourists. For a healthy climate, pure water, good fishing, and an agreeable class of inhabitants, no part of the lake exceeds the above favored localities. From Ashland, the *Wisconsin Central Railroad* runs south through Northern Wisconsin to Menasha, Milwaukee, etc.

On leaving Bayfield, proceeding eastward, the steamer usually runs between Bass Island and Madeline Island, passing Presque Island, toward Michigan Island, where there is a light-house. There are also several other islands in sight, which usually present a most lovely appearance from the deck of the passing steamer. After leaving the Apostle Islands, the steamer runs direct for Ontonagon, 60 miles, passing the *Porcupine Mountains*, and the Silver Region near the mouth of Iron River, a beautiful stream entering the lake about 15 miles west of Ontonagon. Proceeding eastward, the steamer either passes through Portage Ship Canal or sails around Keweenaw Point to Houghton, Marquette, etc.

On leaving Marquette the steamer usually runs direct for White Fish Point, 130 miles, passing Grand Island and the *Pictured Rocks*, all being in sight from the deck of the steamer. Distance from Duluth to Saut Ste. Marie, passing around Keweenaw Point, 560 miles.

SAUT STE. MARIE—FROM AMERICAN SIDE.

LAKE SUPERIOR — NORTH SHORE.

Trip around the North Shore of Lake Superior,

MADE ON BOARD THE CANADIAN STEAMER CHICORA, *July*, 1873.

On leaving the mouth of the *Ship Canal*, above the Rapids of St. Mary's River, a beautiful view is presented from the deck of the steamer. The Rapids and the settlements on both sides of the river appear to advantage, while a lovely stretch of the stream above is passed over before arriving at

POINT AUX PINS, Ont., 6 miles above the Rapids. Here is a convenient steamboat landing, a store and a few dwellings, being handsomely situated among a grove of small pine trees. Fishing parties can here be fitted out with guides and canoes for fishing and exploring the North Shore.

WAISKA BAY, an expanse of the river, is next entered, extending westward to Point Iroquois. Here may be seen an Indian settlement on the South shore.

GROS CAP, 15 miles from the Saut, lies on the Canada side directly opposite Point Iroquois, forming prominent landmarks to the entrance of Lake Superior. This bold headland consists of hills of porphyry rising from 600 to 700 feet above the waters of the lake. "Gros Cap is a name given by the *voyageurs* to almost innumerable projecting headlands; but in this case appropriate, since it is the conspicuous feature at the entrance of the Great Lake."

Immediately north of Gros Cap lies GOULAIS BAY and GOULAIS POINT, another bold highland which is seen in the distance. *Goulais River* enters the bay, affording, in connection with the ad-jacent waters, good fishing-grounds—the brook or speckled trout being mostly taken in the river. Here is an Indian settlement of the Chippewa tribe.

TAQUAMENON BAY, lying to the westward, is a large expanse of water, being about 25 miles long and as many broad, terminating at *White Fish Point*, 40 miles above Saut Ste. Marie.

PARISIEN ISLAND is passed 30 miles from the Saut, lying near the middle of the above bay, being attached to Canada.

MAPLE ISLAND, 20 miles above Gros Cap, is a small wooded island lying near the shore. SANDY ISLANDS, 5 miles farther, are a low group of islands lying off BATCHEEWANAUNG BAY, a large and beautiful sheet of water which receives a stream of the same name, being famed for trout fishing. Here is an Indian settlement, and a village where is a large steam saw-mill. On *Corbay Point*, at the entrance to the above bay, stands a lighthouse.

MAMAINSE POINT, (*Little Sturgeon*,) lying opposite White Fish Point, is another bold headland, where is a fishing station and a few dwellings. It is about 55 miles from the Saut, and was formerly a copper mining location, but has been abandoned. Here is a good harbor and steamboat landing.

Some 12 or 15 miles north of Mamainse Point are located, on MICA BAY, the old Quebec Copper Mining Company's Works, at present abandoned, owing to their being found unproductive. Still

181

farther north, skirting Lake Superior, is to be found a vast *Mineral Region*, as yet only partially explored.

MONTREAL ISLAND and RIVER, 25 miles north of Mamainse, afford good fishing-grounds. Here is a harbor exposed to the west winds from off the lake, which can safely be approached when the winds are not boisterous.

Aguawa River, 30 miles from Mamainse Point, affords a good harbor, where is a trading post and good fishing.

LIZARD ISLAND and LEECH ISLAND, some 10 miles farther northward, are next passed, lying contiguous to the mainland.

CAPE GARGANTUA, 45 miles north of Mamainse, is a bold headland. On the south side is a harbor protected by a small island. From this cape to the island of Michipicoten the distance is about 30 miles.

The steamers bound for Prince Arthur's Landing usually run direct for *Michipicoten Island*, lying 70 miles northwest of Mamainse Point; but when bound for *Michipicoten Harbor* they sail along the eastern end of the lake, passing several small islands and headlands.

This portion of Canada is as yet an almost unbroken wilderness for several hundred miles around Lake Superior. With the exception of a few scattered Hudson Bay Co's. Posts, there are no settlements until you reach the vicinity of Silver Islet or Thunder Bay, two or three hundred miles distant.

Coasting along the North Shore of Lake Superior.

There are *two* modes of reaching St. Ignace Island and the mouth of Nepigon River, if bound on a pleasure excursion to the far-famed fishing-grounds on the North Shore of Lake Superior, where speckled trout, varying from one to eight pounds, can be taken in large quantities.

The *first* plan and mode of conveyance is that of hiring a birch canoe of from 18 to 20 feet in length, with two Indians, or half-breeds, as voyageurs, for which service there is usually a charge of five dollars a day. Then lay in tents and a supply of provisions for two or more weeks, according to the length of the intended trip. After this is accomplished, embark on one of the Canadian steamers running direct for the mouth of the Nepigon River; then land and commence your canoe sailing or paddling until you reach the desired fishing-grounds. Here land and spread your tents for a time, alternately fishing, cooking, eating and sleeping, as you may desire.

This is a truly independent style of travelling and living, combining health, pleasure and economy. The party may consist of two, four, or more persons besides the two guides. You have nothing to annoy you except the rainy weather or mosquitoes, which are at times very numerous and hungry, often attacking the sportsman in a fierce manner. The best mode of avoiding these pests, is to camp on a point of land and clear away all the underbrush; then build a smudge-fire and take your comfort, regardless of the buzzing outside world.

The *second* mode is to hire a Mackinac sailing-boat with oars, and lay in your provisions; hire your crew of two Indians, and set sail from the mouth of the Ship Canal above the Saut Ste. Marie, or any other point from which you may choose to depart, if bound for the North Shore or mouth of the Nepigon. The supplies for this excursion are usually obtained on the Canadian side, either at the *Saut* or at *Point aux Pins*, 7 miles above, where is a good landing and camping-ground.

Starting from above the Rapids with a fair wind, *Gros Cap*, 15 miles, is soon reached; opposite lies Point Iroquois, the "Pillars of Hercules" of Lake Superior, being the true entrance to the lake. You next pass *Goulais Bay*, Maple Island, Sandy Island, and enter *Batcheewanaung Bay*, where is found good fishing-grounds and abundance of brook trout in the streams which enter here.

Proceeding northward, *Mamainse Point*, 50 miles distant from the Saut, is passed. From thence the voyager coasts along the North Shore toward Michipicoten Harbor, proceeding westward toward Nepigon Bay and River.

Michipicoten Harbor and RIVER, 120 miles north of the Saut Ste. Marie, situated in N. lat. 47° 56′, W. long. 85° 6′, affords a safe anchorage, being surrounded by high hills. Here is established a Roman Catholic mission, and an old Hudson Bay Company's post, from whence diverges the river and portage route to James's Bay, some 350 miles distant. The shore of the lake here tends westward toward *Otter Head*, about 50 miles distant, presenting a bold and rugged appearance.

MICHIPICOTEN ISLAND (the *Island of Knobs or Hills*), 65 miles from Mamainse Point in a direct course, running in a north-west direction, lies about 40 miles west of Michipicoten Harbor. This island, 15 miles in length and 6 miles wide, may be called the *gem* of Lake Superior, presenting a most beautiful appearance as approached from the southward, where a few picturesque islands may be seen near the entrance to a safe and commodious harbor, which can be entered during all winds. Nature seems to have adapted this island as a place of resort for the seekers of health and pleasure. Within the bay or harbor a beautiful cluster of islands adorns its entrance, where may be found agates and other precious stones; while inland is a most charming body of water, surrounded by wooded hills rising from 300 to 500 feet above the waters of Lake Superior. The shores of the island abound with greenstone and amygdaloid, while copper and silver mines are said to exist in the interior of great value. The fisheries here are also valuable, affording profitable employment to the hardy fisherman of this region. As yet, but few houses are erected on the shores of this romantic island, where, sooner or later, will flock the wealthy and beautiful in search of health and recreation, such as are afforded by pure air, boating, and fishing and hunting.

The fish taken in this part of the lake are mostly white fish, siskowit, Mackinac trout, and speckled trout, the former being taken by gill-nets.

On the mainland are found the carabou, a large species of deer, bears, foxes, otters, beavers, martins, rabbits, partridges, pigeons, and other wild game. The barberry, red raspberry, and whortleberry are also found in different localities.

CARIBOU ISLAND, lying about 25 miles south of Michipicoten, near the middle of the lake, is a small body of land attached to Canada. It is usually passed in sight when the steamers are on their route to Fort William.

Otter Head, 60 miles north-west of Cape Gargantua, is one of the best natural harbors on Lake Superior, where is to be seen grand and romantic scenery. *Otter Island*, lying to the north of the harbor, is a bold piece of land. In this vicinity are supposed to be rich deposits of iron, tin, and other minerals.

Pic River, 40 miles farther, is a fine stream, entering the lake from the north. Here is an old Hudson Bay Company's Post and a settlement of Indians. *Pic Island*, lying 20 miles to the westward, is next passed, also the mouths of several small rivers.

Slate Island, 25 miles farther west, lies out in the lake, about 12 miles from shore. Here is the widest part of Lake Superior, being about 160 miles from shore to shore, being nearly due north of Marquette.

Wilson's Island, 20 miles west, lies within a few miles of the mainland, which is bold and rugged.

Salter Island, a few miles farther west, is a small island, affording fine scenery.

Simpson's Island, 40 miles west of Slate Island, is an important body of land, which, no doubt, contains valuable minerals.

St. Ignace Island, the largest and most important of the Canadian Islands, has long been famed for its supposed mineral wealth. Copper, silver, and lead are said to be found on this island. Here the land rises to a great height in some places, while the shores are bold and rugged. On the west end there is a convenient steamboat landing, being distant 30 miles from the mouth of Nepigon River.

On approaching *Nepigon Bay* the steamer runs between Simpson's and Salter Island, the former being a large and rugged body of land, while the highlands on the main coast are of an equally high and rugged character, presenting a mountainous appearance.

NEPIGON BAY is about 30 miles long and 20 miles wide, containing several beautiful wooded islands, the largest being *Verte Island*, lying near the middle of the bay, with lesser islands as the mouth of the river is approached.

Nepigon River, the "Ultima Thule" of the angler, is a bold and dashing stream, falling 300 or 400 feet within the distance of 22 miles. There are alternate rapids and expansions, with high, rugged hills for most of the distance to the foot of Lake Nepigon. Those who are in the habit of visiting this stream from year to year represent it as one of the most wild and charming retreats on the Continent, where health and sport can be obtained during the summer months, affording almost endless enjoyment to the angler and sportsman.

Red Rock is the name of the settlement at the mouth of the river, where is a convenient steamboat landing, a store and storehouse, a few dwellings and Indian wigwams, surrounded by cleared lands. This is an Hudson Bay Company's Post, where goods and Indian curiosities are sold. Here parties intending to fish for trout in the river disembark, and by means of birch canoes, paddled by Indian guides, they ascend the stream, often proceeding to Lake Nepigon, passing over several portages on their route.

No words can faithfully describe the wild beauties of this region of country, surrounded by rugged hills, dashing streams, placid lakes, water-falls, and rapids. Here brook trout of a large size are taken in fabulous numbers, and of a delicious flavor. No place on the Continent of America exceeds this romantic spot, which will ever afford sport to the angler and seekers of pleasure.

Nepigon River — Its Rapids and Lakes.

COPIED FROM REPORT OF ROBERT BELL, C. E.

The Nepigon River empties into the head of Nepigon Bay, which is the most northern point of Lake Superior, (48° 45′ N. lat.) The water is remarkable for its coldness and purity, and is the largest stream flowing into the lake. The general upward course of the Nepigon is due north, the length of the river being 31 miles. Four lakes occur in its course, between which are rapids and falls. The lowest, Lake Helen, is only one mile from *Red Rock*, a Hudson Bay Company's post, at the head of Nepigon Harbor. At the outlet of this lake the river is very narrow, apparently about 100 yards wide, with a very swift current, flowing between banks from thirty to fifty feet high. Lake Helen, which runs due north, is about eight miles long and one mile wide. For six miles above the lake the river has a width of about five chains, with deep water, and a moderately strong current. Here the river makes a sharp bend to the right, and is broken by a slight chute at Camp Alexander. At a quarter of a mile above this point the Long Rapids begin, and continue for two miles. Here is a portage that leads to the foot of Lake Jessie. This lake, which is three miles long, and studded with islands, is separated from Lake Maria, immediately above it, and two and a half miles in length, by "The Narrows," six or eight chains wide, in which there is a strong current.

A very high west-facing cliff of columnar trap approaches the river from the south-westward, at the head of Lake Maria, and runs from this point, in a tolerably straight course, all along the east side of the river to Lake Nepigon. Cedar Portage, two miles above the lake, is 250 yards long. A mile and a quarter above it there is another portage of fifty yards over an island in the middle of the river. Three-quarters of a mile above Island Portage the One-Mile Portage begins. At rather more than one mile from the head of this portage the river breaks in a white, foaming chute, which separates Lake Emma from the lower level. This lake is nearly four miles long. Between it and the point at which the river leaves Lake Nepigon, a distance of six miles, four principal rapids occur, the lowest of which is seen where the river enters the northern extremity of Lake Emma. Four miles more, in a north-westerly course, brings us to the head of Lake Hannah, from which Flat Rock Portage, one mile in length, carries us to the shore of Lake Nepigon.

The following list shows the levels in ascending the river, and the height of Lake Nepigon above Lake Superior:

RAPIDS, PORTAGES, ETC.	Feet.
Current between Red Rock and Lake Helen	2
Current in river from Lake Helen to Camp Alexander	6
Chute at Camp Alexander	4
From the last to Long Portage	8
Rapids at Long Portage	137
Current in the Narrows, between Lakes Jessie and Maria	1
Current from last lake to Cedar Portage	1
Cedar Chute	10
Current from Cedar Chute to Island Chute	1
Island Chute	7
Current from the Island to the One-Mile Portage	2
Rapids of One-Mile Portage	45
Current from One-Mile Portage to White Chute	1
White Chute	6
Current between Lakes Emma and Hannah	1
Rise from last lake to Lake Nepigon (Flat Rock Portage)	81
Lake Nepigon* above Lake Superior	313

WATERS OF THE NEPIGON.

"Good water is a luxury that cannot be over-estimated. In this respect, the sportsman angling for trout on the Nepigon is favored above all others. He has always before him a never-failing supply, so remarkable for its coldness and its purity that even those who have drunk of it habitually for years cannot but speak of it whenever they taste its refreshing waters. Its temperature is too cold for bathing, and a morning ablution brings a rosy glow to the cheeks that the fairest belle might envy."

---

* *Nepigon*, the name by which the lake is known, is a contraction of an Indian word signifying "Deep Clear-water Lake." It is about 70 miles long and 50 miles wide, being deeply indented by large bays, having a coast-line of upwards of 500 miles, and numerous islands, numbering upwards of one thousand.

Nepigon River, Rapids, and Lake.

"The Nepigon," says HALLOCK in his *Fishing Tourist*, "is a noble stream, with waters cold and clear as crystal, flowing, with a volume five hundred feet wide, into a magnificent bay of great extent. This bay is surrounded by long undulating ranges of hills, rugged precipices, huge bluffs, and lofty hills more or less wooded with evergreens, interspersed with deciduous trees, and filled with islands of all sizes and every variety of outline.

"The first rapids on the river occur about a quarter of a mile above the landing at Red Rock, and are a mile in length. They can be run by canoes with safety, while a portage road passes along the shore. There are fifteen rapids or chutes in all, and at each there is the best of trout fishing. Three miles below the head of the river are the Virgin Falls, 25 feet high. Altogether the scenery is the wildest and most diversified imaginable, and constantly presents changes of the most enchanting character. Above the falls the river widens gradually, inclosing within its area dozens of small islands variegated with evergreens, birch, poplar, larch, tamarack, etc., and then expands into a vast inland sea, whose shores gradually recede beyond the limit of vision. In the far distance, horizon, sky, and water meet, and the waves roll up on shore with a volume and dash as turbulent in storm as those of Erie or Superior. Its bays are numerous; some of them very deep, extending inland for a number of miles — all teeming with speckled trout, lake trout, pike, and pickerel in great quantities. Into it flows several rivers that have their sources in the heights of land which constitute the water-shed that divides the waters of the St. Lawrence chain from those of Hudson Bay."

Lake Nepigon Explored.

For many generations the existence of LAKE NEPIGON has been known, but only partially, to the traders, trappers, and Indians in the employ of the Hudson's Bay Fur Company. The general impression has been, that Lake Nepigon was located about 80 miles north of Lake Superior, and that it was as large as Lake Erie. But during the past three years the Government of Canada has been extending its surveys to the public lands along the North Shore of Lake Superior, past Fort William toward the Red River country. Mr. Herrick, one of the provincial surveyors, made a traverse of the Nepigon River in 1866, and found the distance to the lake to be in a right line less than thirty miles. The lake, instead of lying east and west for two miles in length, has its longest diameter north and south, seventy miles; its breadth east and west being fifty miles. Its height above Lake Superior is determined approximately at 313 feet, or 913 above the ocean.

In 1869, Robert Bell, an assistant geologist and engineer of the Canada survey, in company with surveyor McKellar, made the entire circuit of Lake Nepigon, exploring it topographically and geologically.

From the reports and maps of these bold explorers it is learned that Lake Nepigon is accessible and has around it a valuable country. It has a shore line of about five hundred and eighty miles of navigable water, and probably one thousand islands. Its waters are clear, cold, and pure, and its outlet is the largest river of Lake Superior. A large part of the rocks are such as produce copper.

Near the south-west angle of the lake at Grand Bay, with only a few feet of elevation between them, the Black Sturgeon River has its rise. The space between them is so narrow and so low that in a high stage of the lake the waters appear to flow over the intervening ground from the lake into the head of this river. *Nepigon*, in Chippewa, signifies deep and clear water.

When the party reached the head of the river at the lake, they divided, McKellar following the right hand or eastern shore, and Bell the western.

They travelled in canoes, Indian fashion, taking the courses with a theodolite, and the distances with a micrometer. They also took observations for latitude and longitude. At the end of eight weeks they met at the northern extremity of the lake, having had a pleasant and prosperous journey.

The latitude of the northern extremity is 50° 15′ north. Its surroundings are less mountainous than Lake Superior. To the north, between Nepigon and Hudson's Bay, the country is represented by the Indians to be comparatively level.

The *Ombabika River* coming in from that direction heads with the Albany River, which discharges into salt water at James' Bay, and between them is supposed to be a flat limestone region. This is one of the routes of the Hudson Bay Company to their posts on Hudson Bay.

South-west of Lake Nepigon, between it and Thunder Bay of Lake Superior, the country is represented as capable of cultivation, and the climate the counterpart of Quebec.

By referring to the map can be seen the route usually pursued by the Canadian steamers; also, the islands and headlands along the North Shore.

On leaving Nepigon River and Bay, proceeding westward toward Thunder Bay, the steamer usually passes through Nepigon Straits, lying between St. Ignace and the mainland. Then the broad waters of Lake Superior are again entered, and you steam past numerous islands before reaching *Point Porphyry*, where is located a light-house to guide the mariner.

BLACK BAY, 15 miles east of Thunder Cape, presents a large expanse of water, being about 40 miles long and 10 or 12 miles wide, surrounded by a rugged and wild section of country. Towards the north are two peaked eminences, termed the *Mamelons* or *Paps*, from their singular formation, resembling a female's breast, when seen at a distance.

On approaching the famed *Silver Islet* from the east, the view presented is of the grandest and most interesting character, —the Island, with its treasures of wealth, and Silver Islet Settlement, are seen in the foreground, while bold Thunder Cape and romantic Pie Island are seen in the distance. Also, southward, may be seen Isle Royale and Passage Island, both being attached to the State of Michigan.

Trip from Marquette to Silver Islet,

NORTH SHORE, LAKE SUPERIOR.

Leave MARQUETTE at 2 P. M. by steamer, with a large party of tourists on board, bound for Silver Islet and Duluth. Arrive at Houghton, 85 miles, at 9 P. M.; leave at 4 A. M. next morning for Silver Islet, distant 150 miles from Marquette by direct route. Pass Keweenaw Point and Copper Harbor at 11 A. M. Fresh wind from the N.W., steering northward; thermometer 42° Fahr. Pass

Passage Island at 4 P. M., running direct for Thunder Cape. The N.E. part of Isle Royale presents a rugged appearance; the highest part being elevated 500 feet. Arrive at SILVER ISLET or RYANTON at 6 P. M.; distant 24 miles E. of Fort William. Take a tug-boat and visit *Silver Islet Mine*, lying 3,000 feet from the mainland; descend down the shaft, within the coffer dam, about

100 feet, by means of ladders, the passage being lighted by candles; length opened, about 70 feet from south to north; vein about 8 feet wide with a dip of 75°. On arriving at the end of the opening a rich sight was presented to our view, the face of the mine sparkling with silver as viewed by candle-light. The ore yields from $2,000 to $4,000 a ton.

Silver Islet or Ryanton lies on the mainland, having a secure harbor and easy of access, with a convenient wharf for the shipment of silver ore, which is mostly put up in barrels for the purpose of transportation.

Mining for silver has been commenced on the mainland near the water's edge, which bids fair to be very productive. The vein has been traced for several miles northward in a straight line from Silver Islet. Back of the settlement lies Surprise Lake, while Thunder Cape, to the westward, rises almost perpendicular from the waters of Lake Superior, here containing several small rocky islands scattered along the coast.

Silver Island, Lake Superior.

"Silver Islet, a desolate rock some 70 or 80 feet square, projecting from the stormy waters of Lake Superior, lies some 3,000 feet away from the Canadian Shore of Lake Superior, south of Thunder Bay, off Thunder Cape, 24 miles east of Fort William, and until recently was regarded merely as a danger to navigation, an object to be scrupulously avoided. Now, it is known that the rocks of the island and its vicinity are rich in silver, and arrangements have been made for mining in them, even as far as the mainland, if found desirable. The island is so low, and such a mere speck in the lake, that in a sea the waves were wont to wash entirely over it, rendering it altogether uninhabitable, and presenting great difficulties in the way of opening a vein. It became necessary that breakwaters should be built all about it, and large expenses incurred before the location could be worked at all. The present owners bought the property from a Montreal Company, and commenced building breakwaters Sept. 1, 1870. Before the close of navigation in Nov., they had completed their erection, and had mined 22 days, sending to the smelting works in New Jersey some $100,000 worth of ore.

"The miners are now boring and blasting 100 or 200 feet below the surface, and, though they will deepen the passages as they go forward with the mine, it is thought that no very deep work will be required, even in getting safely beneath the surface of the Lake and working toward the mainland, 3,000 feet distant. All indications point to the great success of the enterprise, and incline one to the belief that this recent addition to the mineral wealth of Lake Superior is one of the most brilliant discoveries that have been made in that marvellous region." The treasure shipped in 1871 amounted to $820,000; in 1872, about $1,000,000.

Silver Islet Settlement is a flourishing village near Thunder Cape, opposite Silver Islet, on the main shore. Some 60 or 80 dwellings are already erected for the accommodation of the miners, and bids fair to be the nucleus of a large mining town. A church, a hotel, and school-house have been erected by the Silver Mining Company.

From a map issued by the Crown Land Department of Canada, showing a proposed route around the North Shore of Lake Superior, it gives the course of numerous streams and locates some good timbered lands. The trees are mostly birch, balsam, cedar, spruce, pine, and hard maple on the high ridges. Iron, copper, and silver ore are also found for a distance of 400 or 500 miles from near Saut Ste. Marie to Prince Arthur's Landing, situated near the mouth of the Kaministiquia River.

THUNDER BAY presents a large expanse of water, being about 25 miles in length and from 10 to 15 miles wide, into which flow several small streams, abounding in speckled trout. *Thunder Cape*, on the east, is a most remarkable and bold highland, being elevated 1,350 feet above Lake Superior. It rises in some places almost perpendicular, presenting a basaltic appearance, having on its summit an extinct volcano. From the elevated portions of this cape a grand and imposing view is obtained of surrounding mountains, headlands, and islands — overlooking *Isle Royale* to the south, and the north shore from McKay's Mountain to the mouth of Pigeon River, near Grand Portage, Minn.

Thunder Bay and its vicinity has long been the favorite residence of Indian tribes who now roam over this vast section of country, from Lake Superior to Hudson Bay on the north. The mountain peaks they look upon with awe and veneration, often ascribing some fabulous legend to prominent localities. A learned Missionary, in describing this interesting portion of Lake Superior and its inhabitants, remarked, that "the old Indians were of the opinion that *thunder clouds* are large gigantic birds, having their nests on high hills or mountains, and who make themselves heard and seen very far off. The head they described as resembling that of a huge eagle, having on one side a wing and one paw, on the other side an arm and one foot. The lightning is supposed to issue from the extremity of the beak through the paw, with which they launch it forth in fiery darts over the surrounding country."

PIE ISLAND, in the Indian dialect called "*Mahkeneeng*" or *Tortoise*, bounding Thunder Bay on the south, is about 8 miles long and 5 miles wide, and presents a most singular appearance, being elevated at one point 850 feet above the lake. This bold eminence is shaped like an enormous *slouched hat*, or inverted pie,

giving name to the island by the French or English explorers, while the Indians gave it the name of *tortoise* from its singular shape. This elevated point is basaltic, rising perpendicular near the top, like the *Palisades* of the Hudson River.

Prince Arthur's Landing, District of Thunder Bay, Ont., is favorably situated on the west side of Thunder Bay, $3\frac{1}{2}$ miles north of Fort William. Here is a government wharf, a court-house and register's office, a town hall, 2 churches, 4 hotels, 12 or 15 stores, about 200 dwelling houses, and 800 inhabitants. A government road extends from this Landing westward toward Fort Garry, Manitoba. There are also in the vicinity several rich silver mines, as yet but partially developed.

CURRENT RIVER, 3 miles above Prince Arthur's Landing, is a dashing trout stream, where is a beautiful fall near the shore of Thunder Bay. On *McKenzie's River*, about 20 miles north-east, are found rich deposits of amethyst.

The view presented on approaching Thunder Bay is perhaps the grandest sight on the shores of Lake Superior. Here may be seen from the deck of the steamer *Thunder Cape*, with its Sleeping Giant; *Pie Island*, assuming the singular shape of a slouched hat; *McKay's Mountain*, lying westward on the main land; the *Welcome Islands*, and numerous other islands and headlands in the distance, altogether presenting a most magnificent view of land and water.

The *Military Road*, running from Prince Arthur's Landing to Manitoba, is 443 miles in length, of which 138 is by land (eleven portages) and 305 miles by water. This route passes through Lake Shebandowan and the new gold region, some 60 or 70 miles west of Thunder Bay, Lake Superior, continuing westward through Rainy Lake and the Lake of the Woods to Lake Winnipeg.

Silver Mining Companies on the North Shore, L. S.

Algoma Mine, Thunder Bay.
3 A Mine, Thunder Bay.
3 B Mine, Thunder Bay.
Cornish Mine, Thunder Bay.
Dawson Mine, Thunder Bay.
Howland Mine, near Pigeon River.
Jarvis Island Mine.
Ontario Mine, Thunder Bay.
Shuniah, Thunder Bay.
Silver Harbor or Beck, Thunder Bay.
Silver Islet, near North Shore.
Thunder Bay Mine.
Trowbridge Mine.

Fort William, an important Hudson Bay Company's Post, is advantageously situated at the mouth of the Kaministiquia River, in north latitude 48 degrees 23 minutes, west longitude 89 degrees 27 minutes. Here is a convenient wharf and safe harbor, the bar off the mouth of the river affording 7 or 8 feet of water, which can easily be increased by dredging. The Company's buildings consist of a spacious dwelling-house, a store, and 3 storehouses, besides some 10 or 12 houses for the accommodation of the *attachés* and servants in the employ of the above Company. The land is cleared for a considerable distance on both sides of the river, presenting a thrifty and fertile appearance. Wheat, rye, oats, barley, potatoes, and most kinds of vegetables can here be raised; also grass and clover of different kinds. The early frosts are the great hindrance to this whole section of country, which is rich in minerals, timber, furs, and fish; altogether producing a great source of wealth to the community. Pine, spruce, hemlock, cypress, and balsam trees are common, also white birch, sugar-maple, elm, and ash, together with some hardy fruit - bearing trees and shrubs.

The *Roman Catholic Mission*, situated 2 miles above the Company's post, on the opposite side of the river, is an interesting locality. Here is a Roman Catholic church and some 50 or 60 houses, being mostly inhabited by half-breeds and civilized Indians, numbering about 300 souls. The good influence of the Roman Catholic priests along the shores of Lake Superior are generally admitted by all unprejudiced visitors — the poor and often degraded Indian being instructed in agriculture and industrial pursuits, tending to elevate the human species in every clime.

McKay's Mountain, lying 3 miles west of Fort William, near the Roman Catholic Mission, presents an abrupt and grand appearance from the water, being elevated 1,000 feet. Far inland are seen other high ranges of hills and mountains, presenting altogether, in connection with the islands, a most interesting and sub-lime view.

KAMINISTIQUIA, or "*Gah-mahnatek-waiahk*" River, signifying, in the Chippewa language, the "*place where there are many currents,*" empties its waters into Thunder Bay. This beautiful stream affords navigation for about 12 miles, when rapids are encountered by the ascending *voyageur*. Some 30 miles above its mouth is a fall of about 200 feet perpendicular descent.

Extracts from Report on the Geology of the Lake Superior Country.

NORTHERN SHORE. — " Beginning at *Pigeon Bay*, the boundary between the United States and the British Possessions (N. lat. 48°), we find the eastern portion of the peninsula abounds with bold rocky cliffs, consisting of trap and red granite.

"The Falls of Pigeon River, eighty or ninety feet in height, are occasioned by a trap dyke which cuts through a series of slate rocks highly indurated, and very similar in mineralogical characters to the old graywacke group. Trap dykes and interlaminated masses of traps were observed in the slate near the falls.

"The base of nearly all the ridges and cliffs between Pigeon River and Fort William (situated at the mouth of Kaministiquia River), is made up of these slates and the overlaying trap. Some of the low islands exhibit only the gray grits and slates. *Welcome Islands*, in Thunder Bay, display no traps, although in the distance they resemble igneous products, the joints being more obvious than the planes of stratification, thus giving a rude semi-columnar aspect to the cliffs.

"At *Prince's Bay*, and also along the chain of Islands which line the coast, including Spar, Victoria, Jarvis, and Pie Islands, the slates with the crowning traps are admirably displayed. At the British and North American Company's works the slates are traversed by a heavy vein of calc-spar and amethystine quartz, yielding gray sulphuret and pyritous copper and galena. From the vein where it cuts the overlaying trap on the main shore, considerable *silver* has been extracted.

"At *Thunder Cape*, the slates form one of the most picturesque headlands on the whole coast of Lake Superior. They are made up of variously-colored beds, such as compose the upper group of SIR WM. LOGAN, and repose in a nearly horizontal position. These detrital rocks attain a thickness of nearly a thousand feet, and are crowned with a sheet of trappean rock three hundred feet in thickness."

SPECKLED TROUT FROM NEPIGON RIVER— Weighing 5½ lbs. each.

ISLE ROYALE — LAKE SUPERIOR.

As this large and important Island, attached to Michigan, is attracting much attention, owing to its mineral deposits, we copy the following account from a late number of the *Ontonagon Miner*, May, 1874.

" The Island is situated near the North Shore, but at the time the treaty was made with Great Britain, Benjamin Franklin secured it for the United States. Its general course is north-east and south-west; its length about 40 miles. The western end is about 10 miles wide, narrowing toward the east; distant 60 miles north from Ontonagon.

" The surface of the island is rough and uninviting, although healthy and cool during the summer months. It is almost entirely destitute of a large growth of timber, its chief supply being cedar, spruce and tamarack. Its shores are indented with numerous bays and harbors, many of them being large enough for the entrance of steamers and sailing craft. In the early days of mining on Lake Superior, Isle Royale attracted much attention, — say from 1850 to 1860. It was then abandoned, and the few dwellings erected allowed to go to decay.

" During the summer of 1872 an extensive system of exploration was commenced in different parts of the Island. Numerous pits and Indian diggings were found and opened, and the veins proved to be of so much promise that work was continued throughout the season of 1873, and belts of rich copper-bearing con-glomerates were found and opened, being mostly situated in Sections 29 and 30, Town 64 North of Range 37 West. A company was formed, called the *Island Mining Company*, and a large force of miners set to work last fall.

" A vein of silver, apparently of great promise, has also been discovered on the western end of the Island. This will, no doubt, be followed by other important discoveries, as it lies on the same range with Silver Islet, some 20 or 30 miles north, situated on the Canada shore."

Lake Desor and Siskowit Lake are considerable bodies of water, lying near the middle of the Island. There are also a number of islands along the coast, where are several good fishing stations.

Blake's Point lies on the extreme north end of the Island. To the N.E., distant 3 miles, lies *Passage Island*, and 5 miles farther north are the *Gull Islands*, all being attached to the United States; the boundary line of Canada running immediately north, in latitude 48° 15′, being the most northern bounds of the United States east of the Lake of the Woods.

Steamers now run from ports on the South Shore and from Duluth to Siskowit Bay, Washington Harbor, and Rock Island Harbor, Isle Royale; also, to Pigeon River, Thunder Bay, and Silver Islet, forming a deeply interesting excursion during the summer months.

LATITUDE, ELEVATION ABOVE SEA-LEVEL AND MEAN ANNUAL TEMPERATURE OF SIGNAL STATIONS ON THE UPPER LAKES.

ALPENA, (Lake Huron,) Mich.
Latitude, 45° 05′.
Longitude, 83° 30′.
Elevation of Barometer, 608 feet.
Mean Annual Temp., 43° Fahr.

BUFFALO, (Lake Erie,) N. Y.
Latitude, 42° 53′.
Longitude, 78° 55′.
Elevation of Barometer, 660 feet.
Mean Annual Temp., 46¼° Fahr.

CHICAGO, (L. Michigan,) Ill.
Latitude, 41° 52′.
Longitude, 87° 38′.
Elevation of Barometer, 650 feet.
Mean Annual Temp., 47° Fahr.

COPPER HARBOR,* (L. S.,) Mich.
Latitude, 47° 30′.
Longitude, 88° 00′.
Elevation of Barometer, 620 feet.
Mean Annual Temp., 40° Fahr.

CLEVELAND, (Lake Erie,) Ohio.
Latitude, 41° 30′.
Longitude, 81° 74′.
Elevation of Barometer, 682 feet.
Mean Annual Temp., 49° Fahr.

DETROIT, (Detroit River,) Mich.
Latitude, 42° 21′.
Longitude, 83° 07′.
Elevation of Barometer, 656 feet.
Mean Annual Temp., 47° Fahr.

DULUTH, (L. S.,) Minn.
Latitude, 46° 48′.
Longitude, 92° 06′.
Elevation of Barometer, 642 feet.
Mean Annual Temp., 40° Fahr.

ERIE, (Lake Erie,) Penna.
Latitude, 42° 07′.
Longitude, 80° 03′.
Elevation of Barometer, 671 feet.
Mean Annual Temp., 48° Fahr.

ESCANABA, (Green Bay,) Wis.
Latitude, 46° 44′.
Longitude, 87° 16′
Elevation of Barometer, 600 feet.
Mean Annual Temp., 41° Fahr.

GRAND HAVEN, (L. M.,) Mich.
Latitude, 43° 05′.
Longitude, 86° 15′.
Elevation of Barometer, 616 feet.
Mean Annual Temp., 47° Fahr.

MACKINAC,* (Lake Huron,) Mich.
Latitude, 45° 51′.
Longitude, 84° 33′.
Elevation of Barometer, 700 feet.
Mean Annual Temp., 41° Fahr.

MARQUETTE, (L. S.,) Mich.
Latitude, 46° 33′.
Longitude, 87° 36′.
Elevation of Barometer, 666 feet.
Mean Annual Temp., 40° Fahr.

MILWAUKEE, (L. M.,) Wis.
Latitude, 43° 03′.
Longitude, 87° 54′.
Elevation of Barometer, 661 feet.
Mean Annual Temp., 46½° Fahr.

TOLEDO, (Lake Erie,) Ohio.
Latitude, 41° 40′.
Longitude, 83° 32′.
Elevation of Barometer, 649 feet.
Mean Annual Temp., 50° Fahr.

NOTE.—Variation of Mean Annual Temperature, from 40° to 50° Fahr.
* Not Signal Stations.

LAKE SUPERIOR,

Its Bays, Harbors, Islands, Tributaries, &c.

OBJECTS OF INTEREST ON THE SOUTH SHORE.

	Miles.
Saut Ste. Marie, Mich. N.	
lat. 46°30′, W. long. 84°43′.	
Head of Ship Canal and Rapids,	1
Round Island and Light,........	6——7
Waiska Bay,..	2——9
Iroquois Point and Light,.......	7—16
Toquamenon Bay and River,....	9—25
White Fish Point and Light, ...	15—40
Carp River, ...	20—60
Two-Heart River, ...	6—66
Sucker River, ...	12—78
Grand Marais River, ...	12—90
Grand Sauble, 300 feet high,....	10–100
Cascade Falls, 100 feet, ...	18–118
The Chapel,..........	2–120
Arched Rock, or Grand	
Portail, ... Pictured	4–124
Sail Rock, ... Rocks.	2–126
The Amphitheatre, ...	2–128
Miner's Castle, & River,	3–131
Grand Island and Harbor, ...	3–134
Munising, and Iron Works......	2–136
ONOTA, Schoolcraft Co........	4–140
Au Train Island, and River	9–149
Au Sauble River,.............	6–155
Harvey, Mouth Chocolat River,..	12–167
MARQUETTE * AND LIGHT, N. lat.	
46°32′, W. long. 87°41′.	3–170
Dead River, ...	2–172
Presque Isle, ...	1–173
Middle Island, ...	2–175
Granite Point and Island, ...	5–180
Garlic River, ...	8–188
Little Iron River, and Lake Inde-	
pendence	10–198
Salmon Trout River,..........	8–206
Pine River and Lake, ...	6–212
Huron Islands and Light, ...	10–222
Huron Bay, ...	6–228
Point Abbaye, ...	4–232
Keweenaw Bay (13 miles wide)	
L'Ance, and Settlement (South), .	13–245

	Miles.
Portage Entry and Light (West),.	245
Portage Lake,................	6–251
HOUGHTON, † N. lat. 46°40′, W.	
long. 88°30′................	8–259
Hancock,†	1–260
Head of Portage Lake (8 miles),	268
Traverse Island from Portage	
Entry, ...	10–255
Tobacco River,..............	12–267
Lac La Belle, † and Bete Grise Bay	14–281
Mount Houghton (900 feet high).	
Montreal River (Fishing Station),	6–287
Keweenaw Point,.............	8–295
Manitou Island and Light	3–298
COPPER HARBOR,† N. lat. 47°30′,	
W. long. 88°, ...	10–308
Agate Harbor, ...	10–318
Eagle Harbor † and Light, ...	6–324
EAGLE RIVER, † N. lat. 47°25′,	
W. long. 88°20′,.............	9–333
Entrance to Ship Canal & Portage	
Lake Route, ...	20–353
Salmon Trout River,..........	8–366
Graveraet River, ...	5–366
Elm River, ...	8–374
Misery River, ...	5–379
Sleeping River,..............	6–385
Flint Steel River,.............	6–391
ONTONAGON,† N. lat. 46°52′, W.	
long. 89°30′	9–400
Iron River, ...	11–411
Porcupine Mount'ns, 1,300 ft. high,	8–419
Carp River, ...	10–429
Presque Isle River,...........	6–435
Black River, ...	6–441
Montreal River boundary between	
Michigan and Wisconsin,.....	20–461
Maskeg, or Bad River.........	10–471
Chaquamegon Point, Bay, & Light,	11–482
Ashland, Wisconsin, head Chaqua-	
megon Bay.	

194

	Miles.		Miles.
La Pointe, Madeline Island,.....	3–485	Burnt Wood River,...........	10–545
(The Twelve Apostle Islands).		Poplar River, ...	9–554
BAYFIELD, Wisconsin, N. lat.		Cotton Wood River,...........	4–558
46°45′, W. long. 91°3′, ...	3–488	Amican River, ...	2–560
Basswood Island, ...	7–495	Mouth Nemadje, or Left Hand	
Raspberry Island, and Light,....	6–501	River, (Alloues Bay).	
York Island, ...	4–505	Wisconsin Point,.............	7–567
Fishing Island,................	4–509	Minnesota Point and Light, ...	1–568
Bark Bay and Point, ...	10–519	St. Louis Bay and River.	
Apakwa, or Cranberry River, ...	6–525	**Superior City,** Wisconsin,	1–569
Iron River, ...	10–535	**Duluth,** Minnesota,.......	7–576

* Shipping Port for Iron. † Shipping Ports for Copper.

Objects of Interest on the North Shore.

	Miles.		Miles.
Superior City, Wisconsin, N. lat. 46° 40′, W. long. 92°, (*Superior Bay.*)		FORT WILLIAM, Canada, N. lat. 48°23′, W. long. 89°27′,... ...	3–150
Duluth, Minn.	7	*Prince Arthur's Landing*, 3½ miles,	
Lester River, ...	5—12	Thunder Cape, 1,350 feet high,..	15–165
Kassabika River,............	7—19	*Silver Islet*, Canada, ...	5–170
French River (Copper Mines), ...	3—22	Black Bay,..	5–175
Buchanon, ...	4—26	Point Porphyry,.............	5–180
Knife River (Copper Mines), ...	1—27	Entrance to Neepigon Bay & River	30–210
Burlington,...................	10—37	Ste. Ignace Island,..........	10–220
Stewart's River,..............	3—40	Slate Islands,................	30–250
Encampment River and Island ..	3—43	Pic Island, ...	15–265
Split Rock River,............	7—50	Peninsula Harbor,...........	8–273
Beaver Bay, N. lat. 47°12′, W. long. 91°.	10—60	Pic River and Harbor, ...	10–283
Palisades, ...	5—65	Otter Island, Head and Cove, ...	30–313
Baptism River,...............	1—66	Michipicoten Island, 800 ft. high,	25–338
Little Marais, ...	6—72	*Michipicoten*, Harbor and River, N. lat. 47°56′, W. long. 85°06′.	45–383
Manitou River, ...	3—75	Cape Gargantua, ...	25–408
Two-Island River,............	8—83	Leach Island, ...	12–420
Temperance River, ...	5—88	Lizard Islands, ...	6–426
Poplar River, ...	6—94	Montreal Island and River, ...	14–440
Grand Portage, Indian trading post, N. lat. 47°50′, W. long. 90°......	16—110	Mica Bay (Copper Mine), ...	20–460
Isle Royal, attached to Michigan,		Mamainse Point, ...	6–466
Pigeon River, boundary between the United States and Canada,	10—120	Batchewanaung Bay (Fishing St'n)	10–476
Pie Island, 700 feet high,.......	20—140	Sandy Islands, ...	4–480
Welcome Island, ...	7—147	Maple Island, ...	7–484
Mouth Kaministiquia River.		Goulois Bay and Point,.........	8–495
		Parisien Island, ...	5–500
		Gros Cap, 700 feet high,........	10–510
		Point Aux Pins,..............	7–517
		Saut Ste. Marie, Canada, ..	8–525

RAILROAD AND STEAMBOAT ROUTE,

FROM CHICAGO TO GREEN BAY AND MARQUETTE, LAKE SUPERIOR.

Chicago and North-Western Railroad.

MILES.	STATIONS,		MILES.
242	**CHICAGO**.............		0
230	Canfield...........................		12
225	Des Plaines...................	5	17
220	Dunton	5	22
216	Palatine	4	26
210	Barrington.....................	6	32
199	CRYSTAL LAKE.............11		43
196	Ridgefield.....................	3	46
191	Woodstock.....................	5	51
179	Harvard Junction.........12		63
177	Lawrence......................	2	65
164	Clinton Junction...........13		78
160	Shopiere.......................	4	82
151	**Janesville**.............	9	91
143	Milton Junction...........	8	99
131	Fort Atkinson...............12		111
125	Jefferson......................	6	117
120	Johnson's Creek...........	5	122
112	WATERTOWN.................	8	130
97	Juneau 15		145
94	Minn. Junction*...........	3	148
90	Burnett.......................	4	152

MILES.	STATIONS.		MILES.
82	Chester.......................	8	160
74	Oakfield........................	8	168
65	**Fond du Lac**............	9	177
49	OSHKOSH	6	193
35	Neenah........................14		207
29	APPLETON	6	213
24	Little Chute.................	5	218
21	Kaukauna.	3	221
16	Wrightstown	5	226
6	De Pere.......................10		236
0	FORT HOWARD............	6	242
	Green Bay............		0
189	**GREEN BAY,**.....		242
161	Oconto.........................28		270
140	Marinette 21		291
138	MENOMONEE	2	293
74	**Escanaba** 64		357
	Peninsula Division.		
61	Day's River.................13		370
54	Centerville..................17		387
12	NEGAUNEE..................32		419
	Marquette, Houghton & Ontonagon R. R.		
0	**MARQUETTE** 12		431

Steamers run daily from MARQUETTE to the Saut Ste. Marie on the East, 170 miles; and to Houghton, Ontonagon, Bayfield, and Duluth on the West; a total distance of 400 miles,—passing around Keweenaw Point. This distance can be shortened about 100 miles by passing the Portage and Lake Superior Ship Canal.

GRAND PORTAIL—Pictured Rocks, L. S.

The Pictured Rocks, of which almost fabulous accounts are given by travellers, are one of the wonders of this "Inland Sea." Here are to be seen the *Cascade Falls* and other objects of great interest. The Amphitheatre, Miners' Castle, Chapel, Grand Portal, and Sail Rock, are points of great picturesque beauty, which require to be seen to be justly appreciated.

Extract from Foster and Whitney's Report of the Geology of the Lake Superior Land District:

Pictured Rocks.—

" The range of cliffs to which the name of the Pictured Rocks has been given, may be regarded as among the most striking and beautiful features of the scenery of the Northwest, and are well worthy the attention of the artist, the lover of the grand and beautiful, and the observer of geological phenomena.

"Although occasionally visited by travellers, a full and accurate description of this extraordinary locality has not as yet been communicated to the public.*

"The *Pictured Rocks* may be described, in general terms, as a series of sandstone bluffs extending along the shore of Lake Superior for about five miles, and rising, in most places, vertically from the water, without any beach at the base, to a height varying from fifty to nearly two hundred feet. Were they simply a line of cliffs, they might not, so far as relates to height or extent, be worthy of a rank among great natural curiosities, although such an assemblage of rocky strata, washed by the waves of the great lake, would not, under any circumstances, be destitute of grandeur. To the voyager coasting along their base in his frail canoe they would, at all times, be an object of dread; the recoil of the surf, the rockbound coast, affording for miles no place of refuge; the lowering sky, the rising wind; all these would excite his apprehension, and induce him to ply a vigorous oar until the dreaded wall was passed. But in the Pictured Rocks there are two features which communicate to the scenery a wonderful and almost unique character. These are, first, the curious manner in which the cliffs have been excavated and worn away by the action of the lake, which for centuries has dashed an ocean-like surf against their base; and, second, the equally curious manner in which large portions of the surface have been colored by bands of brilliant hues.

"It is from the latter circumstance that the name by which these cliffs are known to the American traveller is derived; while that applied to them by the French *voyageurs* ('Les Portails'*) is derived from the former, and by far the most striking peculiarity.

" The term *Pictured Rocks* has been in use for a great length of time, but when it was first applied we have been unable to discover.

"The Indian name applied to these cliffs, according to our *voyageurs,* is *Schkuee-archibi-kung,* or 'The end of the rocks,'

* Schoolcraft has undertaken to describe this range of cliffs, and illustrate the scenery. The sketches do not appear to have been made on the spot, or finished by one who was acquainted with the scenery, as they bear no resemblance, so far as we observed, to any of the prominent features of the Pictured Rocks.

"It is a matter of surprise that, so far as we know, none of our artists have visited this region, and given to the world representations of scenery so striking, and so different from any which can be found elsewhere. We can hardly conceive of any thing more worthy of the artist's pencil; and if the tide of pleasure-travel should once be turned in this direction, it seems not unreasonable to suppose that a fashionable hotel may yet be built under the shade of the pine groves near the Chapel, and a trip thither become as common as one to Niagara now is."

* Le Portail is a French term, signifying the principal entrance of a church or a portal, and this name was given to the Pictured Rocks by the *voyageurs,* evidently in allusion to the arched entrances which constitute the most characteristic feature. Le Grand Portail is the great archway, or Grand Portal.

which seems to refer to the fact that, in descending the lake, after having passed them, no more rocks are seen along the shore. Our *voyageurs* had many legends to relate of the pranks of the *Menni-boujou* in these caverns, and in answer to our inquiries seemed disposed to fabricate stories without end of the achievements of this Indian deity.

"We will describe the most interesting points in the series, proceeding from west to east. On leaving Grand Island harbor,* high cliffs are seen to the east, which form the commencement of the series of rocky promontories, which rise vertically from the water to the height of from one hundred to one hundred and twenty-five feet, covered with a dense canopy of foliage. Occasionally a small cascade may be seen falling from the verge to the base in an unbroken curve, or gliding down the inclined face of the cliff in a sheet of white foam. The rocks at this point begin to assume fantastic shapes; but it is not until having reached Miners' River that their striking peculiarities are observed.

* The traveller desirous of visiting this scene should take advantage of one of the steamers or propellers which navigate the lake and land at Grand Island, from which he can proceed to make the tour of the interesting points in a small boat. The large vessels on the lake do not approach sufficiently near the cliffs to allow the traveller to gather more than a general idea of their position and outlines. To be able to appreciate and understand their extraordinary character, it is indispensable to coast along in close proximity to the cliffs and pass beneath the Grand Portal, which is only accessible from the lake, and to land and enter within the precincts of the Chapel. At Grand Island, boats, men, and provisions may be procured. The traveller should lay in a good supply, if it is intended to be absent long enough to make a thorough examination of the whole series. In fact, an old voyager will not readily trust himself to the mercy of the winds and waves of the lake without them, as he may not unfrequently, however auspicious the weather when starting, find himself weather-bound for days together. It is possible, however, in one day, to start from Grand Island, see the most interesting points, and return. The distance from William's to the Chapel—the farthest point of interest—is about fifteen miles.

Here the coast makes an abrupt turn to the eastward, and just at the point where the rocks break off and the friendly sand-beach begins, is seen one of the grandest works of nature in her rock-built architecture. We gave it the name of 'Miners' Castle,' from its singular resemblance to the turreted entrance and arched portal of some old castle—for instance, that of Dumbarton. The height of the advancing mass, in which the form of the Gothic gateway may be recognized, is about seventy feet, while that of the main wall forming the background is about one hundred and forty. The appearance of the openings at the base changes rapidly with each change in the position of the spectator. On taking a position a little farther to the right of that occupied by the sketcher, the central opening appears more distinctly flanked on either side by two lateral passages, making the resemblance to an artificial work still more striking.

"A little farther east, Miners' River enters the lake close under the brow of the cliff, which here sinks down and gives place to a sand-bank nearly a third of a mile in extent. The river is so narrow that it requires no little skill on the part of the voyager to enter its mouth when a heavy sea is rolling in from the north. On the right bank, a sandy drift plain, covered with Norway and Banksian pine, spreads out, affording good camping-ground —the only place of refuge to the voyager until he reaches Chapel River, five miles distant, if we except a small sand-beach about midway between the two points, where, in case of necessity, a boat may be beached.

"Beyond the sand-beach at Miners' River the cliffs attain an altitude of one hundred and seventy-three feet, and maintain a nearly uniform height for a considerable distance. Here one of those cascades of which we have before spoken is seen foaming down the rock.

"The cliffs do not form straight lines, but rather arcs of circles, the space between the projecting points having been worn out in symmetrical curves, some of which are of large dimensions. To one of the grandest and most regularly formed we gave the name of 'The Amphitheatre.' Looking to the west, another projecting point—its base worn into cavelike forms—and a portion of the concave surface of the intervening space are seen.

" It is in this portion of the series that the phenomena of colors are most beautifully and conspicuously displayed. These cannot be illustrated by a mere crayon sketch, but would require, to reproduce the natural effect, an elaborate drawing on a large scale, in which the various combinations of color should be carefully represented. These colors do not by any means cover the whole surface of the cliff even where they are most conspicuously displayed, but are confined to certain portions of the cliffs in the vicinity of the Amphitheatre ; the great mass of the surface presenting the natural light-yellow or raw sienna color of the rock. The colors are also limited in their vertical range, rarely extending more than thirty or forty feet above the water, or a quarter or a third of the vertical height of the cliff. The prevailing tints consist of deep-brown, yellow, and gray—burnt sienna and French gray predominating.

" There are also bright blues and greens though less frequent. All of the tints are fresh, brilliant, and distinct, and harmonize admirably with one another, which, taken in connection with the grandeur of the arched and caverned surfaces on which they are laid, and the deep and pure green of the water which heaves and swells at the base, and the rich foliage which waves above, produce an effect truly wonderful.

"They are not scattered indiscriminately over the surface of the rock, but are arranged in vertical and parallel bands, extending to the water's edge. The mode of their production is undoubtedly as follows: Between the bands or strata of thick-bedded sandstone there are thin seams of shaly materials, which are more or less charged with the metallic oxides, iron largely predominating, with here and there a trace of copper As the surface-water permeates through the porous strata it comes in contact with these shaly bands, and, oozing out from the exposed edges, trickles down the face of the cliffs, and leaves behind a sediment, colored according to the oxide which is contained in the band in which it originated. It cannot, however, be denied that there are some peculiarities which it is difficult to explain by any hypothesis.

" On first examining the Pictured Rocks, we were forcibly struck with the brilliancy and beauty of the colors, and wondered why some of our predecessors, in their descriptions, had hardly adverted to what we regarded as their most characteristic feature. At a subsequent visit we were surprised to find that the effect of the colors was much less striking than before ; they seemed faded out, leaving only traces of their former brilliancy, so that the traveller might regard this as an unimportant feature in the scenery. It is difficult to account for this change, but it may be due to the dryness or humidity of the season. If the colors are produced by the percolation of the water through the strata, taking up and depositing the colored sediments, as before suggested, it is evident that a long period of drought would cut off the supply of moisture, and the colors, being no longer renewed, would fade, and finally disappear. This explanation seems reasonable, for at the time of our second visit the beds of the streams on the summit of the table-land were dry.

"It is a curious fact, that the colors are so firmly attached to the surface that they are very little affected by rains or the dashing of the surf, since they were, in numerous instances, observed extending in all their freshness to the very water's edge.

"Proceeding to the eastward of the Amphitheatre, we find the cliffs scooped out into caverns and grotesque openings, of the most striking and beautiful variety of forms. In some places huge blocks of sandstone have become dislodged and accumulated at the base of the cliff, where they are ground up and the fragments borne away by the ceaseless action of the surge.

"To a striking group of detached blocks the name of 'Sail Rock' has been given, from its striking resemblance to the jib and mainsail of a sloop when spread—so much so that when viewed from a distance, with a full glare of light upon it, while the cliff in the rear is left in the shade, the illusion is perfect. The height of the block is about forty feet.

"Masses of rock are frequently dislodged from the cliff, if we may judge from the freshness of the fracture and the appearance of the trees involved in the descent. The rapidity with which this undermining process is carried on, at many points, will be readily appreciated when we consider that the cliffs do not form a single unbroken line of wall; but, on the contrary, they present numerous salient angles to the full force of the waves. A projecting corner is undermined until the superincumbent weight becomes too great, the overhanging mass cracks, and, aided perhaps by the power of frost, gradually becomes loosened and finally topples with a crash into the lake.

"The same general arched and broken line of cliffs borders the coast for a mile to the eastward of Sail Rock, where the most imposing feature in the series is reached. This is the Grand Portal—*Le Grand Portail* of the *voyageurs*. The general disposition of the arched openings which traverse this great quadrilateral mass may, perhaps, be made intelligible without the aid of a ground-plan. The main body of the structure consists of a vast mass of a rectilinear shape, projecting out into the lake about six hundred feet, and presenting a front of three hundred or four hundred feet, and rising to a height of about two hundred feet. An entrance has been excavated from one side to the other, opening out into large vaulted passages which communicate with the great dome, some three hundred feet from the front of the cliff. The Grand Portal, which opens out on the lake, is of magnificent dimensions, being about one hundred feet in height, and one hundred and sixty-eight feet broad at the water-level. The distance from the verge of the cliff over the arch to the water is one hundred and thirty-three feet, leaving thirty-three feet for the thickness of the rock above the arch itself. The extreme height of the cliff is about fifty feet more, making in all one hundred and eighty-three feet.

"It is impossible, by any arrangement of words, or by any combination of colors, to convey an adequate idea of this wonderful scene. The vast dimensions of the cavern, the vaulted passages, the varied effects of the light, as it streams through the great arch and falls on the different objects, the deep emerald green of the water, the unvarying swell of the lake, keeping up a succession of musical echoes, the reverberations of one's own voice coming back with startling effect, all these must be seen, and heard, and felt, to be fully appreciated.

"Beyond the Grand Portal the cliffs gradually diminish in height, and the general trend of the coast is more to the southeast; hence the rock, being less exposed to the force of the waves, bears fewer marks of their destructive action. The entrance to Chapel River is at the most easterly extremity of a sandy beach which extends for a quarter of a mile, and affords a convenient landing-place, while the drift-terrace, elevated about thirty feet above the lake-level, being an open pine plain, affords excellent camping-ground, and is the most central and convenient spot for the traveller to pitch his tent, while he examines the most interesting localities in the series which occur in this vicinity—to wit, the Grand Portal and the Chapel. (*See Engraving.*)

"The Chapel—*La Chapelle* of the *voyageurs*—if not the grandest, is among the most grotesque of Nature's architecture here displayed. Unlike the excavations before described, which occur at the water's edge, this has been made in the rock, at a height of thirty or forty feet above the lake. The interior consists of a vaulted apartment, which has not inaptly received the name it bears. An arched roof of sandstone, from ten to twenty feet in thickness, rests on four gigantic columns of rock, so as to leave a vaulted apartment of irregular shape, about forty feet in diameter, and about the same in height. The columns consist of finely stratified rock, and have been worn into curious shapes. At the base of one of them an arched cavity or niche has been cut, to which access is had by a flight of steps formed by the projecting strata. The disposition of the whole is such as to resemble very much the pulpit of a church; since there is overhead an arched canopy, and in front an opening out toward the vaulted interior of the chapel, with a flat tabular mass in front, rising to a convenient height for a desk, while on the right is an isolated block, which not·inaptly represents an altar; so that if the whole had been adapted expressly for a place of worship, and fashioned by the hand of man, it could hardly have been arranged more appropriately. It is

hardly possible to describe the singular and unique effect of this extraordinary structure; it is truly a temple of nature— 'a house not made with hands.'

"On the west side, and in close proximity, Chapel River enters the lake, precipitating itself over a rocky ledge ten or fifteen feet in height.*

"It is surprising to see how little the action of the stream has worn away the rocks which form its bed. There appears to have been hardly any recession of the cascade, and the rocky bed has been excavated only a foot or two since the stream assumed its present direction.

"It seems therefore impossible that the river could have had any influence in excavating the Chapel itself, but its excavation must be referred to a period when the waters of the lake stood at a higher level.

"Near the Grand Portal the cliffs are covered, in places, with an efflorescence of sulphate of lime, in delicate crystallizations; this substance not only incrusts the walls, but is found deposited on the moss which lines them, forming singular and interesting specimens, which however cannot be transported without losing their beauty.

"At the same place we found numerous traces of organic life in the form of obscure fucoidal markings, which seem to be the impressions of plants, similar to those described by Prof. Hall as occurring in the Potsdam sandstone of New York. These were first noticed at this place by Dr. Locke, in 1847."

* "At this fall, according to immemorial usage among the *voyageurs* in ascending the lake, the *mangeurs de lard*, who make their first trip, receive baptism; which consists in giving them a severe ducking—a ceremony somewhat similar to that practised on green-horns when crossing the line.

Lake Superior Region.

The following verses were written by J. G. WHITTIER, on receiving an *eagle's quill*, when on a visit to Lake Superior in 1846.

THE SEER.

I hear the far-off voyager's horn,
  I see the Yankee's trail—
His foot on every mountain pass,
  On every stream his sail.

He's whistling round St. Mary's Falls,
  Upon his loaded train;
He's leaving on the Pictured Rocks
  His fresh tobacco stain.

I see the mattock in the mine,
  The axe-stroke in the dell,
The clamor from the Indian lodge,
  The Jesuit's chapel bell!

I see the swarthy trappers come
  From Mississippi's Springs;
And war-chiefs with their painted brows,
  And crests of eagle wings.

Behind the scared squaw's birch canoe,
  The steamer smokes and raves;
And city lots are staked for sale
  Above old Indian graves.

By forest, lake and water-fall,
  I see the peddler's show;
The mighty mingling with the mean,
  The lofty with the low.

I hear the tread of pioneers
  Of nations yet to be;
The first low wash of waves where soon
  Shall roll a human sea.

The rudiments of empire here
  Are plastic yet and warm;
The chaos of a mighty world
  Is rounding into form!

Each rude and jostling fragment soon
  Its fitting place shall find—
The raw materials of a state,
  Its muscle and its mind!

And, westering still, the star which leads
  The new world in its train,
Has tipped with fire the icy spears
  Of many a mountain chain.

GRAND ISLAND, 125 miles distant from the Saut, is about 10 miles long and 5 wide, lying close in to the south shore.

This is a wild and romantic island; the cliffs of sandstone, irregular and broken into by the waves, form picturesque caverns, pillars, and arches of immense dimensions. There are several romantic bays and inlets protected from storms, which are frequent on this great lake, where the brook trout of a large size can be caught in quantities. The forests also afford a delightful retreat, while all nature seems hushed—save by the moaning winds and billowy surges of the surrounding waters.

A few families reside on the south shore, facing the mainland, where is a clearing of considerable extent. The main-shore in full sight, and the Pictured Rocks, visible from its eastern shore, altogether add a charm to this truly Grand Island, unsurpassed by no other spot in this interesting region.

MUNISING, formerly called Grand Island City, lies on the south side of Grand Island Bay, here about 3 miles in width. Here is a steamboat wharf and hotel, together with a few dwellings, being, no doubt, destined to become a favorite place of resort, as from this place the Pictured Rocks can be easily reached by canoes or small boats during calm weather. Trout fishing is also good in Ann's River, which enters Grand Island Bay, and in Miner's River, near the Pictured Rocks.

The bay or harbor is capacious, deep, and easy of access from the east or west, being 6 miles in length by from 2 to 4 in width, with a depth of water of 100 feet and upwards. It is perfectly landlocked by hills rising from 100 to 300 feet high, and capacious enough to contain the entire fleet of the lakes.

The *Schoolcraft Iron Works*, near Munising, have recently been erected close to the water's edge, for the manufacture of pig iron, where is a landing for teamers.

MINER'S POINT, a most remarkable headland, lies 6 miles east of Munising, at the mouth of a small stream of the same name.

The action of the waters has here disintegrated portions of the sand-stone formation, forming romantic caverns and grottoes where the waters of the lake penetrate, making strange music in the subterranean passages.

MONUMENT ROCK.

MONUMENT ROCK, about one mile west of Miner's Point, is another strange freak of nature, being an upright column standing in full view, near the water's edge, elevated some 80 or 100 feet above the lake. (*See Engraving.*) All these points can easily be reached from Munising, or Grand Island, by a sail or row boat.

Remarkable Phenomena on Lake Superior.

The sudden and singular changes of the weather on Lake Superior, in connection with its healthy influence, during the summer and fall months, present one of the phenomena of nature which seems almost unaccountable. The sun frequently rises clear and cloudless, giving indications of continued sunshine, when suddenly the sky becomes overcast with white, fleecy clouds, scudding low and giving out a chilly atmosphere, not unfrequently accompanied with rain,—the clouds as suddenly disappear, and a pleasant afternoon usually follows, with light winds. This influence, causing a fluctuation of several degrees of the thermometer, seems to have an injurious effect on most kinds of fruit and vegetables requiring a warm sun throughout the day in order to arrive at maturity; the country a few miles inland, however, being less subject to these frequent changes.

On the 6th of August, 1860, there occurred a remarkable phenomenon, as witnessed on Grand Island Bay, near the Pictured Rocks—Lake Superior being here about 170 miles wide. During the forenoon of a pleasant summer's day, the water was observed suddenly to fall some three or four feet perpendicularly on the south shore, then rise in about half an hour, as suddenly again to recede and rise several times; exposing the bed of the lake for a considerable distance where ·

the water was shallow, affording a fine opportunity to collect pebbles of different hues, and precious stones.

At noon the wind blew moderately from the southward, while the thermometer ranged at about 74° Fahr. This apparently calm and pleasant weather was taken advantage of by a party of pleasure to cross the bay in a sail-boat from Munising to Grand Island, 3 miles distant, affording a delightful excursion. On looking eastward at about 4 o'clock, P. M., a dense fog or low cloud was seen rapidly to enter the east channel of the bay, from the northward, rolling on in majestic grandeur, and presenting apparently the smoke caused by the discharge of a park of artillery, obscuring every object in the far distance, while the headlands within one or two miles were distinctly visible. As it approached, the thermometer fell several degrees, and rain followed, attended with lightning and thunder. Soon, however, the wind lulled, or entirely ceased, while the rain poured down in torrents. The mist or fog seemed mostly to ascend as it passed over the high lands on the main land, and assumed the appearance of clouds, while portions remained, in low and wet localities, above the forest-trees, —presenting altogether a most magnificent appearance. The rain-storm and cloud effect, after continuing some two hours, as suddenly ceased, followed by a splendid rainbow,—being the harbinger of a pleasant evening and calm weather for a time.

Mackenzie, who wrote in 1789, relates a very similar phenomenon, which occurred at Grand Portage, on Lake Superior, and for which no obvious cause could be assigned. He says: "The water withdrew, leaving the ground dry which had never before been visible, the fall being equal to four perpendicular feet, and rushing back with great velocity above the common mark. It continued thus rising and falling for several hours, gradually decreasing until it stopped at its usual height."

To the mariner these sudden storms and fluctuations, accompanied by fog, are attended with much danger, more particularly if near the land, when the sun and all objects in sight suddenly disappear as if in darkest night, the terrific noise of the waves and wind alone being heard. When followed by snow the danger is still more increased, frequently causing the most disastrous shipwrecks. In this high latitude a perfect calm seldom continues but for a short time; the wind will occasionally lull, when fitful gusts disturb the waters, to be followed by a breeze or storm from some quarter of the compass.

On examining the meteorological record kept at Fort Mackinac, about 100 miles distant in a southeast direction from Grand Island, it was found that the thermometer ranged at 78° Fahrenheit at 2 P. M. on the above day; the wind being from the south. A. 7 P. M. a heavy rain and thunder storm commenced, which lasted two hours, the same as on Lake Superior, terminating with a gorgeous sunset view, exceeded only by the magnificent aurora, which frequently illuminates the northern heavens in this high latitude, or the beautiful mirage of mid-day, which reflects with remarkable distinctness the invisible landscape, and vessels floating on the bosom of this vast inland sea.

How far the receding of the waters had to do with the above coming storm, must be left to conjecture or further investigation—no doubt, however, it caused a displacement of water at some remote parts of the lake, which was almost immediately felt at other and far distant points. So with the vapory clouds which suddenly rise over Lake Superior; they, no doubt, being caused by cold currents of air from the higher regions or northwest, passing over warmer portions along the south shore, when immediately a mist or fog is created, which ascends in the form of clouds into the upper regions; not, however, at first very far above the lake level—thus giving out the cold influence above referred to as peculiar to the south shore of the lake when the northwest winds prevail: this cold influence being most probably wafted far to the east and southward, producing, no doubt, an effect on the weather along the Atlantic coast several hundred miles to the southeast. The northwest winds which mostly prevail in the States of New York and Pennsylvania have a modified character, similar to the winds from the same quarter passing over the upper lakes of North America—affording a cool and bracing influence on the human system.

Another remarkable feature in the climate of Lake Superior, is its healthy and invigorating influence on residents and invalids suffering from incipient pulmonary and throat complaints—the sudden changes of hot and cold, or wet weather, seem to brace the constitution, without producing any other injurious effects than rheumatism, when too much exposure is endured.

While the balmy southern clime too often disappoints the invalid, this northern climate, its influence extending westward toward the Rocky Mountains, seems to give strength to the respiratory and digestive organs—thereby often effecting most miraculous and permanent cures.

without the aid of medicine, other than that afforded by nature—pure air and water. The intense colds of winter are here represented as being far more endurable than in more southern latitudes, along the Atlantic coast, where damp northeast storms prevail.

In *Foster and Whitney's Report* on the Geology of Lake Superior, the phenomena of these fluctuations are elaborately discussed; and, for the most part, they are found to be the premonition of an approaching gale. They remark, that the earth may be regarded as surrounded by two oceans—one aërial, the other liquid. By the laws which regulate two fluids thus relatively situated, a local disturbance in the one would produce a corresponding disturbance in the other.

Every rise or fall of one-twentieth of an inch in the mercurial column, would be attended with an elevation or depression in the surface of the water equal to one inch. A sudden change of the atmospheric pressure over a large body of water would cause a perpendicular rise or fall, in the manner of waves, greater than the mere weight itself, which would propagate themselves in a series of undulations from the centre of disturbance. These undulations result from an unusual disturbance of the atmosphere occurring around the margin of the storm, and its effects are perceived before the storm actually breaks.

Lake Superior Region.—

PHENOMENA OF THE SEASONS, By J. W. FOSTER, LL.D. "That portion of the lake region occupied by the great coniferous forest, has but two seasons—Summer and Winter. About the middle of September, heavy gales sweep over the lake, and hoar-frosts fall, nipping the leaves of the deciduous trees, of which the maple is conspicuous, and dyeing them with many colored tints. This equinoctial term is followed by a delightful Indian Summer of several day's duration. By the middle of October, the snow begins to fall, and to this succeeds an interval of calm, lasting two or three weeks, when winter sets in in earnest. The interior lakes are closed with a covering of ice, and the land and water are wrapped in a mantle of snow, so that the ground becomes frozen to no great depth. The dense forests prevents the drifting of the snows, and the warmth of the soil is retained until the opening of Spring. The thermometer accasionally drops to 30°, followed by a dry, cold, and elastic north-west wind, which seems to rob the temperature of its intensity, so far as relates to its effects upon the human system. The trapper, amid these intense colds, and shod with snow shoes, pursues his accustomed round; camping at night with his feet towards the log-built fire, with no covering than a Mackinac blanket.

"During the long winter nights the northern sky is frequently illumined with brilliant streaks of variously-colored light, which reach to the zenith, and then dissolve in luminous waves. So intense are they, at times, as to communicate a crimson tint to the snow, and clothe every object with an unnatural hue. The Northern Lights increase in number and intensity in September and March, as though there was an intimate connection between these phenomena and the changes of the equinoxes.

"Towards the end of April the streams become released from their icy fetters. When the weather has become so far mollified, as it ordinarily does by the middle of March, to thaw at mid-day and freeze at night, the sap of the maple begins to flow, and then commences the sugar harvest. This tree, as far north as the shores of Lake Superior, clothes most of the ridges and high grounds, and the bird's-eye and curled varieties are the most abundant. By the beginning of May, when the sun's rays have acquired sufficient power to dissolve the snows, the trees start from their winter's sleep, and commence the process of foliation, with an activity unknown in lower latitudes; the air becomes vocal with the hum of insects; the birds resume their accustomed haunts; and all nature seems roused from a lethargic sleep.

"In June, the thermometer often rises to 90° in the shade, and the sun's rays have a scalding effect, but the pressure of so large a body of water as is contained in the Great Lakes, modifies the range of the temperature, lessening the Winter's cold and the Summer's heat.* In freezing, the water evolves a large amount of heat, and during the Summer the winds are tempered in passing over its surface, often giving out a chilling influence. When the unclouded sun sinks behind the western horizon, leaving a long twilight, a cool breeze commences blowing toward the heated surface of the land; so that, however hot it may have been, the night is rarely sultry."

* A singular phenomena occurs at Copper Harbor, on Keweenaw Point; here, the mean annual temperature of 40° Fahr. and Summer temperature of 60° come together.

North Shore of Lake Superior.—*Extract from the Chicago Tribune of August,* 1870. "If you ever go to Lake Superior, and have the opportunity, don't fail to visit the North Shore and Isle Royale. On Monday, the 25th, we left Duluth, and at five o'clock the next morning we were in sight of *Bayfield,* quietly sleeping on its hill-side, its neat, trim houses and old churches snugly nestled among the pines, cow-bells tinkling among the trees, crows lazily sailing through the air, over the woods full of birds, whose songs come floating over the water like a morning benison. It is a lovely place, and as quiet as a dream, and soon thereafter we were steaming among the Apostle Islands, which, for an hour, appeared and disappeared, each more beautiful than the other, sometimes but one in sight, and, as we round it, a whole cluster coming into view, each overlapping the other with its curving outlines, the sun gilding the tree tops, a little boat here and there dancing on the waves, beauty everywhere. By seven o'clock the outermost Apostle had disappeared, and at half-past one we were passing Isle Royale, sleeping in the blue distance. We are out of the world, beyond the pickets of civilization, and heading for the great wilderness, which stretches towards the pole, without a single habitation. The North Shore is in view, an unending range of hills and mountains, indented here and there with beetling crags and frowning precipices on their summits, with solid walls of porphyry and red granite at their bases, worn by the water into caves and all manner of fanciful shapes, and the whole bathed in that deep, dreamy glamor of purple which artists love so much, giving you no detail, but the most suggested outlines. Here are caverns which might shelter the Titans, gorges which seem fathomless. Sometimes a bold headland projected out into space, with magnificent relief, and sometimes a solid wall of granite plunging down into the Lake from a mountain spur. The beauty of this North Shore is the beauty of sublimity. Nature here is not in her pretty moods, toying with water, playing with flowers, or decorating herself in any bridal attire of gossamers, blossoms and shells. She is in her stern moods. She has piled up Ossa on Pelion. She frowns at you from stupendous crags. Her music is the thunder. Her attire is the sombre green of the pine. Her play is the everlasting wash of the waves against solid granite walls.

"About nightfall, we had reached Pigeon river, the boundary line between the United States and Canada, and came to anchor in its still waters, shut in by frowning mountains of rock towering above us on either hand. There was time for us to land in small boats before dark, and walk up to the Pigeon River Falls, a sight which amply repaid us for our scramble up hill and down dale. Minnehaha must yield the palm of beauty—or, at least, of grandeur—to these. They make a descent of about one hundred and fifty feet, in two distinct streams, over a table of trap, with projecting spurs, which break up and divide the fall in the wildest manner. Standing upon the rocks which overhang the fall, you look down through a rocky gorge, its crevices bursting into beauty with vines and flowers, glistening in the spray, to a bend in the river, which is spanned from side to side with a clear, well-defined rainbow. The river is not very wide at this point, scarcely wider than the brook which feeds Minnehaha: but its sudden and headlong plunge, the immensity of its rocky surroundings, and their fantastic shapes and forms, especially of one huge, towering rock, which stands upon the very edge of the fall and divides the water before it goes over, the beauty of the mosses, and ferns, and vines, and the utter solitude of nature around it, give to it not only beauty but an element of solemnity and sublimity Minnehaha does not possess.

"The next morning at daybreak we were off for Thunder Bay, rightly enough named withal, skirting among towering islands of rock rising hundreds of feet out of the Lake. It was fair when we started, but we had hardly been under way half an hour when, as if by magic, the heavens clouded over. We were near the entrance to the bay which runs between two headlands of solid rock, Pie Island, rising 850 feet, and Thunder Cape, 1,300 feet, massive, threatening walls of granite, standing there in an awful silence, with nothing around them but smaller islands of naked rock, except in one instance, where a perfect little gem of an island, hardly fifty feet square, rose from the water, covered with grass and shrubs, with a solitary pine in the centre, which, like Heine's, seemed to be dreaming of the palm, which 'lonely and silent sorrows 'mid burning rocks and sands.' In the midst of this sublime scene, these terrible rock-shapes around us—which looked like rocky monsters just risen from the submarine caverns—this terrible solitude and silence, which seemed like the solitude of another world, the clouds closed blackly around the steamer as if they would crush it. It seemed as if you could raise your hand and touch their dark, fleecy whirls. The storm burst forth in awful fury. For a half hour the air was apparently all ablaze, not with flashes, but absolute sheets of lightning, which lit up the rocks and cliffs with a supurb rose color, while the thunder crashed from crag to crag like the roar of a thousand batteries. Nearer and nearer we came to the frowning monsters guarding the entrance to the bay, showing their lofty walls in the glare of the lightning. The Captain stood, anxiously, on the forward deck, with a silent group around him, peering ahead through the darkness. The sea was now running heavily, and the brave steamer plunged wildly about. As we came abreast of the entrance, a sheet of lightning, more vivid than the rest, lit up the bay, and showed a dense bank of fog inside. To enter would be folly; the Captain gave the signal to the pilot to about ship, and in a few minutes the steamer was heading for the South Shore, pelted with rain, girt about with lightning, and running at full speed to outstrip the fog, which was chasing us at headlong force across the mad waves of Superior. It overtook us, however, like magic, and shut us in; we flanked it, in turn, by running into a harbor in Isle Royale, being the first Steamer which ever entered those waters; and, before night, we were in sight of the South Shore again, and dancing over waves which were ruddy with sunlight. The Indians have a legend, that, the thunder was a huge bird, which laid its eggs and hatched the lightnings about Thunder Bay, having their nests on the high hills or surrounding mountains."

Meteorological Table,

SHOWING THE LATITUDE, LONGITUDE, ALTITUDE, TEMPERATURE, ETC., OF THE PRINCIPAL CITIES AND PORTS ON THE AMERICAN SIDE OF THE GREAT LAKES.

CITIES, etc.	Latitude.	Longitude.	Altitude.	Yearly Mean.	Four Seasons.			
					Spring.	Summer.	Autumn.	Winter.
LAKE ONTARIO.			Ft.	°Fahr.	°Fahr.	°Fahr.	°Fahr.	°Fahr.
Sacket's Harbor, N. Y.....	43° 55'	76° 00'	250	46.40	42.49	67.82	50.58	24.80
Oswego, "	43° 20'	76° 40'	250	46.44	43.70	67.00	50.40	24.72
Charlotte, "	43° 12'	77° 51'	250	47.88	43.72	68.46	50.77	28.56
Fort Niagara, "	43° 15'	79° 00'	250	46.60	41.38	67.20	50.00	27.86
LAKE ERIE.								
Buffalo, N. Y.............	42° 53'	78° 50'	600	47.26	43.60	67.56	50.14	27.80
Cleveland, Ohio...........	41° 30'	81° 47'	600	49.70	46.84	69.86	51.97	30.00
Toledo, "	41° 45'	83° 36'	565	50.00	47.00	71.00	52.00	29.00
Monroe City, Mich........	41° 43'	83° 24'	565	49.23	46.22	71.00	51.33	28.62
Detroit (Detroit R.), Mich..	42° 20'	83° 00'	580	48.00	45.94	69.20	49.81	28.17
LAKE HURON.								
Port Huron (St.Clair R),Mich	42° 53'	82° 24'	590	47.00	43.68	67.00	49.00	25.60
Tawas City, Mich........	44° 15'	......	590	44.33	37.22	65.15	47.06	24.61
Fort Mackinac, "	45° 51'	84° 33'	700	41.00	38.70	62.00	43.54	18.30
Green Bay (Green B.), Wis.	44° 30'	88° 05'	600	44.50	43.52	68.50	46.00	20.00
LAKE MICHIGAN.								
Grand Haven, Mich.......	43° 05'	86° 10'	580	47.36	44.59	68.62	49.56	26.62
Milwaukee, Wis...........	43° 03'	87° 55'	600	46.00	42.89	67.08	48.34	25.00
Chicago, Ill..............	41° 52'	87° 35'	590	47.00	45.00	68.50	49.00	26.00
Michigan City, Ind........	41° 40'	86° 53'	590	49.00	46.00	70.00	50.00	28.00
LAKE SUPERIOR.								
Saut Ste. Marie, Mich.....	46° 30'	84° 43'	600	40.50	37.60	62.00	43.54	20.00
Marquette, "	46° 32'	87° 41'	630	41.50	38.30	63.10	43.84	20.00
Copper Harbor, "	47° 30'	88° 00'	620	41.00	38.47	60.80	42.96	21.78
Ontonagon, "	46° 52'	89° 30'	600	40.00	37.00	62.60	42.86	17.85
Bayfield, Wis.............	46° 45'	91° 00'	620	40.00	38.00	62.00	43.00	15.60
Superior City, Wis........	46° 40'	92° 03'	600	41.00	38.00	63.00	43.50	14.60

THE GEOGRAPHY OF CONSUMPTION.

"Consumption originates in all latitudes, from the Equator, where the mean temperature is 80° Fahrenheit, with slight variations, to the higher position of the Temperate Zone, where the mean temperature is 40°, with sudden and violent changes. The opinion, long entertained, that Consumption is peculiar to cold and humid climates, is founded in error. Far from this being the case, the tables of mortality warrant the conclusion, that Consumption is sometimes more prevalent in tropical than in temperate countries. Consumption is rare in the Arctic Regions, in Siberia, Iceland, the Orkneys, and Hebrides, also in the north-western portions of the United States, and western portions of Canada.

"In North America, 'the disease of the respiratory organs, of which Consumption is the chief, have their maximum in New England, in latitude about 42° where north-east winds prevail, and diminish in all directions from this point inland. The diminution is quite as rapid westward as southward, and a large district near the fortieth parallel is quite uniform at twelve to fifteen per cent. of deaths from Consumption, while Massachusetts varies from twenty to twenty-five.* At the border of the dry climate of the plains, in Minnesota, a minimum is obtained as low as that occurring in Florida, and not exceeding five per cent. of the entire mortality. It is still lower in Texas, and the absolute minimum for the continent in temperate latitudes is Southern California.'"

The Upper Peninsula of Michigan, embracing the whole of the Lake Superior region, Minnesota, Dakota, Montana, and Washington Territory, are all alike exempt, in a remarkable degree, from the above fatal disease. Invalids, suffering from pulmonary complaints and throat diseases, are almost uniformly benefitted by the climate of the above northern region; having a mean annual temperature of from 40° to 50° Fahrenheit, with moderately cold winters and a dry atmosphere.

DEATHS IN THE UNITED STATES, ACCORDING TO THE CENSUS OF 1860.

STATES.	CONSUMP- TION.	RATIO.	FEVERS.	RATIO.	TOTAL DEATHS.
Eastern States	10,792	25.0	2,596	6.2	45,361
Middle States	16,213	18.4	4,439	5.0	94,612
Southern States	11,201	7.4	19,431	15.4	165,800
Western States	10,308	13.0	7,882	10.6	82,643
Pacific States	614	10.2	550	8.8	5,734
Total,	49,118	14.8	35,898	9.1	394,150

Making 23.9 per cent. of deaths caused by the two above prevalent diseases.

* The mortality is even greater in New Hampshire and Maine.

DULUTH TO ST. PAUL,

VIA LAKE SUPERIOR AND MISSISSIPPI RAILROAD.

MILES.	STATIONS.	MILES.	MILES.	STATIONS.	MILES.
155	**Duluth**......................	0	54	Rush City....................11	101
154	Rice's Point..................	1	42	North Branch.............12	113
151	Oneota....................... 3	4	30	Wyoming...................12	125
141	**Fond du Lac**..........10	14	25	Forrest Lake................. 5	130
	(*Dalles of the St. Louis.*)		17	Centreville................. 8	138
			12	White Bear Lake.......... 5	143
133	**Thomson**.............. 8	22			
132	Northern Pacific R. R. J. 1	23		STILLWATER BRANCH....12	155
110	Moose Lake..................22	45			
95	Kettle River................15	60		MINNEAPOLIS BRANCH...13	156
77	**Hinckley,** (Din.Sta.).18	78			
65	Pine City...................12	90	0	**St. Paul**..................12	155

Stages run from Pine City to Chengwatana, 4 miles. *Stages* run from North Branch to Sunrise, 8 miles.

RAILROAD ROUTE FROM DULUTH TO ST. PAUL,

VIA LAKE SUPERIOR AND MISSISSIPPI RAILROAD.

ON leaving DULUTH, the line of the Railroad passes Rice's Point, which separates the Bay of Superior from St. Louis Bay; the former is a beautiful sheet of water, about one mile wide and seven in length, being an expanse of St. Louis river. St. Louis Bay is a smaller body of water, being about two miles in length and breadth.

ONEOTA, 4 miles from Duluth, is a village situated on the west bank of St. Louis Bay, where is a good harbor. Here are 2 saw-mills, a church, a hotel, 2 stores, and about 300 inhabitants. After leaving St. Louis Bay the channel of the river is winding—passing along low banks and islands with a luxuriant growth of reeds and pond lilies.

DEVIL'S BEND, 11 miles, is seen to the best advantage from the deck of the Steamer; it is very circuitous, and is surrounded by beautiful, bold scenery.

SPIRIT ISLAND and LAKE, 12 miles; here is presented a most interesting and romantic view of land and water. The island rises about fifty feet above the surface of the river, and is clothed with a thick growth of trees. According to the Indian tradition, here died a celebrated chief of an unknown disease, and ever since it has been avoided by the children of the forest as haunted ground. As you approach Fond du Lac by water the scenery is wild and beautiful, with high and abrupt hills in sight, which seems to intercept the river in its course to Lake Superior.

FOND DU LAC, 14 miles by Railroad and 20 miles by water, is an old and interesting settlement, mostly inhabited by half-breeds, who, until recently, were almost excluded from the outside world. Here are about 250 inhabitants, 30 dwelling houses, a church, 2 public houses, a warehouse and steamboat landing.

The DALLES OF ST. LOUIS, immediately above Fond du Lac, are the next great object of interest in connection with the tressle-work and railroad bridge which here spans the river. The scenery is thus glowingly described by a recent tourist, and copied from the *St. Paul Press:*—

DALLES OF ST. LOUIS,
July 26, 1870.

Editors St. Paul Press.

"It is a matter of some surprise how little is known of the topography and general characteristics of the country pierced by the *Lake Superior and Mississippi Railroad.* Some denominate it a cold, damp, uninviting pine forest, and others a barren, almost God-forsaken region. Now, while we cannot agree wholly with either class of complainants, we allow them all due consideration for their opinions.

"The *Dalles of St. Louis* make up a scene of attractions rarely equalled and never excelled. There Nature is seen in her original, pristine wildness. Great towering forests of pine; grey, rugged dykes of slate rock; roaring cascades; the never ceasing lullaby of echoing hillside or far-reaching ravine; not one sign of the footstep of the coming march of progress, save the narrow, uninviting, thread-like line of the Lake Superior and Mississippi Railroad. There it goes—through hillside, great rock formations, over deep ravines, across roaring streams—all, all bespeaking a tribute of praise to the enlightened, indomitable will of man, and the almost perfection of engineering skill.

"Let us quietly start from the bridge that spans the St. Louis river near the head of the Dalles, and while walking slowly along, note the difficulties encountered, overcome, and made subservient to scientific skill.

"Great slate dykes rear their tesselated fronts like some ancient ruin, defiant and frowning. Soon the regular, almost measured reverberation of "click," "click," as the iron drill slowly but surely enters the deep bosom of each rock, tells the war has commenced. Soon the deep ravine, the forest, the hillside give back, in measured tones, the echoing voice of that element, called into requisition by human skill and will. The tall rock is shivered, and then comes the great loud blast. Twelve hundred pounds of powder is placed deep beneath the mass of rock. The fuse burns slowly, but surely. At last the rugged mass pulsates, heaves, tosses, and away go tons of ragged, broken fragments, separated, never to be reunited again; a huge material type of far too many episodes in the great world's history.

"But on we go. Deep cuts, huge embankments lend enchantment to the view. Stop by this trickling rill, and as the rude tin-cup bears the pure water to our parched lips, note the peculiarities of the deeply cloven bank, strata upon strata, clay, sand, interseamed alternately. Up some ten feet is seen a soapy, slimy composite, that looks unstable, in fact dangerous. This singular, deceitful strata has cost the Railroad Company hundreds of thousands of dollars, and yet is not wholly subdued. In the cleft hill-side slides occur in the embankment; it glides away with sin-like defiance. Go along still farther and the ruins of great huge stone walls betray its instability, as the deep gulch tells of its slippery characteristics. Engineering skill was baffled, capitalists grew restive, and on, on went the untiring task of labor.

"The nine miles of Railroad that reaches from the *head* to the *foot* of the Dalles of St. Louis river is a lasting monument to scientific progress and untiring human toil. Generals gain bloody victories and are immortalized on the page of history. Their fame is tarnished by the tears, wails, and sufferings of unappreciated thousands. But here nature bows to human ingenuity, channels for commerce are being opened; unthinking crowds glide idly along; and yet scarce a thought or a word in praise of those 'who have brought the hills low and made the rough places even.' Is not fame an idle breath and deserving praise a myth?

"But on we go, till wearied nature warns us that rest is the natural panacea for toil. A grassy dell, a pure trickling rill, invites repose. Here seated, we may survey the roaring cascade, the tumbling, tossed, frothy floods, and wonder when man, endowed with engineering skill, will utilize the sixty-eight thousand horse-power opposed by the Dalles of the St. Louis river to the interests of manufacturing demands. That day, in our humble opinion, is not far distant. The untold wealth of eastern capitalists will soon render its tribute, and what to-day is the quiet haunt of the slowly sailing fish-hawk, will become resonant with the music of the shrill cleaving saw, or the music of untold thousands of busy spindles. This is no idle phantasy, nor yet the dream of poetic enthusiasm.

"The men who have wielded the Lake Superior and Minnesota Railroad are earnest, determined men.

Foresight suggested the first enterprise as only a stepping-stone to the vestibule of greater, mightier consummations. The Dalles of St. Louis river will become speedily utilized, and new and important manufacturing centres created. Nature has here lavished untold facilities, subservient to and awaiting man's behests.

"The almost endless slate quarries —enough to supply the world—will be opened, and render tribute to wide-awake capitalists.

"The iron ore treasures of Lake Superior can be landed at the foot of the Dalles from steamers or barges.

"The superior quality of the lumber to be there manufactured, the facilities for reaching an almost boundless market, ample capital and cheap production, will soon render the Dalles of St. Louis the greatest lumber producing centre of the entire North-west. There can be no doubt that the City of Thomson must soon become an important point. Railroad shops alone built up Altoona, amid the Alleghany Mountains, to a City of many thousands of inhabitants. Then, why doubt the future of Thomson, with railroad shops, manufacturing facilities, slate quarries, and the illimitable pine forests contiguous thereto."

At the Dalles of St. Louis, on the line of the railroad, is one of the largest and best slate formations in America, and the only available one in the United States west of Pennsylvania. It is twenty miles long, six miles wide, and hundreds of feet deep.

In connection with the slate deposit at this point, and the mines and forests of Lake Superior, and the relation of both to the agricultural and commercial development of this section of the North-west, no point possesses more interest than the Dalles of St. Louis. The succession of cataracts, within a distance of eight miles, packs the accumulated waters of the numerous tributaries of the St. Louis, from the plateau of the Upper Mississippi to the level of Lake Superior at Fond du Lac, whence the river is navigable for 22 miles to its termination in the bays of St. Louis and Superior, fronting the city of Duluth. The future utilization of the extensive water-power at and near the Dalles of the St. Louis river, (ascertained to be 288,000 cubic feet per minute), gives assurance of the rapid growth of manufactures at this point. At the upper end of the Dalles the *Northern Pacific Railroad* forms a junction with the *Lake Superior and Mississippi Railroad*, thus increasing in an immense degree the importance of this locality, where the commerce of the Pacific Ocean is destined, ere long, to add its wealth to the already local advantages of this highly favored section, embracing both the cities of Thomson and Duluth.

Thomson, 23 miles from Duluth, on the line of the Lake Superior and Mississippi Railroad near its junction with the Northern Pacific Railroad, has natural advantages of the most commanding character,—being situated near the head of the Dalles of the St. Louis river, where is afforded the most reliable water-power in the country. Here is a substantial bridge crossing the river, an hotel, store and several dwellings. It is, no doubt, destined to be a favorite resort for pleasure seekers and admirers of sublime river scenery,—the St. Louis having a fall of about four hundred feet in the distance of eight miles,—jumping from precipice to precipice in wild grandeur.

The *Northern Pacific Railroad* has its present terminus at *Thomson Junction*, 24 miles from Duluth, where is a small settlement. From thence it runs westward to Brainerd, 115 miles from Duluth; crosses the Mississippi River and extends to Moorhead, 254 miles, forming in part a line of travel to Fort Garry, Manitoba; thence westward across the Red River of the north, through Dakota, to *Bismarck*, 450 miles, situated on the east bank of the Missouri River, here connecting with steamers running on the Upper Missouri for several hundred miles.

MOOSE LAKE STATION, 45 miles from Duluth, is a small settlement; in the vicinity are several small lakes and saw mills.

KETTLE RIVER STATION, 15 miles further, is situated on the banks of the above stream, which here affords good water-power. Here is a hotel, a store, and several dwellings.

Hinckley, 78 miles from Duluth, and 77 from St. Paul, is a new town erected since the commencement of the Railroad. It is situated on Grindstone Creek, a small stream, the outlet of Grindstone Lake. The Railroad Company have erected machine shops at this place, where passengers stop for refreshments. Here is a good hotel, a saw mill, a shingle mill, and a number of dwellings.

PINE CITY, 12 miles further, is situated on Snake river, where are several saw mills, a stave factory, a hotel, 2 stores, a number of dwellings, and about 300 inhabitants. *Cross Lake*, in the immediate vicinity, is navigable for large-size boats, as well as Snake river which enters into St. Croix river on the east.

CHENGWATANA, 4 miles east of Pine City, is the capital of Pine County. This is a thriving place of business, where is good water-power on Snake river; being surrounded by a fine lumbering region, and much good land.

RUSH CITY, 54 miles from St. Paul, is a small settlement situated on the outlet of Rush Lake, in the midst of a fertile region. Here are 2 saw mills, a hotel, a store, and a number of dwellings.

NORTH BRANCH, 12 miles further, is situated in the midst of a good farming country. The village of *Sunrise*, 8 miles east of the station, is a flourishing place of business, where are located several mills propelled by water-power. Trout fishing is excellent in Sunrise river and other streams flowing into the St. Croix river, which here divides the State of Minnesota from Wisconsin.

WYOMING, 30 miles from St. Paul, is pleasantly situated on the west banks of Sunrise river. Here is a small village surrounded by a fine section of country. This is a great resort for sportsmen in Autumn, when deer and other game are abundant.

FORREST LAKE, 5 miles, is another popular place of resort during the Summer months. This Lake is almost four miles long and from one to two miles wide. Bass and pickerel abound in this Lake, and afford fine sport for the angler. Deer and other wild game are found in great plenty. Cranberries, raspberries, blackberries, and whortleberries, grow here in great abundance.

CENTERVILLE STATION, 17 miles from St. Paul, is three miles east of the village of *Centerville*, situated on one of Rice Lakes. These Lakes, and the streams connecting them, abound with wild rice, which is much prized by the inhabitants for food. Wild ducks and other game frequent these Lakes in the Spring and Fall months in immense numbers.

WHITE BEAR LAKE STATION, 12 miles from St. Paul, situated near the Lake, is a place of great resort during the Summer months. Here are several hotels for the accommodation of visitors. The Lake is a beautiful sheet of water, of great depth and purity, covering a surface of about 3,500 acres. The hotels are very popular resorts during the warm weather, for boating, fishing and hunting. From this station a Branch Railroad, 12 miles in length, extends to *Stillwater* on the St. Croix river, and another branch road to *Minneapolis*, 13 miles.

A late writer in speaking of *White Bear Lake*, remarks:—

"This Lake is certainly a most beautiful one; it is surrounded by splendid forests, and its waters are as limpid and transparent as the clearest of crystal. Its circumference exceeds twenty miles, and there is a lovely island in its midst, covering an area of over eighty-five acres. Having enjoyed a delightful sail of about an hour, followed by a ramble among the trees on the margin of the Lake, we resumed our carriages and drove back to the City, delighted beyond measure with the excursion, and glad at having had an opportunity of seeing what is confessedly one of the most interesting spots in the vicinity of St. Paul. The climate at White Bear Lake has quite a reputation as a sanatorium for invalids, and persons suffering from diseases of the chest and lungs; consequently, many people have taken up their residence in the neighborhood, one and all of whom concur as to the health-restoring influence of its bracing atmosphere; which physicians believe possesses an excess of oxygen, together with a peculiar dryness and lightness which admirably adapts it to the purposes of free respiration. The lowness of the temperature during the Summer months, (ranging from 40° to 70° Fahrenheit,) too far from tending to produce colds or coughs, is found to afford direct relief to those affected with asthma, bronchitis, and other pulmonary complaints. Although the Winters are long, they are considered the most healthy period of the year; the extreme cold not being greater than is sometimes felt in the Northern or Eastern States, near the sea-board."

On approaching St. Paul from White Bear Lake, or Stillwater, the railroad runs near LAKE PHALEN, which supplies the City with pure and wholesome water, being three miles distant. Several other small Lakes are passed lying to the west, their waters all flowing into Lake Phalen, being tributary to St. Paul; the following are their area in acres:—

Lake Phalen....................237 Acres.
Lake Gervais...................210   "
Lake Pleasant..................730   "
Vadnais Lake...................500   "
Lambert's Lake................750   "
Three lesser Lakes............560   "

In addition to these Lakes a number of others are readily available, which will swell the aggregate water surface from which St. Paul is to be supplied, to several thousand acres. The quality of the water is superior in all respects to any in the West, and is probably, in regard to purity, unsurpassed in the country.

RAILROAD AND STEAMBOAT CONNECTIONS.

The *Lake Superior and Mississippi Railroad* connects at St. Paul with Steamers running on the Mississippi river, and with the St. Paul and Pacific Railroad, the St. Paul and Sioux City Railroad, the Milwaukee and St. Paul Railroad, and with the West Wisconsin Railroad. Steamers also run from St. Paul up the Minnesota river as far as navigable, and up the St. Croix river to Taylor's Falls—the latter trip affording a most delightful excursion. *For a description of St. Paul, see page* 202.

THE UPPER MISSISSIPPI.

THE vast range of country drained by the Mississippi river proper, independent of its great tributary, the Missouri river, embraces most of the State of Illinois, and a great portion of the States of Missouri, Iowa, Wisconsin, and Minnesota; a small part of the waters of Illinois, on its northeast border, flows into Lake Michigan, while nearly one half of the waters of Wisconsin flow in the same direction, finding their outlet through the Great Lakes and the St. Lawrence river into the Atlantic Ocean. All the waters of Missouri and Iowa find their way into the Missouri or Mississippi river, and thence into the Gulf of Mexico. The waters of Minnesota in part flow northward, through the Red river of the North, into Lake Winnipeg, and thence into Hudson's Bay. A portion flows eastward into Lake Superior, whilst its most important streams are the Upper Mississippi, fed by numerous lakes, and the St. Peter's or Minnesota river, falling into the Mississippi a few miles below the Falls of St. Anthony.

The Mississippi river is navigable for steamers of a large class, during a good stage of water from St. Paul to St. Louis, a distance of about 800 miles, and from St. Louis to New Orleans at all seasons of the year, except when interrupted by ice, a further distance of about 1,200 miles; making an uninterrupted navigation, during most of the year, of upward of 2,000 miles, from the Falls of St. Anthony, to the Gulf of Mexico. It is also navigable for steamers of a small class for about 150 miles above the Falls of St. Anthony. The entire navigation of this great river and its numerous tributaries being estimated at 16,000 miles.

The Area and Population of the *five* States mostly drained by the Mississippi, are as follows:

	Area sq. miles.	Population, 1860.	Population, 1870.
Illinois	55,400	1,711,951	2,539,638
Missouri	65,000	1,182,012	1,714,102
Iowa	55,000	674,913	1,182,933
Wisconsin	53,924	775,881	1,032,880
Minnesota	83,500	172,023	460,037
Total	312,824	4,516,780	6,948,590

This rich and fertile portion of the Union, when as densely populated as the State of New York, will contain about 25,000,000 inhabitants, and be capable of raising annually an immense amount of bread stuffs, meats, and other agricultural products for home consumption and foreign markets.

The North-Western States, proper, including Ohio, Indiana, Illinois, Michigan, Wisconsin, Iowa and Minnesota, contains an area of 378,920 square miles; and, in 1860, contained a population of 7,775,678 souls, which increased in 1870, by the new Census, to 10,757,944 inhabitants—being about *one-fourth* the entire population of the United States. The agricultural products and wealth having increased in about the same rapid ratio.

The principal cities and centres of trade for the above States, lying on navigable waters, and from which Railroads diverge to different sections of the country, are St. Louis, Burlington, Davenport, Chicago, Milwaukee, Dubuque, and St. Paul. Between these different cities a healthy rivalry exists for the trade of this great North-Western region, which is annually increasing in population and wealth.

A large number of Steamers run between St. Louis, Dubuque, and St. Paul, stopping at intermediate landings, affording daily opportunities for travellers visiting the Upper Mississippi, now annually thronged with pleasure-seekers and invalids in search of health.

Steamers, propellers, and sailing vessels run from Chicago, Milwaukee, and other lake ports on Lake Michigan, to Green Bay, Mackinac, Lake Superior; also, to Detroit, Buffalo, and Lake Ontario, *via* the Welland Canal. These steamers and propellers are usually thronged with passengers during the Summer months. Mackinac, Saut Ste. Marie, Marquette, and the different ports on Lake Superior being delightful and healthy places of resort.

A Railroad and Steamboat Route is now in operation, running from Chicago to Green Bay, and thence to Marquette, on Lake Superior, affording a speedy conveyance to this health-restoring region.

The *Lake Superior and Mississippi Railroad*, finished in August 1870, forms the most desirable route to and from Lake Superior, running from St. Paul to Duluth; uniting the travel on the Mississippi, with the Great Lakes or Inland Seas of America; forming a line of travel from New Orleans to Lake Superior, and from thence to Montreal and Quebec, a distance of about 3,800 miles; or, in other words, from the Gulf of Mexico to the Gulf of St. Lawrence.

TABLE OF DISTANCES.

FROM NEW ORLEANS TO QUEBEC, VIA LAKE SUPERIOR.

CITIES, &C.	MILES.	CITIES, &C.	MILES.
New Orleans	0	LA CROSSE, Wis	1,853
BATON ROUGE, La.	135	ST. PAUL, Minn.	2,060
VICKSBURG, Miss.	387	DULUTH.	2,215
HELENA, Ark	715	SAUT STE. MARIE	2,715
MEMPHIS, Tenn	800	DETROIT, Mich.	3,088
CAIRO, Ill	1,020	TORONTO, Can.	3,312
ST. LOUIS, Mo	1,247	MONTREAL	3,645
DUBUQUE, Iowa	1,707	**Quebec**	3,815

Mississippi River and its Principal Tributaries.

	NAVIGABLE.
UPPER MISSISSIPPI.—St. Louis to St. Paul, Minn..................	810 miles.
Minnesota River, St. Paul to Mankato........	148 "
St. Croix, Prescott, to St. Croix Falls, Wis....	54 "
Chippewa, Wisconsin......................	50 "
Wisconsin River.........................	
Fevre River, from Mouth to Galena, Ill.......	8 "
Rock River, Illinois.......................	
Iowa River, to Iowa City..................	
Des Moines River, Mouth to Ottumwa, Iowa..	90 "
Illinois River, Mouth to La Salle, Ill.........	270 "
MISSOURI RIVER — Mouth to Fort Benton, Montana..............	3,090 "
Yellow Stone River, Montana................	
Platte River,* Nebraska...................	
Kansas River, Mouth to Junction City.........	225 "
Osage River, Mouth to Osceola, Mo...........	
OHIO RIVER—Cairo, Ill., to Pittsburgh, Penn..................	1,000 "
Kentucky River, Mouth to Frankfort, Ken.........	
Cumberland, Mouth to Burkesville, Tenn............	370 "
Tennessee, Mouth to Muscle Shoals, Ala.............	600 "
LOWER MISSISSIPPI—St. Louis to Gulf of Mexico..................	1,350 '
St. Francis River, Mouth to Wittsburg, Ar.....	80 '
Arkansas River, Mouth to Fort Gibson, In. Ter.	740 "
White River, Mouth to Batesville, Arkansas...	175 "
Red River, Mouth to Shreveport, Lou.........	500 "
Yazoo River, Mouth to Greenwood, Miss......	240 "

The Mississippi River is the continental stream of North America. It forms a line of unbroken navigation from the Gulf of Mexico to Fort Snelling, Minn., a distance of 2,160 miles. No stream has ever served such valuable purposes to commerce and civilization; and no city upon its banks has ever, or can ever, share so largely in the commerce that floats upon its waters as St. Louis.

The Platte is an important tributary of the Missouri, which, like the Arkansas river, reaches to the base of the Rocky Mountains, and spreads over a wide space, so that it is totally unfit for navigation.

Steamboat Route from St. Louis to Dubuque and St. Paul

USUAL TIME, to DUBUQUE, 2½ days; to ST. PAUL, 4½ days.   THROUGH FARE, $18.

LANDINGS.	Miles.	LANDINGS.	Miles.
St. LOUIS.............	0	DUNLEITH, Ill................	1–461
Mouth Missouri River.........	20	Potosi Landing, Wis...........	14–475
Alton, Ill..................	5—25	Buena Vista, Iowa............	15–490
Mouth Illinois River........		Cassville, Wis................	4–494
Cap au Gris.................	40—65	Guttenburg, Iowa.............	10–504
Clarksville, Mo..............	37–102	Clayton, Iowa................	12–516
Louisiana, Mo................	12–114	McGREGOR, Iowa.............	11–527
HANNIBAL, Mo...............	30–144	**Prairie du Chien,** Wis ..	3–530
QUINCY, Ill..................	20–164	☞ To Chicago, 229 Miles.	
Lagrange, Mo................	12–176	Lynxville, Wis................	14–544
Canton.....................	8–184	LANSING, Iowa................	16–560
Alexandria, Mo..............	20–204	De Soto, Wis.................	6–566
Warsaw, Ill.................		Victory, Wis.................	10–576
Keokuk, Iowa............	4–208	Bad Ax City, Wis....	10–586
Montrose, Iowa..	12–220	Brownsville, Minn.............	16–602
Nauvoo, Ill.................	3–223	**La Crosse,** Wis...........	12–614
Fort Madison, Iowa..........	9–232	☞ To Milwaukee, 195 Miles.	
Pontoosuc, Ill...............	6–238	La Crescent, Minn............	2–616
BURLINGTON, Iowa...........	17–255	Richmond, Minn..............	16–632
OQUAWKA, Ill...............	15–270	Trempeleau, Wis..............	5–637
Keithsburg, Ill..............	12–282	**Winona,** Minn.............	17–654
New Boston, Ill..............	7–289	Fountain City, Wis...........	12–666
MUSCATINE, Iowa...........	18–307	Mount Vernon, Minn.........	14–680
ROCK ISLAND, Ill. ⎱		Minneiska, Minn............	4–684
DAVENPORT, Iowa ⎰	30–337	Alma, Wis..................	14–698
Le Claire, Iowa..............	18–355	WABASHA, Minn..............	10–708
Princeton, Iowa..............	6–361	Reed's Landing..............	6–714
Camanche, Iowa..............	10–371	Foot Lake Pepin.............	2–716
Albany, Ill..................	3–374	North Pepin, Wis............	6–722
Clinton, Iowa................	6–380	LAKE CITY, Minn............	5–727
FULTON, Ill. ⎱		Maiden Rock, Wis.........	8–735
LYONS, Iowa ⎰	2–382	Frontenac, Minn.............	3–738
Sabula, Iowa................	20–402	RED WING, Minn.............	18–756
SAVANNA, Ill................	3–405	PRESCOTT, Wis..............	28–784
Bellevue, Iowa...............	23–428	Mouth St. Croix River.	
GALENA, Ill.................	12–440	Point Douglass, Minn.........	1–785
Dubuque, Iowa...........	20–460	HASTINGS, Minn..............	3–788
☞ To Chicago, 189 Miles.		**St. PAUL,** Minn...........	32–820

Steamboat Route from St. Paul to Dubuque and St. Louis,

CONNECTING with RAILROADS RUNNING to MILWAUKEE and CHICAGO.

LANDINGS.	Miles.	LANDINGS.	Miles.
St. PAUL	0	**Dubuque,** Iowa	1–360
HASTINGS, Minn	32	☞ To Chicago, 189 Miles.	
Point Douglass, Minn	3—35	GALENA, Ill	20–380
Mouth St. Croix River.		Bellevue, Iowa	12–392
PRESCOTT, Wis	1—36	Savanna, Ill	23–415
RED WING, Minn	28—64	Sabula, Iowa	3–418
Head Lake Pepin	2—66	LYONS, Iowa }	20–438
Frontenac, Minn	16—82	FULTON, Ill. }	
Maiden Rock, Wis	3—85	Clinton, Iowa	2–440
LAKE CITY, Minn	8—93	Albany, Ill	6–446
North Pepin, Wis	5—98	Camanche, Iowa	3–449
Reed's Landing, Minn	8–106	Princeton, Iowa	10–459
WABASHAW, Minn	6–112	Le Claire, Iowa	6–465
Alma, Wis	10–122	DAVENPORT, Iowa }	18–483
Minneiska, Minn	14–136	ROCK ISLAND, Ill. }	
Mount Vernon, Minn	4–140	MUSCATINE, Iowa	30–513
Fountain City, Wis	14–154	New Boston, Ill	18–531
Winona, Minn	12–166	Keithsburg, Ill	7–538
Trempeleau, Wis	17–183	OQUAWKA, Ill	12–550
Richmond, Minn	5–188	BURLINGTON, Iowa	15–565
La Crescent, Minn	16–204	Pontoosuc, Ill	17–582
La Crosse, Wis	2–206	Fort Madison, Iowa	6–588
☞ To Milwaukee, 195 Miles.		Nauvoo, Ill	9–597
Brownsville, Minn	12–218	Montrose, Iowa	3–600
Bad Ax City, Wis	16–234	**Keokuk,** Iowa	12–612
Victory, Wis	10–244	Warsaw, Ill	4–616
De Soto, Wis	10–254	Alexandria, Mo	
LANSING, Iowa	6–260	Canton, Mo	20–636
Lynxville, Wis	16–276	Lagrange, Mo	8–644
Prairie du Chien, Wis	14–290	QUINCY, Ill	12–656
☞ To Milwaukee. 194 Miles.		HANNIBAL, Mo	20–676
McGREGOR, Iowa	3–293	Louisiana, Mo	30–706
Clayton, Iowa	11–304	Clarksville, Mo	12–718
Guttenburg, Iowa	12–316	Cap au Gris	37–755
Cassville, Wis	10–326	Mouth Illinois River	
Buena Vista, Iowa	4–303	**Alton,** Ill	40–795
Potosi Landing, Wis	15–345	Mouth Missouri River	5–800
DUNLEITH, Ill	14–359	**St. LOUIS**	20–820

TABLE OF DISTANCES
FROM ST. LOUIS to NEW ORLEANS.

LANDINGS.	Miles.	LANDINGS.	Miles.
St. LOUIS, Mo............	00	Commerce, Miss..............	40–487
Jefferson Barracks............	12	HELENA, Ark.................	45–532
Herculaneum.................	18—30	Mouth of White River.........	75–607
Selma, Mo...................	6—36	NAPOLEON....................	35–642
Ste. Genevieve...............	23—59	Gaines' Landing..............	40–682
Kaskaskia Landing, Ill.........	6—65	Columbia, Ark...............	20–702
Mouth Kaskaskia River........	15—80	Greenville	12–714
Chester, Mo.................	4—84	Port Worthington............	30–744
Grand Tower.................	46–130	Grand Lake, Ark.............	5–749
Bainbridge..................	10–140	Ashton.....................	15–764
Cape Girardeau..............	15–156	Lake Providence, La..........	10–774
Commerce, Mo...............	16–172	Miliken's Bend..............	50–824
Cairo, Ill.................	35–207	**Vicksburg,** Miss..........	26–850
COLUMBUS, Ken..............	18–225	Grand Gulf, Miss............	50–900
Hickman, Ken...............	25–250	Rodney, Miss...............	17–917
New Madrid, Mo.............	32–282	**Natchez,** Miss............	60–977
Island No. 11...............	5–287	Mouth Red River...........	60–1,037
Needham's Cut-off............	54–341	Bayou Sara, La.............	40–1,077
Plumb Point	20–361	Port Hudson, La............	11–1,088
Fulton, Tenn................	10–371	**Baton Rouge,** La.......	24–1,112
Mouth of Hatchee River.......	6–377	Plaquemine, La.............	25–1,137
Randolph...................	5–382	DONALDSONVILLE, La.........	30–1,167
Memphis, Tenn............	65–447	**NEW ORLEANS,** La...	80–1,247

Steamboat Route from St. Paul to Mankato, Minn.

LANDINGS.	Miles.	LANDINGS.	Miles.
St. PAUL.	0	**MANKATO**..............	0
Mendota....................	5	**St. Peter.**.............	30
Fort Snelling...............	1—6	Ottawa....................	16—46
Credit River...............	10—16	Le Sueur..................	12—58
Bloomington................	4—20	Henderson.................	10—68
SHAKOPEE..................	12—32	Belle Plaine...............	11—79
Chaska....................	6—38	St. Lawrence..............	6—85
Carver....................	4—42	Strait's Landing............	7—92
Louisville..................	4—46	Louisville.................	10—102
Strait's Landing.............	10—56	Carver...................	4—106
St. Lawrence...............	7—63	Chaska...................	4—110
Belle Plaine...............	6—69	SHAKOPEE.................	6—116
Henderson.................	11—80	Bloomington..............	12—128
Le Sueur..................	10—90	Credit River..............	4—132
Ottawa...................	12—102	Fort Snelling..............	10—142
St. Peter...............	16—118	Mendota..................	1—143
MANKATO, ...	30–148	**St. PAUL.**..............	5–148

The Maiden's Rock—Lake Pepin.

(*Copied from Harper's Magazine, July, 1853.*)

THE MAIDEN'S ROCK.

"Toward noon we entered that grand expansion of the Mississippi, called LAKE PEPIN. Its width is from three to five miles, and its length about twenty-five. It is destitute of islands, and all along its shores are high bluffs of picturesque forms, crowned with shrubbery, and commingled with dense forests. The white man has not yet made his mark upon Lake Pepin and its surroundings; and there lay its calm water, and yonder uprose its mighty watch-towers in all their primal beauty and grandeur. High above all the rest loomed the bare front of the Maiden's Rock, grand in nature, and interesting in its romantic associations. It has a sad story to tell to each passer-by; and as each passer-by always repeats it, I will not be an exception, It is a true tale of Indian life, and will forever hallow the *Maiden's Rock*, or Lover's Leap.

"*Winona*, a beautiful girl of Wapasha's tribe, loved a young hunter and promised to become his bride. Her parents, like too many in Christian lands, were ambitious, and promised her to a distinguished young warrior, who had smitten manfully the hostile Chippewas. The maiden refused the hand of the brave, and clung to the fortunes of the hunter, who had been driven to the wilderness by menaces of death. The indignant father declared his determination to wed her to the warrior that very day. The family were encamped on Lake Pepin, in the shadow of the great rock. Starting like a frightened fawn at the cruel announcement, she swiftly climbed to the summit of the cliff, and there, with bitter words, reproached her friends for their cruelty to the hunter and her own heart. She then commenced singing her dirge. The relenting parents, seeing the peril of their child, besought her to come down, and take her hunter lover for a husband. But the maiden too well knew the treachery that was hidden in their promises, and, when her dirge was ended, she leaped from the lofty pinnacle, and fell among the rocks and shrubbery at its base, a martyr to true affection. Superstition invests that rock with a voice; and oftentimes, as the birch canoe glides near it at twilight, the dusky paddler fancies he hears the soft, low music of the dirge of Winona."

SCENERY ABOVE WINONA.

THE **City of St. Paul,** a port of entry, capital of Minnesota, and seat of justice of Ramsey county, most advantageously situated on the left bank of the Mississippi, 2,100 miles from its mouth, 820 miles above St. Louis, and 10 miles by land below the Falls of St. Anthony; being elevated 690 feet above the Gulf of Mexico; in lat. 44° 52' north, long. 93° 5' west from Greenwich. It is situated on a bluff, 60 or 70 feet high, rising to 100 feet, and presents a grand view from the river. It is near the head of steamboat navigation on the Mississippi, 5 miles below the mouth of the Minnesota river, which enters from the west at *Fort Snelling*, the river here being about a quarter of a mile in width. No place on the continent of America has a more commanding position or healthy location than this most favored city. Steamers of a large class, during a good stage of water, can descend to New Orleans, 2,060 miles distant; above the Falls of St. Anthony, navigation is afforded, for steamers of a small class, for about 150 miles, while the St. Peter's or Minnesota river affords about a like extent of navigation, flowing through a very fertile section of country.

Saint Paul is one of the oldest settlements in the State. Father Hennepin visited and speaks of its site (1680). Jonathan Carver made a treaty in 1766 with the Dakotas in Carver's Cave, which is still in existence under Dayton's Bluff, within the present limits of the city. The site of the city was known to the Dakotas from time immemorial as *"Im-min-i-jas-ka,"* or *"White Rock,"* from its high bluff of white sandstone, a prominent landmark.

The first actual settlement was made in 1838 (just after the Indian title to the land east of the Mississippi had been extinguished) by one Parrant, a Canadian, who built a cabin on Bench Street. In 1840, a little log chapel was built by Father Gaultier, a Catholic missionary, on the present site of "Catholic Block." The church, or mission was called "St. Paul's," which henceforth became the name of the settlement.

Its growth in population for a few years was perhaps unsurpassed by any city in the Union. In 1838 it had only 3 inhabitants; in 1846, 10; in 1848 about 50 (white); in 1849, 400; 1850 (census), 1,112; 1854, 4,500; 1857, 9,973; 1860 (census), 10,277; 1870 (census), 20,031.

The public buildings in St. Paul are a State House; which is a brick edifice, standing on elevated ground; a courthouse, jail, and city hall; a public market building, five public-school edifices, an opera house; 20 church edifices, many of them being fine structures; also, 4 national banks, besides several firms engaged in the banking business; three insurance companies, a gas company, several large and well-kept hotels, the *Metropolitan Hotel*, the *Merchants' Hotel*, and *Park Place* being the most frequented; numerous stores and storehouses, several extensive breweries, flouring mills, and printing offices, besides numerous other manufacturing establishments. A wooden bridge here spans the Mississippi River, being a quarter of a mile in length; cost, $150,000. A Government custom-house and post-office building has lately been erected. Several railroads are constructed running west, north, and south from St. Paul, making it the centre of an extensive system of railways.

The arrival and departure of steamers are numerous during the season of navigation, there being daily lines from St. Louis, Dubuque, La Crosse, Winona, and up the Minnesota River.

Progress of Minnesota in Population and Wealth.

The following Table shows the general increase of population and assessed property valuation in the State at large, from the date of its Territorial organization, and the superficial expansion of settlement as indicated by the number of counties assessed. The census enumerations of population are given for the years 1850, 1860, 1865, and 1870, the population for the remaining years being estimated from the popular vote:

Year.	No. Assessed Counties.	Val. of Pers. and Real Estate.	Population.
1855	18	$10,424,157	40,000
1856	24	24,394,395	100,000
1857	31	49,336,673	150,037
1858	37	41,846,778	156,000
1859	40	35,564,492	162,000
1860	41	36,753,408	172,022
1861	44	39,077,531	190,000
1862	...	29,832,719	200,000
1863	...	32,211,324	225,000
1865	48		250,099
1870	...		460,037

POPULATION OF ST. PAUL.

The following Table will indicate the growth of population since 1850:

TABLE SHOWING THE GROWTH OF THE STATE SINCE 1850.

Year.	No. Assessed Counties.	Val. of Pers. and Real Estate.	Population.
1850	6	$806,447	6,077
1851	3	1,282,123	7,000
1852	8	1,715,835	10,000
1853	6	2,701,437	14,000
1854	13	3,508,518	32,000

Year.	Population.	Year.	Population.
1850	840	1857	9,973
1852	1,800	1858	10,000
1853	2,500	1860	10,600
1854	4,500	1865	13,176
1856	8,500	1870	20,031

Railroads Diverging from St. Paul, Minnesota.

NAME.	FROM	TO	MILES.
St. Paul and Pacific, Main Line	St. Paul,	Breckinridge	217
" " " Branch Line*,	St. Anthony,	Junc. Northern Pacific.	
St. Paul and Sioux City	St. Paul,	Sioux City, Iowa,	270
Lake Superior and Mississippi,	St. Paul,	Duluth,	155
Stillwater Branch,	White Bear Lake,	Stillwater,	12
Milwaukee and St. Paul,	St. Paul,	State Line,	134
Chicago and St. Paul	St. Paul,	Winona,	105
Dubuque and St. Paul*,	St. Paul,	Cedar Falls, Iowa,	
West Wisconsin	St. Paul,	Chicago, Ill.,	410

Early History of St. Paul.

The history of what is now St. Paul, divides itself into three distinct periods, marked by corresponding changes of names.

· 1. The period of Indian occupancy till 1838, when it was known as *Imnijaska*, or "White Rock."

2. The period of squatter settlement, from 1838 to 1849, when it was known by the Indians as "the place where they sell whiskey," and by the whites as "*Pig's Eye*."

3. Since 1849, when it was selected as the Capital of the Territory of Minnesota by the name of *St. Paul*, which had been bestowed upon it two years before.

* Unfinished.

FIRST WHITE MAN IN ST. PAUL.—
Louis Hennepin, whose name is immortally associated with the history of Minnesota as the first white man who ascended the Mississippi within its borders, and as the discoverer of the Falls of St. Anthony, was undoubtedly the first white man who ever set foot upon the site of St. Paul. On April 30th, 1680, over one hundred and eighty-four years ago, Hennepin, a captive in the hands of a war party of Dakotas on their way to Mille Lacs, "landed in a bay, five leagues below the Falls of St. Anthony," a description of which, with other circumstances, fixes the locality under Dayton's Pluff, at the mouth of Trout Brook—about three quarters of a mile below the Steamboat landing.

THE FIRST AMERICAN IN ST. PAUL.—
Eighty-seven years have passed since the arrival of Hennepin. Perrot has built and abandoned a fort on Lake Pepin, and planted the arms of France in Minnesota. Le Seuer has explored the Minnesota and given it the name of his gallant friend, Capt. St. Pierre. The Dakotas have been driven from the northern lakes by the Chippewas, and Minnesota, by the treaty of Marseilles, has just passed from the dominion of France to the flag of England, when on one fine morning in November, 1766, a keen, practical Yankee, the forerunner of all the Yankees in this part of the world, stepped into St. Paul near where Hennepin had landed three generations before. It was Brother JONATHAN CARVER, fresh from Connecticut, come to trade—Carver, great progenitor of the land speculators of Minnesota, first and greatest of the race.

CARVER'S CAVE.—Jonathan's landing was at the foot of Dayton's Bluff, and his account of the discovery made there is the first memorial which links St. Paul with the traditions of the Dakotas:—

"About thirteen miles below the Falls of St. Anthony * * is a remarkable cave, of amazing depth. The Indians term it Wakan teebe, that is, *the dwelling of the Gods.*

"The arch within is near fifteen feet high and about thirty broad; the bottom consists of clear sand. About thirty feet from the entrance begins a lake, the water of which is transparent, and extends to an unsearchable distance, for the darkness of the cave prevents all attempts to acquire a knowledge of it. * * * * I found in this cave many Indian hieroglyphics, which appeared very ancient, for they were so covered with moss that it was with difficulty I could trace them. They were cut in a rude manner upon the inside of the wall, which was composed of a stone so extremely soft that it might be easily penetrated with a knife. * * * At a little distance from this dreary cavern is the burying-place of several bands of Naudowessie [Dakota] Indians. Though these people have no fixed residence, being in tents, and seldom but a few months in one spot, yet they always bring the bones of their dead to this place, which they take the opportunity of doing when the *chiefs meet to hold their councils and to settle public affairs for the ensuing summer.*"

These ancient burial mounds still exist on Dayton's Bluff, and, a few years ago, Mr. Neill had one of them opened. In this, which was 218 feet in circumference and 18 feet high, he found the remains of skulls and teeth at the depth of three or four feet.

In 1807, Major Long was obliged to creep through the sandstone *débris* at its mouth on all fours. In 1837, Nicollet worked for two days to effect an entrance, and confirmed the accuracy of Carver's description.

"A Chippewa warrior made a long

harangue on the occasion, threw his knife into the lake as an offering to Wakan tibi." Indian pictographs still remain, gray with age, upon portions of the wall still standing.

After a voyage to what is now Anoka, and up the Minnesota river for 200 miles, Carver, on the 1st of May, 1767, returned to the "Great Cave," where he officiated as the first representative of the whites in the great Annual Legislative Session of the Dakota bands, and made the first speech ever delivered by a Yankee in St. Paul.

"At this season," says Carver, "these bands go annually to the Great Cave before mentioned *to hold a grand council with all the other bands, wherein they settle all their operations for the ensuing summer.*" Thus early was St. Paul the *Capital of Minnesota.*

Nothing could be more significant of the geographical centrality of St. Paul than this fact, that from immemorial time it had, at that date, been the political centre of the scattered bands of the Dakota nation.

THE FIRST LAND SPECULATOR IN ST. PAUL.—It was here, too, at this "Great Cave," that *the first conveyance of land* was made and the *first deed signed* in Minnesota. This was the instrument by which the heirs of Carver founded their title to Carver's tract, which contained St. Anthony, St. Paul, and a large part of Wisconsin. The document is curious, and runs in this wise:

"To Jonathan Carver, a chief under the most mighty and potent George the Third, King of the English and other nations, the fame of whose warriors has reached our ears, has been now fully told us by our good *brother Jonathan*, aforesaid, whom we rejoice to have come among us and bring us good news from his country.

"We, the chiefs of the Naudowessies, who have hereunto set our seals, do, by these presents for ourselves and our heirs forever, in return for the aid and other good services done by the said Jonathan to ourselves and our allies, give, grant, and convey to him, the said Jonathan, and to his heirs and assigns forever, the whole of a certain tract of territory or land, bounded as follows, viz.: From the Falls of St. Anthony, running on the east side of the Mississippi, nearly south-east, as far as Lake Pepin where the Chippewa joins the Mississippi, and from thence eastward five days' travel, accounting twenty English miles per day, and from thence again to the Falls of St. Anthony. We do, for ourselves, heirs, and assigns forever give unto the said Jonathan, his heirs and assigns, with all the trees, rocks, and rivers therein, reserving the sole liberty of hunting and fishing on land not planted or improved by the said Jonathan, his heirs and assigns, to which we have affixed our respective seals, at the Great Cave, May 1st, 1767.

"[Signed]

"HAW-NO-PAW-A-TON.

"O-TOH-TON-GOOM-LISH-RAW."

———

It was here, too, nearly a century ago, that Carver anticipated that splendid scheme of commercial intercommunication whose realization in our day is to make St. Paul the focus of the internal commerce of the continent. With the Delphic *numen* of the cave upon him, he foresaw that in the fat soil and laughing waters of Minnesota the elements were ripening for the sustenance of future populations, who, he says, will be "able to convey their produce to the seaports with great facility. * * This might also in time be facilitated by canals or shorter cuts, and *a communication opened by water with New York, by way of the lakes.*"

Here, too, Carver conceived the project of a Northern Pacific route by the way of the Minnesota and Oregon rivers, which, he says, "would open a passage for conveying intelligence to China and the English settlements in the East Indies"—an idea which will doubtless be consummated in our day.

THE ORACLE OF THE CAVE DUMB.—After Carver robs the "Great Cave" of its mighty secret that has throbbed for ages at its heart, the "Dwelling of the gods" is henceforth shut to all the world.

Henceforth, for seventy years, the oracle is dumb, silent, stony, impenetrable as the Sphinx, its white face turned in speechless prophecy toward the ter raced slopes which lay there before its closed mouth.

History rolled over "White Rock" and past it, but took no notice of it. The brave Pike goes past it in 1805, and ignores it. Long besieges the unutterable oracle in vain in 1807. Fort Snelling is established in 1819. Mendota becomes the depot of the fur trade. Events are clustering around it, but all look past it, till 1837, when the Dakotas were persuaded to cede their lands on the east side of the river to the United States, on account of the valuable pine lands and water power thereon. The treaty was ratified at Washington in 1838, and *Imnijaska* ceased to be Indian territory

Drive from St. Paul to the Falls of St. Anthony, returning via Fort Snelling.

This excursion affords one of the most interesting drives in any part of the country. On leaving St. Paul, by private conveyance, you pass through Madison avenue to the open plains which skirt the city, and then follow the direct road to St. Anthony, 10 miles. One or two beautiful cascades are passed near the roadside, as you approach the great Falls.

The *State University*, another object of interest, situated east of the road, overlooking the Falls, is a flourishing institution of learning. The town of ST. ANTHONY, with its saw mills and factories, propelled by water power, extends for near a mile above and below the Falls. Here is a *Suspension Bridge* of fine proportions, spanning the stream above the cascade.

MINNEOPOLIS, a large and flourishing place, is situated on the west bank, surrounding the Falls, where are very extensive saw mills, grist mills, paper mills, and other factories, all being propelled by water-power, and all well worthy of a visit. Here is a good hotel, where visitors usually stop for refreshments.

On returning, the road runs along the west bank of the Mississippi for four miles, when the *Falls of Minne-ha-ha* are reached. This beautiful fall of water, made famous by poetry and romantic scenery, is almost beyond description, as seen at different seasons. It has a perpendicular fall of about 40 feet, and can be viewed from the rear, as the rocks recede so as to allow a passage from side to side under the fall of water. About half a mile below, this pure stream enters into the Mississippi.

FORT SNELLING, two or three miles farther, and six miles above St. Paul, is an old Government post, where are usually quartered more or less troops; at the present time (1865) there are two regiments. Standing at the junction of the Minnesota and Mississippi Rivers, on elevated ground, it has a very picturesque appearance. Here is a rope ferry across the river, leading toward St. Paul, it being reached by a circuitous road running under the bluffs, affording highly romantic views. Here the *Milwaukee and St. Paul Railroad* crosses the Minnesota River.

The next object of interest is a Cave, 2 miles above St. Paul, which will well repay a visit to its subterranean caverns, from whence issues a lovely sheet of pure water.

MENDOTA, Minn., is situated on the right bank of the Mississippi River, at the mouth of the Minnesota, 5 miles above St. Paul. This is one of the earliest settled places in the State, being formerly the head-quarters of the American Fur Company. Here are two churches, an hotel, and several stores. Population, 600. The *Milwaukee and St. Paul Railroad* and the *Minnesota Valley Railroad* form a junction at Mendota, both rivers being crossed by a drawbridge.

FORT SNELLING, 6 miles above St. Paul, is an important United States post and rendezvous, situated on a commanding eminence at the junction of the Minnesota and Mississippi Rivers, 6 miles below the Falls of St. Anthony.

PLACES AND OBJECTS OF INTEREST

TWENTY-FOUR MILES AROUND ST. PAUL.

West Side Mississippi River.	Miles.	*East Side Mississippi River.*	Miles.
MENDOTA	5	Carver's Cave	1
Fort Snelling and Ferry	6	Fountain Cave	2
Falls of Minnehaha	8	Lake Como	3
Diamond Lake	9	Phalon's Lake	3
Rice Lake	10	*Little Canada*	6
Lake Amelia	10	Gervais Lake	6
Mother Lake	10	Mazaska Falls	6
Wood Lake	11	Bass Lake	6
Grass Lake	11	Vadnois Lake	7
MINNEAPOLIS	11	Black-Bass Lake	9
Cedar Lake	13	Fawn's Leap and Silver Cascade	9
Crystal Lake	14	ST. ANTHONY'S FALLS	10
Lake Calhoun	14	White-Bear Lake	12
Lake Harriet	14	Bald-Eagle Lake	14
Medicine Lake	16	*Stillwater* (St. Croix River)	18
Shakopee	22	*Hudson*, Wis	20
Lake Minnetonka	24	Forest Lake	24

Remarks.

At LAKE COMO, 3 miles from St. Paul, there is good fishing and two well-kept public-houses.

WHITE-BEAR LAKE, 12 miles by railroad; here is good boating and fishing, and several well-kept public-houses.

At LAKE HARRIET, 14 miles, there is a well-kept hotel, boating, and fishing.

LAKE MINNETONKA, 24 miles from St. Paul, by railroad route, is one of the largest sheets of water in the State. Its shores are indented with beautiful bays, fertile lands, and sloping bluffs, crowned with forest trees, coming down to the water's edge. Islands, covered with the gorgeous green of Minnesota's foliage, are scattered liberally over its surface of pure sparkling waters, abounding with the finest of fish, affording great sport to the angler. Sail and row boats can be procured at the hotels for pleasure parties, and those desiring to see the extended beauties of land and water scenery can do so by taking a trip on the miniature steamer, "Lady of the Lake." This is a charming place of resort for invalids and seekers of pleasure.

The fish which are mostly taken in these lakes are bass, pike, and pickerel of a fine quality.

MINNEHAHA RIVER, the outlet of some of the small lakes in this vicinity, is a shallow, sparkling stream, dashing over its pebbly bed and around its little islands in the most gleeful manner. Without a warning, without even any preliminary rapids, it makes the leap which is called the *Falls of Minnehaha*. A graceful leap it is. The stream springs over in one sheet of sparkling foam, landing in a basin which for centuries it has been busily hollowing out for itself—a basin much like that into which the Kaaterskill Fall leaps, and like that, too, in presenting behind the sheet of water a smooth concave recess, around which it is possible for a man to pass, coming out at the opposite side of the cataract. The foliage in the vicinity is as gracefully disposed by nature as the artist could wish, and in itself and all its surroundings Minnehaha is a type of perfection of its class—a

model for all ambitious young waterfalls who may wish to win the poet's as well as the public's regard, and be ever associated with the fate of some dear maiden, as beautiful as itself, who in her delirium would rave about it, as did the old arrow-maker's daughter:

> "Hark! she said, I hear a rushing,
> Hear a roaring and a rushing,
> Hear the Falls of Minnehaha
> Calling to me from the distance.
> No, my child, said old Nokomis,
> 'Tis the wind among the pine-trees."

At CHICAGO CITY, near the line of the Lake Superior Railroad, there is a well-kept hotel and accommodations for parties visiting the several beautiful lakes in this vicinity. Here is afforded good fishing, boating, and hunting.

In addition to the lakes which surround St. Paul, a visit to the *Dalles* on the St. Croix River and *Taylor's Falls*, 40 miles distant by stage, is recommended to all travelers fond of sublime river scenery. Farther up the St. Croix good trout fishing can be found—many of the small streams flowing into the river on the Wisconsin side being almost alive with speckled trout.

The Falls of St. Anthony,

once the *Ultima Thule* of the northwestern traveler, are not so striking or grand as one might expect from the description given by the early explorers of Minnesota. "There is no prodigious height for the water to leap from, as at Niagara, but the rapids are grander and quite as extensive, while their power is shown by the large slabs of stone which lie in distorted piles along the shore, some standing up on end like giant tombstones, others piled irregularly as if trying to crowd away from the fearful force of the water. The retrogression of the falls has been very slow until the

Spring of 1867 when the great freshet, which proved so disastrous to log-owners and lumber-merchants, told to an unprecedented extent on the cataract itself.

"The reason of this is clearly understood. For a quarter of a mile above the main fall the bed of the river is composed of a thin stratum of limestone, supported by sandstone. This latter, being soft and crumbling, is worn away under the constant action of the water, thus forming a sort of cave, with the slab of limestone from which the water falls overhanging it. Of course, as this excavation grows deeper, the limestone having nothing to support it, breaks away, and thus St. Anthony's Falls recede.

"As a mere spectacle, St. Anthony's Falls is grander by moonlight than at any other time, for then the unpoetical and unsightly buildings around it do not obtrude themselves, while the noise and dash of the rapids are heard and seen to perfection. At such a time St. Anthony's waters present an overpowering idea of furious strength, and one worthy to be remembered along with the recollections of Niagara itself. A greater contrast to the gentle beauty of Minnehaha it would be impossible to find; and yet these two cataracts are within a short walking distance of each other, and to tourists both will be, for ages to come, among the greatest attractions of the Northwest."

Here is a perpendicular fall of about 18 feet, and a rapid descent of 46 feet, within a distance of one mile.

Minnesota, or St. Peter's River,

one of the largest streams that rises in the State, is navigable for steamers, at most seasons of the year, from St. Paul to Mankato, 148 miles, passing St. Peter and other important towns on its banks. In good stages of water, small boats run to the mouth of the Yellow Medicine, 238 miles from its mouth. Beyond this, at a slight expense, it might be rendered navigable to Big Stone Lake, where a portage of about three miles in length separates it from the equally navigable waters of the Sioux Wood, which empties into the Red River of the North. The Red River gives over 300 miles of navigable water on the western boundary of the State, before entering into British America, above Lake Winnipeg.

Railroads have been constructed from St. Paul, Hastings and Winona, which run up the Valley of the Minnesota River, superseding in a great measure steamboat traffic.

BRIDGING THE MISSISSIPPI RIVER.

There are now some ten or twelve Railroad Bridges spanning the Mississippi River between St. Louis and St. Paul, a distance of 820 miles.

These bridges, although provided with draws, or an opening for steamers, more or less obstruct navigation on this noble stream. They vary from 1,000 to 2,000 feet in length, from shore to shore.

The principal structure at St. Louis, recently finished, is thus described:—

THE GREAT ST. LOUIS BRIDGE.

"The iron work on the Keystone bridge over the Mississippi, at St. Louis, has been completed amid general rejoicing. The entire work will be speedily finished and that important structure thrown open for business. The bridge connects St. Louis with East St. Louis, in the State of Illinois. The river at that point is 1,500 feet wide. It is spanned by three arches of 500 feet length inside the piers on which they rest. In its construction chrome steel has been used. It has a tensile strain double that of ordinary steel. The bridge is arranged for railway and carriage tracks. It enters St. Louis near its business centre. There is no draw in it for the passage of boats, and as it is but about 60 feet above high water, only the smaller class of steamers can pass under it without lowering their chimneys. It is an enormous structure. It is claimed to be most important, or at least the most notable railroad bridge in the world. Its cost, including approaches and tunnel, will not be less than $10,000,000. As a feat of engineering skill it takes the very first rank. But great and remarkable as it is the cry is raised against it that it obstructs navigation. The largest steamers find trouble in passing under it, and plans have been suggested for a remedy. But so long as the bridge stands—and that is likely to be many years—there seems to be no other course than for the steamers to be built and fitted with machinery for the easy lowering of their smoke-stacks. It is evident they must acknowledge obeisance by a graceful bow."

BRIDGING DETROIT RIVER.

This proposed impediment to the outlet of the Great Lakes or Inland Seas should never be allowed. A *graceful bow*, or fatal collision with railroad piers are both objectionable, alike to owners of steamers and sail vessels navigating the Detroit or St. Clair Rivers. Here vessels have to pass at all hours of the day and night,—in storm and in calm weather,—and be subject to the loss of both cargo and vessel, as well as the life of passengers.

231

FALLS OF ST. ANTHONY.

232

MINNE-HA-HA.

"Here the Falls of Minne-ha-ha
Flash and gleam among the oak trees,
Laugh and leap into the valley."

The City of **St. Anthony**, situated 10 miles north of St. Paul, by railroad, now forming part of Minneapolis, is a favored locality. Incorporated in 1855, and in 1870 contained 5,000 inhabitants. Here are nine churches, two banks, three hotels, several stores, and numerous manufacturing establishments, propelled by water power. The "St. Anthony Falls Water Power Company" is capable of sawing 40,000,000 feet of lumber annually. There are also three flouring mills, a paper mill, five saw-mills, machine shop, two breweries, and other extensive manufacturing establishments. The University of the State of Minnesota is located here, on an eminence overlooking the falls and the two towns. An elegant suspension bridge, erected in 1855, 620 feet long, spanning the main branch of the river above the Falls, connects the city with Minneapolis. The *St. Paul and Pacific Railroad*, completed to a point 108 miles northward, now extends from St. Paul to Melrose.

Minneapolis, Minn., the capital of Hennepin county, is delightfully situated on the west side of the Mississippi, at the Falls of St. Anthony, where is afforded one of the most magnificent water powers on the continent. Here are 14 flouring mills, 1 cotton mill, 2 woollen mills, 1 paper mill, 10 extensive saw-mills, 8 planing mills, 6 sash and blind factories, 2 iron foundries, machine shops and lath and shingle mills all using water power. In addition to these are manufactured ploughs, wagons, furniture, &c.; 2 foundries and the machine shops and car factory of the Milwaukee and St. Paul Railway Company. The Minneapolis Water Power Company, and the St. Anthony Company, have combined properties of quantity and availability unsurpassed in the United States. The lineal frontage along which the power can be carried and applied at a trifling cost, so as to supply a mill with power in every hundred feet of its course, is over 15,000 feet. The value of such a power, as well as the amount of machinery it is destined to propel, as the vast and fertile region north and west of it becomes settled, can hardly be estimated.

Minneapolis has become a fashionable place of resort for invalids and seekers of pleasure. The *Nicollet House* is a popular and well-kept hotel. Besides the county buildings, Minneapolis contains 3 national banks, 8 churches, 4 hotels, numerous stores and store-houses, an opera house, with many fine private residences. Population in 1870, 13,066.

The picturesque scenery in and around these two cities at the Falls, their topographical beauty, the fine hard roads leading in all directions, the charming lakes in the vicinity, the celebrated *Minne-ha-ha Falls*, being a few miles below Minneapolis on the Fort Snelling road, taken together with the dry, bracing atmosphere that distinguishes Minnesota from all other Western States, have contributed to draw crowds of pleasure-seekers, travelers, and invalids to this locality. Two beautiful lakes, Harriet and Calhoun, lying within a half hour's drive, and Lake Minnetonka, 12 miles westward, are places of constant resort in summer. These lakes, and about thirty others in the country, abound with sunfish, bass, and pickerel, as also the woods and prairies with the usual varieties of game. The old *Fort Snelling*, and its reservation of 10,000 acres, is situated in this county, at the confluence of the Minnesota and Mississippi rivers. The Fort is now used as a rendezvous for troops and recruits. Although once abandoned by the Government, the prospect now is that it will be permanently retained for military purposes. The *St. Paul and Pacific Railroad* runs west from Minneapolis to Breckinridge, 217 miles, and the *Chicago, Milwaukee and St. Paul Railroad* runs south through Minnesota and Wisconsin to Milwaukee and Chicago, connecting with railroads in northern Iowa. A railroad also extends to White Bear Lake, 14 miles.

ST. PAUL AND PACIFIC RAILROAD ROUTE.

MANOMIN, the capital of Manomin county, is a small village on the east bank of the Mississippi river, 17 miles north of St. Paul by railroad route.

ANOKA, Minn., 25 miles north of St. Paul, by railroad, is the county-seat of Anoka county, being handsomely situated on the east bank of the Mississippi river, at the mouth of Mille Lac, Rum river lying on both sides of the latter stream. The surface of the country is here diversified, and the climate highly salubrious; the soil being well adapted to agriculture. The natural meadows are an important feature, and, taken in connection with other facilities which the place affords, make it particularly adapted to the raising of cattle and sheep. Here are three church edifices, two hotels, several stores, and about 1,000 inhabitants.

ITASKA, Anoka County, Minn., is a small settlement on the east bank of the Mississippi river, 35 miles from St. Paul by railroad route.

ELK RIVER, Minn., is the name of a village situated on a stream of the same name, half a mile east of the Mississippi river, distant 40 miles from St. Paul, by railroad route.

BIG LAKE, Minn., the county-seat of Sherburne county, 50 miles north of St. Paul by railroad route, is situated about two miles east of the Mississippi river, containing a population of 200 or 300.

ST. CLOUD, Minn., lying on the west side of the Mississippi river, at the foot of the Sauk Rapids, is the capital of Stearns county, 74 miles north of St. Paul by railroad route. This may be called the head of navigation for the river above the Falls of St. Anthony, being on the direct route from St. Paul to the

Red River settlement of the North. A railroad is being constructed to run from Sauk Rapids to Pembina, connecting with the Northern Pacific Railroad, benefiting this whole section of country. The village now contains about 2,000 inhabitants, and is fast increasing in wealth and importance. There are a fine court-house and jail, one bank, United States land-office, five churches, three hotels, twelve stores, and two printing-offices.

From St. Cloud to the Red River is about 200 miles, the distance being about 200 more miles to *Fort Gary*, British America. A large trade is carried on, by means of ox-carts passing over the prairie, including the furs and other articles belonging to the Hudson Bay Company.

SAUK RAPIDS, Minn., lying on the east side of the Mississippi River, at the head of the rapids, two miles above St. Cloud, is the capital of Benton County. It contains about 700 inhabitants, 2 churches, 2 hotels, 2 stores, and manufacturing establishments. The *St. Paul and Pacific Railroad* runs to this place along the east bank of the river. Here is an immense water-power, created by the *Sauk Rapids*, having a descent in half a mile of about 15 feet, where a dam is constructed.

The *Mississippi River*, above the Sauk Rapids, flows through a level country, interspersed with groves of timber of different kinds, having a width of about 100 yards, to Crow Wing, 40 miles above. North of the latter place, pine timber of a large growth is found in abundance, the lumbering business being the principal source of profit.

WATAB, Benton County, Minn., 80 miles above St. Paul, lying on the east side of the Mississippi River, is a small post settlement, containing about 150 inhabitants.

LITTLE FALLS, Minn., 100 miles north of St. Paul, is the capital of Morrison county, where are a fine water-power and saw-mills, it being in the region of a good lumbering section of country.

CROW WING, Minn., is the capital of Crow Wing county, situated on the east bank of the Mississippi, 120 miles north of St. Paul. This is an important post, where is located the Government agency for the Chippewa Indians, and commands a considerable Indian trade. A Branch Railroad is being constructed to extend from Sauk Rapids to Brainerd, forming a connection with the Northern Pacific Railroad.

Northern Minnesota.

The distance from ST. PAUL to CROW WING, Minn., is about 120 miles, the *Chippewa Agency* being seven miles above Crow Wing, on Crow Wing river, a stream larger than the Mississippi proper; it is the outlet of Otter Tail and other numerous lakes, some sixty miles westward. The Indian agent for the Chippewa, Pembina, and Pillager Indians resides at the above agency. The agent makes a yearly payment to the above Indians, usually leaving the agency about the first of October, travels west to Otter Tail Lake, thence north, over the old Red river trail, to Douglas, Polk county, Minn., situated on Red Lake river, emptying into the Red river of the North, about forty miles west. In this vicinity the payments are made.

OTTER TAIL LAKE and the surrounding chain of lakes are of the purest water, abounding in delicious fish of different kinds. The shores are pebbly, surrounded by hard-wood timber, the sugar maple tree here predominating, from which large quanties of maple sugar are annually manufactured. The soil is unusually rich, producing wild grass three or four feet in height. The principal game left is wild fowl of different kinds, among which may be named the prairie chicken, grouse, partridges, ducks, and wild geese. Deer, elk, bear, foxes, badgers, and other fur-bearing animals, heretofore numerous, are now sparse, being nearly exterminated by the Indians, who are expert huntsmen. The healthy influence of this section of the country is unrivaled, it being a luxury to breathe the pure air of this region.

A few years since a resident of Milwaukee, Wis., who had been suffering from ill health, tending to consumption, started for St. Paul and journeyed toward Crow Wing, along the east side of the Mississippi river, arriving about the time of the leaving of the United States agent and his party for the interior, the weather being then cool and delightful. Joining said party, and participating in their fare, he made the journey to Otter Tail Lake, and thence to Red Lake river, on horseback, returning with said party.

During this trip of some four weeks, his health was almost entirely restored, being able to bear almost any amount of fatigue; camping out in the open air, hunting, and fishing as circumstances would permit.

This is the happy experience of hundreds of invalids who have the resolution to visit this health-restoring section of country, where fevers and consumption are almost entirely unknown. Even the winter months are endurable and healthy in this region, extending north to the British settlement near Lake Winnipeg, 50° north latitude.

Buffalo and other large game may be found west of Red river, affording wholesome food, while wheat and vegetables are raised in great abundance wherever settlements have been made.

Interesting to Consumptives.

WHO SHOULD GO TO MINNESOTA AND WHO SHOULD NOT.

Extract from a letter, dated, St. Paul, Minn.,

"It is not the object of your correspondent to court any argument upon the relative merits of a northern or southern climate for the cure of that fell destroyer of human life and happiness, consumption, but merely to give his experience as an invalid during a sojourn of several months in a country which is fast becoming one of the most popular resorts for invalids from all parts of the Union. Neither do I wish to be understood as claiming for Minnesota entire immunity from disease, nor that the climate is a sovereign remedy for all cases of consumption; but, from careful observation, I believe I am justified in asserting that there is no locality on this continent so exempt from 'all the ills that flesh is heir to' as this. The dryness of the atmosphere, the peculiar character of the soil, the almost total absence of fogs and moist winds, all contribute to render the climate one of unrivaled salubrity.

In its first stages, consumption appears to yield readily to the peculiar influence of the climate; and, even in the more advanced stages of the disease, the patient, by a continued residence in this country, finds permanent relief and comparative good health. I find that three classes of cases arrive in this country in search of relief: 1. Those slightly affected, who take time by the forelock, get well in a few months, and return to their homes perfectly cured. 2. Those more seriously affected, who never fully recover the use of their lungs, but by a permanent residence in Minnesota enjoy comparative good health. 3. Those who wait until it is too late, and arrive here only to linger a few weeks and die among strangers.

"It is to be regretted that the majority of the invalids who arrive here are not of the first class. Unfortunately, owing to the ignorance of physicians, the disease is seldom detected in its first stages; and it is not until a hemorrhage takes place, or tubercles commence to soften, that they see the necessity for the removal of the patient to a more salubrious climate.

"The second, or predominating class, are scattered all over the entire State, from the Iowa line to the shores of Lake Superior. Go where you will through Minnesota and you will meet persons, apparently in good health, who could not exist two years under the influence of the cold moist winds of the Atlantic States. Many of them arrive here quite low, but, with the help of a good constitution and the peculiar salubrity of the climate, they manage to rally and enjoy tolerably good health. In one or two instances which came under my observation, the patients had to be removed from the steamboat in a carriage, and several months elapsed before any visible improvement could be noted; but finally the patients commenced to mend, and the clear, bracing atmosphere of winter soon restored them to health. A few Sundays ago we buried one of the oldest residents of this city, who had been ill with consumption for fifteen years. He had been sick with the disease three years when he entered the State, and did not expect to live many months; but he rallied, and by a continued residence in the country managed to prolong his existence a dozen years. Some of the leaning business men of this city, men noted for their enterprise and success in life, belong to the second class, and,

although to all appearance in the full possession of health, tell you that it would be impossible for them to exist East.

"Of the third class not much need be said. They never ought to come here, as the fatigue and excitement of the journey only tend to hasten death. Some die on their way up the river, some at the hotels and boarding-houses before they have been domiciled among us a fortnight, and others, feeling that death is inevitable, start for home before they have been a week in the country.

"A very intelligent gentleman from New York, whose acquaintance I made when I first arrived in St. Paul, estimated that about three out of every ten persons who came here afflicted with lung complaints recovered so as to be able to return to their homes, and that over fifty per cent. of the invalids were afforded permanent relief. My informant, who is an invalid himself, has spent three years in the State, and, although in the enjoyment of apparent good health, says he will never be able to live in his native place again. He has therefore sent East for his family, and intends going into business here.

"It would be a difficult task to arrive at anything like the approximate number of invalids in the State, for there are no statistics on the subject, but it is safe to estimate them by thousands. In the summer you find them scattered all over the State, amusing themselves by fishing and hunting. The attractions in this respect are superior to anything of the kind in the United States perhaps. The entire surface of the State is dotted with lakes, varying in circumference from one mile to one hundred, which abound in the largest and choicest kind of fish. Pickerel, weighing from twelve to fifteen pounds, bass, wall-eyed pike and trout in proportion are caught in large quantities in all the lakes and rivulets. Trolling on the lakes is especially recommended by the physicians as the most fitting exercise for invalids who are too reduced to follow the more fatiguing sport of gunning. In the Fall of the year, which is certainly a delightful season, the woods abound with deer, partridges and quail, while the stubble fields furnish the Nimrod with all the prairie chickens he can carry in an ordinary sized wagon. Geese and ducks of the finest flavor frequent the lakes in immense flocks, and afford splendid sport. Occasionally you stumble upon a bear, but invalids are not very partial to Bruin as a general thing, and usually allow him to follow the bent of his inclination unmolested.

"The cost of living in this far off Western country is by no means as expensive as some would imagine. Board at the best hotels in St. Paul can be procured as cheap as at the East, and in the country towns one can live very comfortably for about five or six dollars a week. As winter sets in, the invalids all flock to the towns, where they can spend the season more agreeably than they can in the country. Such places as St. Paul, Minneapolis, St. Anthony, St. Cloud, Farrihault and Winona are crowded with them, and the citizens derive no little profit by the presence of such visitors. The pineries, which extend along the St. Croix river, and run as far north as Lake Superior, are much frequented by consumptives. A belief is prevalent here that the pine emits an odor which is peculiarly healing, and highly beneficial for invalids; hence it is no uncommon thing for small parties to take up their quarters in the wilderness, and spend the winter there with the numerous gangs of lumbermen engaged in felling trees and hauling logs to the banks of the neighboring creeks, with the view of floating them down the St. Croix in the Spring. Those who have the strength and courage to endure this wild mode of life generally experience the most beneficial effects, and in the Spring are enabled to return fat and hearty."

RAILROADS IN MINNESOTA.—1874.

	MILES.		MILES.
St. Paul and Pacific*.—Main Line	217	Minneapolis and St. Louis*	26
Upper Miss. Branch	76	Northern Pacific*.—(Duluth to	
Red River Branch*		Bismarck, Dak.)	448
St. Paul and Sioux City	270		
St. Paul and Stillwater	20	Total miles, finished	1,913
Milwaukee and St. Paul	324		
Chicago and St. Paul	105	Several other Railroads are projected to run from cities and towns on the Mississippi River; which, when completed, will afford great facilities for travelling, and the transportation of merchandise and agricultural products.	
Lake Superior and Mississippi	155		
Stillwater Branch	12		
Minneapolis Branch	13		
Hastings and Dakota*	74		
Winona and St. Peter	326		
Southern Minnesota	170		

* Unfinished.

Lake Superior and Mississippi Railroad and its Branches.

—This much needed connecting link between the navigable waters of the Upper Mississippi and the head of Lake Superior, 155 miles in length, is of the utmost importance to both Canada and the United States. Branch Roads run to Stillwater and Minneapolis. It is now completed, and affords an uninterrupted route of travel from the Gulf of Mexico to the Gulf of St. Lawrence. The distance from St. Paul to New Orleans, by water, is about 2,000 miles, and the distance from the head of Lake Superior to Quebec, *via* Toronto, passing through Georgian Bay, is about 1,400 miles—thus shortening the distance about 400 miles by avoiding Lakes Huron and Erie.

This international and inland route, passing through the most fertile region of the United States and portions of Canada, is destined to form the shortest and most desirable through line of transit from North America to Europe; carrying the mineral, agricultural, and other products of this wide extended region of country.

This great route, in connection with the Northern Pacific Railroad, for which a very liberal grant of land has been given by the United States Government, ought to encourage capitalists both in America and Europe to invest their surplus means so as to insure their early completion.

St. Paul to Fort Garry.

—Passengers going to Red River from St. Paul can go by way of the Main Line of the St. Paul and Pacific Railroad to Breckinridge, 217 miles, the present terminus of the road, and by team and steamboat from there to Moorhead and Fort Garry (300 miles). Total distance, 500 miles. Passengers going by way of St. Cloud can proceed by railroad to Melrose, 108 miles, and thence by stage to the Junction of the Northern Pacific Railroad at Glyndon. Then proceed by railroad and steamboat to Fort Garry. A Line of Steamers are now running on the Red River of the North, affording a speedy mode of conveyance from St. Paul and Duluth.

The completion of the Branch of the St. Paul and Pacific Railroad to Pembina will afford still greater facilities in order to reach Northern Minnesota and Manitoba.

Climate of the Upper Mississippi Valley.

—"The climate of the plains," says J. W. FOSTER, "exhibits some peculiarities deserving of notice,—such as the purity of the air; the cloudless skies during certain seasons, giving to the landscape such sharply-defined outlines; the dewless nights; the illusive phantoms of the mirage; and the feeling of vastness which impresses every beholder, as he stands on some high swell, and in every direction, sees the surface stretched out like a hemisphere.

"As in the distant north, there is a mingling of Spring and Summer, so here the Summer is protracted far into Autumn. This is the most delightful season of the year, characterized by an absence of severe rainstorms, with a cool, bracing atmosphere, so gratifying to the physical system that man exults 'in the intense consciousness' of his existence.

"That delicious season known as 'Indian Summer,' is often prolonged into December, when a calm, soft, hazy atmosphere fills the sky, through which, day after day, the sun, shorn of his beams, rises and sets like a globe of fire. This peculiarity is observed as far north as Lake Superior, but is more conspicuous and protracted in Iowa and Nebraska, but does not extend south, into the lower latitudes of the United States.

"Thus it will be seen that the Mississippi Valley possesses a great diversity of climate; and this is naturally to be inferred, when we consider that it extends through twenty degrees of latitude, and that its western rim, at many points, rises into the region of perpetual snow, so that the traveller may elect whether to breathe the pure and bracing air of the mountains, or the soft and balmy air wafted from the Tropics."

In reference to the Climatology of the Valley of the Red River of the North, and the surrounding region, Captain PALISER, accompanied by Dr. HECTOR, as geologist, in their report of the Saskatchewan country, and of the region lying between Lake Winnipeg and the Rocky Mountains, which are embraced in a warm belt extending southward, state there is an area of millions of acres of fertile soil, where the Winters are comparatively mild, and where the Hudson Bay Company have produced every crop grown in the north-western States. In the Pembina settlement, on the 49th parallel, we know that for years agriculture has been successfully prosecuted.

"In explanation of these phenomena, it may be said that the Rocky Mountains attain their highest elevation south of the 40th parallel, and as prolonged northward, they drop down and become merged in hills of moderate elevation. Through this depression, it is believed that the warm breath of the Pacific, brought by the southwesterly winds, flows, until it is met by the cooler currents, which prevail over the eastern slope of the great valley. The effect of this flow is to modify the vigor of the climate on the eastern slope of the Rocky Mountains, as far up even as Athabasca and the Assinniboine, and to render subservient to agriculture a region corresponding in latitude, on the Atlantic coast, with the inhospitable wastes of Labrador.

Lakes, Rivers and Water-Falls of Minnesota.

— This health-restoring region owes its purity of atmosphere, in a great degree, to the numerous lakes and lovely streams which abound in almost every section of the State. The Lakes in the vicinity of St. Paul and Minneapolis, as well as those on the line of the Lake Superior and Mississippi, and St. Paul and Pacific Railroads, are easy of access, and deserve a particular description.

LAKE COMO 3 miles north-west of St. Paul, is a most lovely sheet of water, surrounded for the most part by cultivated grounds. It is approached by a good carriage road, being very frequently visited during warm weather. Here are two or three public houses for the accommodation of visitors, where may be obtained row-boats and fishing tackle. Ten pound pickerel are not unfrequently taken from its waters.

LAKE PHALON, 3 miles north of St. Paul, on the line of the Lake Superior and Mississippi Railroad, is another lovely sheet of water from which is obtained a supply of pure water for the city. BALD EAGLE LAKE, a few miles beyond, is another attractive sheet of water. Here is afforded fine amusement for the angler or sportsman.

BASS LAKE, LAKE JOHANNA, VADNAIS LAKE, and several other small Lakes, lying from 6 to 10 miles north of St. Paul, are all worthy of a visit, as they abound with fish and wild game.

WHITE BEAR LAKE, 12 miles north of St. Paul, on the line of the Lake Superior and Mississippi Railroad, is a favorite place of resort for tourists seeking health and pleasure. This is also a most popular Summer drive for the citizens of St. Paul as well as strangers. No idea of the beauty of the Lake, with its graceful and shaded shores, can be given to one who has not witnessed its lovely features. Here are two or three well-kept hotels, and plenty of row and sail boats for the accommodation of visitors. There are lovely groves for pleasure or picnic parties adjoining the Lake.

A cluster of small Lakes, lying west of St. Paul and south of Minneapolis, emptying their pure waters into Minne-ha-ha Creek, are all lovely sheets of water. They are known by the names *Amelia, Mother, Rice, Calhoun, Harriet, Cedar, Christmas, Diamond,* and *Lake of the Isles.* They are all readily reached from St. Paul or Minneapolis by carriage roads. Lakes CALHOUN and HARRIET, 3 or 4 miles from Minneapolis, are the most frequented. Here are well-kept hotels, and several fine country residences.

CEDAR LAKE, 3 miles west of Minneapolis, lies on the line of the St. Paul and Pacific Railroad, and is easily approached by rail from St. Paul or Minneapolis. Here is a public house called the "*Invalid's Home.*"

CRYSTAL LAKE, TWIN LAKES, and other small bodies of water lie north-west of Minneapolis,—all affording delightful excursions during warm weather.

The above Lakes, in connection with *Lake Minnetonka* and the far-famed *Falls of Minne-ha-ha,* situated 5 miles below the Falls of St. Anthony, renders St. Paul and its vicinity, next to Quebec, one of the most interesting portions of America.

All these beauties of Nature, surrounded by a most delightful atmosphere, can be easily reached by rail or private carriage, starting from St. Paul or Minneapolis.

Head Waters of the Mississippi River.

—From a recent survey of the Upper Mississippi, it appears that steamboat navigation is afforded for upwards of six hundred miles above the Falls of St. Anthony, by means of three locks overcoming the Sauk Rapids, the Little Falls, and French Rapids, near the mouth of Rabbit river; terminating at Pokegoma Falls in about 47° 30′ N. lat. Above these falls boat navigation is continued to Leach Lake and Lake Winnebagoshish, one hundred and forty miles further; these being the two largest of the many Lakes from which this noble river takes its rise. About one hundred and fifty miles west of Lake Winnebagoshish is situated Itasca Lake, the head source of the Mississippi.

The river, from the Falls of St. Anthony to *Sauk Rapids*, 80 miles, was successfully navigated before the completion of the St. Paul and Pacific Railroad. There is a company formed to build a steamboat and barges, to be placed upon this reach of the river. From Sauk Rapids to Little Falls there is another navigable reach of water; with a dam and a lock, with a lift of fourteen feet, it would connect the lower with the upper reach. The river through the largest part of this reach, above Little Falls, has a good depth of water and clear channel. French Rapids, near the mouth of Rabbit River, are the most difficult to overcome.

FORT RIPLEY, CROW WING, AND BRAINERD, where crosses the Northern Pacific Railroad, are all landings and places of interest; the latter, no doubt, is destined to become an important railroad station, being situated about 100 miles west of Lake Superior.

There are very few islands between Crow Wing and Pine River, and from thence to Pokegoma, with the exception of Three-Mile Rapids, three miles below Pokegoma. There is no obstruction to steamboat navigation during the entire season, the river being deep and narrow, with a moderate current. There are several places where cut-offs could be made at very slight expense, that would save miles of distance round by the present channel. The entire distance from the Falls of St. Anthony to Pokegoma Falls, by the windings of the river, is estimated at six hundred and seventy-five miles.

Above *Pokegoma Falls*, navigation is afforded to the *Chippewa Agency*, on Leach Lake, and on other tributaries of the Mississippi, extending for 260 miles; thus, altogether, affording about 900 miles of navigation above the Falls of St. Anthony, extending through a deeply interesting section of country.

The numerous Lakes north and west of Crow Wing are described as being most beautiful sheets of water, abounding with fish of a fine flavor. *Otter Tail Lake*, about 50 miles west of Crow Wing, on the line of the Northern Pacific Railroad, is surrounded by a healthy and fruitful section of country, affording a delightful resort during warm weather.

It is said that a canal could be constructed to connect the large Lakes at the source of the Mississippi river with the Red River of the North, passing through Red Lake and Red Lake River—flowing into the latter stream. If this was accomplished, and locks constructed on the Mississippi, an inland navigation would be afforded from Lake Winnipeg to the Gulf of Mexico.

Trip to Pokegoma Falls.

THE HEAD OF NAVIGATION ON THE MISSISSIPPI RIVER.

To the Editors of the St. Paul Pioneer.
July, 1870.

As some of your readers of the PIONEER might be glad to know something more in regard to the first trip of the new steamer Pokegoma, from Little Falls to Pokegoma Falls, a distance of three hundred miles, I will give you a detailed account.

After the steamer left Little Falls, 115 miles above St. Paul, it proceeded along up the river to Belle Prairie. There the people turned out *en masse*, and came to the bank of the river to see the river king, the steamer POKEGOMA.

After leaving Belle Prairie, we passed along up the river through a beautiful country, not much settled. In a short time we arrived in sight of *Fort Ripley*, one of the most beautiful military outposts in the country. The old flag, the stars and stripes, was floating at masthead. As we neared the Fort the commanding officer came to the bank of the river with a six-pounder and fired a salute of five guns. On our arrival at CROW WING, 140 miles above St. Paul, the people assembled on the landing, and fired a salute with small arms, waved flags, and made many demonstrations of joy. On Friday night, the 24th day of June, the people of Crow Wing gave a ball, in honor of the event, at the Whipple House, which was free to the Captain and crew of the steamer.

The night passed pleasantly, and early in the morning, on Saturday, the 25th day of June, we started for the boat to get ready for our trip to Pokegoma Falls. Everything was soon ready, and we passed along up the river to French Rapids, through a country covered with a little white pine, some Norway and black pine, with a poor soil. Further up, towards the mouth of the Big Willow, the bottom land along the river is good and covered with rushes. Just below the mouth of the Big Willow, some two or three miles, is the Northern Pacific Railroad crossing. We found at the Railroad crossing three or four large tents, filled with stores belonging to the Northern Pacific Railroad Company.

On our arrival at Sandy Lake, the citizens and natives fired a salute, with small arms, and the old flag was displayed at the masthead. All seemed glad to see the river king, and many Indians came on board and inspected the boat closely, and said it was the largest canoe they ever saw.

After leaving Sandy Lake, going up the river about seven miles, we stopped to wood, where a gentleman by the name of Vergen was getting out square timber. Here we put on for wood all the score-blocks we could find. They were Norway pine, and burned well.

After getting through wooding we passed along up the river by the mouths of Swan river, Split-Hand and Prairie rivers. Six miles up the Prairie river is the finest iron mines in the world; not worked yet. About one mile above the mouth of Prairie river resides an Indian chief, by the name of Rabbit, who has adopted the habits and customs of civilization, living in a good log house, owning two yoke of oxen, and also farming quite extensively. He too hoisted the flag on the approach of the steamer. On Monday, the 27th day of June, at sunset, we arrived at the foot of Long Rapids, the boat landed, and Captain Charles R. Riggs went on shore with the line, and made fast the second steamboat that ever ascended the Mississippi river to this point.

On Tuesday, the 28th day of June, Captain Houghton and several others, with your correspondent, took a batteau and went up the river some three miles to visit the celebrated Falls of Pokegoma, situated 415 miles above St. Paul.

We found at the Falls a beautiful sheet of water, about sixty feet wide, pouring over a solid ledge of granite rock, at an angle of about 45 degrees. The fall at this place is about eleven feet, with the river falling fast below; so there is about twenty-two feet fall within ten rods. On the west side of the river is a heavy body of pine (white); on the east bank is hard wood timber, but it has been cut off by the claimants in days gone by, so there is about eight or ten acres of land cleared. We found on this cleared land, blueberries, red raspberries, strawberries, and wintergreen berries. After looking at the Falls, and eating what berries we wanted, some of us tried fishing, but without much success, as the fish did not feel inclined to bite. At about eleven o'clock A. M. we all embarked in our batteau and returned to the steamer; and, immediately after dinner, steam was got up, and the steamer was turned round and headed down stream, homeward bound. After two days and a half hard running we arrived at Crow Wing, all tired, and glad to see the city on the border of civilization.

The people living on the Mississippi river, and its tributaries above Little Falls, are anxiously looking forward to the time when they will ship all their grain, by boats, to the crossing of the Northern Pacific Railroad on the Mississippi river, thence by rail to Duluth, and then down the great Lakes to the sea-board.

The country on the Upper Mississippi river above Crow Wing, is a timbered country, rather low and marshy—adapted to raising grass—a fine stock country, on account of its immense rush beds or bottoms, with a fertile soil; vast iron mines; inexhaustible pineries, and white cedar swamps; and, with the water-power at Pokegoma—which is unsurpassed by any in the State—will, when properly drained and dyked, be as rich a country as Holland; and, ere long, it will be filled up and settled with the hardy sons of toil from Norway and Sweden, and be made to bud and blossom like the rose.

A ***** C. R ****.

RIVERS AND LAKES OF CENTRAL NORTH AMERICA.

The River and Lake System of that portion of North America, comprised within the United States and Canada, or the British Possessions, is of the most grand and useful character. It embraces *three* distinct basins or valleys; that of the Mississippi, the St. Lawrence, and the Saskatchewan and Red River of the North. The waters flowing through these different valleys, all take their rise in the immediate neighborhood of each other, between the 46th and 48th parallels of latitude, —the table-lands in Northern Minnesota, elevated from 1,200 to 1,500 feet above the ocean, being the fountainhead of all the above gigantic streams, if we except the Saskatchewan.

The Missouri river—a tributary of the Mississippi—rises far to the west, in the region of the Rocky Mountains, uniting with its mother-stream midway in its flow to the Gulf of Mexico. The above stream drains an immense extent of country, forming, in part, a system of itself.

The *two* great streams of commerce, the Mississippi and the St. Lawrence, —including the Great Lakes, now united by a *band of iron*—afford the longest line of water communication on the globe—altogether amounting to upwards of 16,000 miles of inland navigation, extending from the Gulf of Mexico to the Gulf of St. Lawrence, with numerous tributaries running from the Alleghanies on the east and the Rocky Mountains on the west.

This grand River and Lake System combined, lies almost wholly within the Temperate Zone, embracing an area of over 2,500,000 square miles,—of which 2,000,000 square miles may be said to belong to the Mississippi Valley, extending through twenty-three degrees of latitude and thirty degrees of longitude. Such an extensive region of fertile land is not to found in any other part of the habitable globe.

The Red River of the North, the Saskatchewan and other streams, flowing into Lakes Manitoba and Winnipeg, drain a large region of country, the waters finding their outlet through Nelson River into Hudson Bay. This great valley, running through ten degrees of latitude and twenty degrees of longitude, possesses a *Siberian Summer*,—bringing forth cereals and vegetables in great abundance for the sustenance of man and beast.

The *Mississippi* takes its rise in Lake Itasca, and other small Lakes lying near the 47th parallel. At first its course is circuitous, running eastward through numerous lakes and expansions, when its course inclines southward to the Falls of St. Anthony, affording stretches of navigable waters for several hundred miles. From below the above Falls, or the mouth of the Minnesota river, at Fort Snelling, it affords uninterrupted navigation to the Gulf of Mexico, a distance of upwards of two thousand miles.

Prof. FOSTER says,—"the sources of the Upper Mississippi are among the great forests of conifers, white birches and aspens—subarctic types—which continue north, but dwarfed in stature, until the limits of arborescent vegetation are reached; and its mouth is in the region of the orange, the magnolia, and even the palm—thus approaching the verge of tropical forms. A navigable river, flowing through a region so diversified in climate and productions, can not but become the source of a vast inland commerce."

The *Upper Mississippi*, above its confluence with the Missouri river, for eight hundred miles, is a grand and lovely stream; the channel being about half a mile in width, with numerous islands which spread out the bottom lands often for several miles. Above Dubuque the famed bluffs appear, and, alternately, on both sides of the stream until the mouth of the St. Croix is reached.

Lake Pepin is an expansion of the water, being about twenty-five miles in length and three miles wide. Here is witnessed some of the most grand river scenery imaginable, while the surrounding atmosphere is of the most bracing and healthy character.

The *Lower Mississippi*, "when in flood, extends to a width of twenty or thirty miles, and the surplus waters find their way to the ocean through almost interminable swamps. The ordinary channel, from a half to one mile in width, is marked by an outline of woods, presenting a great sameness of an uninteresting character. As the flood recedes it leaves behind, in the bottom lands, a sediment as fine and fertilizing as the mud of the Nile."

St. Louis River, forming the head waters of Lake Superior, takes its rise on about the 48th parallel of latitude, almost due north of its outlet into Superior Bay. It drains a valley distinct from the above Lake, passing through a wilderness region of country above the Dalles of the River, near Fond du Lac, here having a fall of about 400 feet within the distance of eight or ten miles.

Lakes *Superior, Huron, Erie* and *Ontario,* form a succession of *basins* through which the accumulated waters flow, being increased by Lake Michigan. The Rivers uniting this great chain of Lakes are the St. Mary, St. Clair, Detroit and Niagara. The *St. Lawrence,* proper, commencing at the outlet of Lake Ontario, where commences the "Thousand Islands." On its onward course toward tide-waters, above Quebec, it has a descent of 234 feet, while Lake Superior stands elevated 600 feet above the ocean. In an extended view it may be said to be 2,000 miles in length, before discharging its waters into the Gulf of St. Lawrence.

This noble stream, unlike the Mississippi, never overflows its banks, neither does it pass through any stagnant waters or swamps, but flows onward as a pure and healthy stream, widening to one hundred miles or upwards before passing the island of Anticosti.

The St. Lawrence if thrown open to the free commerce of the world, from the 45th parallel to the ocean, will soon form the main artery of trade between America and Europe.

The *Red River of the North* takes its rise in Lake Travers, 46° N. lat., and flows northward into Lake Winnipeg, a distance of about 400 miles, passing through a rich agricultural section of country. The *Saskatchewan* rises in the foot-hills of the Rocky Mountains of British America, and flows eastward into Lake Winnipeg. This great valley is rich in minerals, with a fruitful soil, susceptible of sustaining a dense population.

The separate system of rivers flowing into Lake Winnipeg, and through Nelson River into Hudson Bay, are, ere .long, destined to be alive with commerce, finding an outlet to the Ocean through Hudson's and Davis' Straits.

RAIN-FALL IN THE UPPER LAKE REGION.

From Rain-Charts recently issued by the Smithsonian Institution, it appears that during the *Winter months,* (December, January and February,) that from 4 to 6 inches of water, (rain or snow) falls in the Lower Peninsula of Michigan; from 6 to 10 inches falls in the Upper Peninsula—the largest amount between Marquette and Keweenaw Point—while but 4 inches falls between Bayfield and Duluth, situated at the west end of Lake Superior, and but 2 inches of moisture falls between St. Paul and Crow Wing, on the Upper Mississippi.

During the *Summer months,* (June, July and August,) 10 inches of rain falls in the Lower Peninsula of Michigan; 8 inches in the Upper Peninsula, near Keweenaw Point, and 10 inches between Bayfield and Duluth, while 12 inches falls in the vicinity of St. Paul.

During the *Year,* from 28 to 30 inches of rain falls in the Lower Peninsula of Michigan; 32 inches in the Upper Peninsula, extending south to Green Bay, Wisconsin, while only from 24 to 28 inches of rain falls during the twelve months in Northern Minnesota, west of Lake Superior.

The above uneven distribution of moisture, in connection with the *Isotheral* and *Isothermal Lines,* when properly understood, afford the key to the climatic influences operating in the Lake Region and Upper Mississippi Valley.

LAKE AND RIVER FISHING.

TROUT FISHING.—Trout, which are usually called brook-trout, are caught in numerous running streams in the northern part of the United States and in Canada. They are also taken in several pure fresh-water lakes where the water is very cold. They generally have red and yellow spots on their sides— concave tail, and belly tinged with orange red—varying in color, appearance and size, with the quality of the soil pertaining to the streams they inhabit. They have large eyes, a wide mouth, sharp teeth, and scaleless skin. The usual weight of brook-trout is from one-half to four pounds, and said to be taken even larger on the North Shore of Lake Superior. The best time for taking them is in the Spring and Summer months.

An experienced angler says:—"You can hook trout in several ways. Some prefer fly fishing, and this is the most interesting mode in Summer. The rod to be used should be light, and the line made of hair, or silk and grass. The fly should be placed on a length of gut, or a single light hair. Do not fish with your back to the sun. Stand as far from the stream as circumstances will allow—always throw your line from you—never whip it out. Fly fishing is only suitable for pleasant weather. The best time of the day is early in the morning or just at sunset. The line should be about half as long as the rod. It should be thrown up stream, and let the fly gradually float down, and, if possible, fall into the eddies where the fish is apt to retreat in case of alarm. Let your line fall into the stream lightly and naturally, and when you raise it, do so gently and by degrees.

"In fishing with the fly, only a small part of the line is allowed to be in the water. If you stand on the bank of the stream, throw your line as far up as possible, as you cannot expect to catch a trout opposite or below where you are standing. If bushes intervene between you and the stream, (which is all the better), do not rustle them or make a noise.

"Worm fishing for trout is practised with similar caution. After a rain, when the water of the brook is a little riley, you can catch trout by this mode sometimes very rapidly. A single split shot will generally be enough to sink your line. The rod should be a bamboo, 16 or 20 feet long, and the line shorter than the rod. When the fish takes the bait, do not let him run with it, but keep a steady hand. Do not jerk, but play gradually with him. If the day be clear, and the stream shallow, the best way is to wade up stream cautiously, throwing your line far up, and let it come gradually towards you. The fish always heads up stream, and you should not fail to remember if he once sees you he vanishes and your sport is spoiled.

"The fin of a trout, or other small fish, is sometimes used as a bait for trout with good success. It is dropped and roved, as with a minnow or fly."

———

MACKINAC TROUT, OR SALMON TROUT.—This is a fish of fine flavor, and some people consider it almost equal to salmon. The flesh is reddish, and hence it is often called the salmon trout. The color is dark, or dusky grey; back and sides sprinkled with spots somewhat lighter; belly light brown or cream color; the teeth, gums and roof of the mouth having a light purple tinge. It usually averages from two to four feet in length, and inhabits all the great Lakes of the West—particularly Lakes Huron, Michigan and Superior.

Sportsmen always take the salmon trout by trolling, or by set lines. For the latter they use the largest size cod-hooks and cod-lines; and for bait, bits of Lake herring or white fish are all that is necessary. In trolling, both tackle and bait are different, the minnow being chiefly used for bait. In trolling use swivels freely to keep your line from getting tangled. When you have a bite, slack the line a little to allow the fish to gorge his bait, then begin to pull steadily, after arranging everything in a proper manner to enable you to play your fish. Trolling for salmon trout is most excellent sport, and amateurs take great delight in this kind of fishing. To insure success you should have an experienced hand to row your boat in trolling for this fish. We have known good success on the Lakes where a small tug boat has been used for this purpose.

THE MUSKELLUNGE, OR LAKE PIKE.—This game fish, found in the Upper Lakes and the St. Lawrence River, grows from one to three feet in length, according to the depth and breadth of the waters that he inhabits. "Built like a pike, he is of a deep greenish brown color, dark back, and pale sides with greenish spots. In fishing for the smaller sizes, your tackle should be similar to that used for pickerel; but for large ones, you want a good sized cod-line, with a cod-hook to match. He will bite greedily at various baits—a bit of fish, a slice of pork, a bundle of worms, or chicken offal, a small fish, or a frog. It requires a good deal of care, caution and physical exertion to land him. He is a most beautiful game fish, and is the best eating fish, next to the white fish, that inhabits the Lakes.

"The Muskellunge, (*long face of the French,*) abound in Niagara River and the St. Lawrence, in the vicinity of the 'Thousand Islands.' He is an enormous pike, with the lower projecting jaw armed with needle teeth clear into the throat, ranging from five to forty pounds weight, agile as lightning and a perfect water-tiger among the smaller fishes. He is game until death—a sharp customer to handle; but a more delicate fish, flesh white as snow and savory as an oyster, well boiled, and served upon the dinner-table with pepper sauces, does not exist."

For *Fish of the Upper Lakes,* see page 19.

PORTAGE ROUTE FROM LAKE SUPERIOR TO LAKE WINNIPEG.

STARTING FROM FORT WILLIAM, C. W.

KAMINISTAQUOIAH RIVER, emptying into Thunder Bay of Lake Superior, forms the west boundary of Canada proper; to the north and west lies the extensive region or country known as the *Hudson Bay Company's Territory.* Here commences the great *Portage Road* to Rainy Lake, Lake of the Woods, and the Red River settlement; also, to Lake Winnipeg, Norway House, and York Factory, situated on Hudson Bay. At the mouth of the Kaministaquoiah stands *Fort William.* "The banks of the river average in height from eight to twenty feet; the soil is alluvial and very rich. The vegetation all along its banks is remarkably thrifty and luxuriant in its appearance. The land is well timbered; there are found in great abundance, the fir-tree, birch, tamarack, poplar, elm, and the spruce, There is also white pine, but not in great plenty. Wild hops and peas are found in abundance, and some bushes and other flowering shrubs, in many places cover the banks down to the very margin of the river, adorning them with beauty, and often filling the air with fragrance. The land on this river up to the Mountain Portage (32 miles), and for a long way back, is unsurpassed in richness and beauty by any lands in British America."

The *Mountain Fall,* situated on this stream, is thus described: "We had great difficulty in finding it at first, but, guided by its thundering roar, through such a thicket of brush, thorns and briars, as I never before thought of, we reached the spot from whence it was visible. The whole river plunged in one broad white sheet, through a space not more than fifty feet wide, and over a precipice higher, by many feet, than the *Niagara* Falls. The concave sheet comes together about three-fourths of the way to the bottom, from whence the spray springs high into the air, bedewing and whitening the precipitous and wild looking crags with which the fall is composed, and clothing with drapery of foam the gloomy pines, that hang about the clefts and fissures of the rocks. The falls and the whole surrounding scenery, for sublimity, wildness, and novel grandeur, exceeds any thing of the kind I ever saw."—*Rev. J. Ryerson's Tour.*

The danger of navigating these mountain streams, in a birch canoe, is greater than many would expect who had never witnessed the force of the current sometimes encountered. Mr. Ryerson remarks: "During the day we passed a large number of strong and some dangerous rapids. Several times the canoe, in spite of the most strenuous exertions of the men, was driven back, such was the violence of the currents. On one occasion such was the force of the stream, that though four strong men were holding the rope, it was wrenched out of their hands in an instant, and we were hurled down the rapids with violent speed, at the mercy of the foaming waves and irresistible torrent, until fortunately in safety we reached an eddy below." (*See Engraving.*)

DOG LAKE is an expansion of the river, distant by its winding course, 76 miles from its mouth. Other lakes and expansions of streams are passed on the route westward.

"The SAVAN, or PRAIRIE PORTAGE, 120 miles from Fort William, by portage route, forms the height of land between Lake Superior and the waters falling into Lake Winnipeg; it is between three and four miles long, and a continuous cedar swamp from one end to the other, and is therefore very properly named the *Savan* or *Swamp*

Portage. It lies seven or eight hundred feet above Lakes Superior and Winnipeg, and 1,483 feet above the sea."

The SAVAN RIVER, which is first formed by the waters of the Swamp, enters into the *Lac Du Mille*, or the Lake of Thousands, so called because of the innumerable islands which are in it. This lake is comparatively narrow, being sixty or seventy miles in length.

The *River Du Mille*, the outlet of the Lake, is a precipitous stream, whereon are several portages, before entering into Lac La Pluie, distant 350 miles from Fort William.

RAINY LAKE, or *Lac la Pluie*, through which runs the boundary between the United States and Canada, is a most beautiful sheet of water; it is forty-eight miles long, and averages about ten miles in breadth. It receives the waters flowing westward from the dividing ridge separating the waters flowing into Lake Superior.

RAINY LAKE RIVER, the outlet of the lake of the same name, is a magnificent stream of water; it has a rapid current and averages about a quarter of a mile in width; its banks are covered with the richest foliage of every hue; the trees in the vicinity are large and varied, consisting of ash, cedar, poplar, oak, birch, and red and white pines; also an abundance of flowers of gaudy and variegated colors. The climate is also very fine, with a rich soil, and well calculated to sustain a dense population as any part of Canada.

The LAKE OF THE WOODS, or *Lac Du Bois*, 68 miles in length, and from fifteen to twenty-five miles wide, is a splendid sheet of water, dotted all over with hundreds of beautiful islands, many of which are covered with a heavy and luxuriant foliage. Warm and frequent showers occur here in May and June bringing forth vegetation at a rapid rate, although situated on the 49th degree of north latitude, from whence extends *westward* to the Pacific

PULLING A CANOE UP THE RAPIDS.

Ocean, the boundary line between the United States and Canada.

"There is nothing, I think, better calculated to awaken the more solemn feelings of our nature, than these noble lakes studded with innumerable islets, suddenly bursting on the traveller's view as he emerges from the sombre forest rivers of the American wilderness. The clear, unruffled water, stretching out on the horizon; here intersecting the heavy and luxuriant foliage of an hundred woody isles; or reflecting the wood-clad mountains on its margin, clothed in all the variegated hues of autumn; and there glittering with dazzling brilliancy in the bright rays of the evening sun, or rippling among the reeds and rushes of some shallow bay, where hundreds of wild fowl chatter as they feed with varied cry, rendering more apparent, rather than disturbing the solemn stillness of the scene: all tend to raise the soul from nature up to nature's God, and remind one of the beautiful passage of Scripture, 'O Lord, how marvellous are thy works, in wisdom hast thou made them all; the earth is full of thy riches.'" —*Ballantyne.*

The WINNIPEG RIVER, the outlet of the Lake of the Woods, is a rapid stream, of large size, falling into Winnipeg Lake, 3 miles below *Fort Alexander*, one of the Hudson Bay Company's Posts. A great number of Indians resort to the Fort every year, besides a number of families who are residents in the vicinity, here being one of their favorite haunts.

Rev. Mr. Ryerson remarks :—"The scenery for many miles around is strikingly beautiful. The climate for Hudson's Bay Territory is here remarkably fine and salubrious, the land amazingly rich and productive. The water in Lakes Lac La Pluie, Lac Du Bois, Winnipeg, &c., is not deep, and because of their wide surface and great shallowness, during the summer season, they become exceedingly warm; this has a wonderful effect on the temperature of the atmosphere in the adjacent neighborhoods, and no doubt makes the great difference in the climate (or at least is one of the principal causes of it), in these parts, to the climate and vegetable productions in the neighborhood of Lake Superior, near Fort William. They grow spring wheat here to perfection, and vegetation is rapid, luxuriant, and comes to maturity before frosts occur."

The whole region of country surrounding Lake Winnipeg, the Red River country, as well as the Assiniboine and Saskatchewan country, are all sooner or later destined to sustain a vigorous and dense population.

LAKE WINNIPEG,

Situated between 50° and 55° north latitude, is about 300 miles long, and in several parts more than 50 miles broad; having an estimated area of 8,500 square miles.* Lake Winnipeg receives the waters of numerous rivers, which, in the aggregate, drain an area of about 400,000 square miles. The *Saskatchewan* (the river that runs fast) is its most important tributary. The Assiniboine, the Red River of the North, and Winnipeg River are its other largest tributaries, altogether discharging an immense amount of water into this great inland lake. It is elevated about 700 feet above Hudson Bay, and discharges its surplus waters through *Nelson River*, a large and magnificent stream, which like the St. Lawrence is filled with islands and numerous rapids,

---

*LAKE BAIKAL, the most extensive body of fresh water on the Eastern Continent, situated in Southern Siberia, between lat. 51° and 55° north, is about 370 miles in length, 45 miles average width, and about 900 miles in circuit; being somewhat larger than Lake Winnipeg in area. Its depth in some places is very great, being in part surrounded by high mountains. The *Yenisei*, its outlet, flows north into the Arctic Ocean.

preventing navigation entirely below Cross Lake.

Lakes *Manitobah* and *Winnipego-sis*, united, are nearly of the same length as Winnipeg, lying 40 or 50 miles westward. Nearly the whole country between Lake Winnipeg and its western rivals is occupied by smaller lakes, so that between the valley of the Assiniboine and the eastern shore of Winnipeg fully one-third is under water. These lakes, both large and small, are shallow, and in the same water area show much uniformity in depth and coast line.

Lakes in the Valley of the Saskatchewan.

	Length in miles.	Breadth in miles.	Elevation in feet.	Area in m's.
Winnipeg, ...	280	57	628	8,500
Manitobah, ...	122	24	670	2,000
Winnipego-sis,..	120	27	692	2,000
St. Martin, ...	30	16	655	350
Cedar, ...	30	25	688	350
Dauphin,........	21	12	700	200

All the smaller lakes lie west of Lake Winnipeg, which receives their surplus waters; the whole volume, with the large streams, flowing into *Nelson River*, discharges into Hudson Bay, near York Factory, in 57° north latitude. The navigation of the latter stream is interrupted by falls and rapids, having a descent of 628 feet in its course of about 350 miles.

"The climate in the region of the above lakes and the Red River Settlement will compare not unfavorably with that of Kingston and Toronto, Canada West. The Spring generally opens somewhat earlier, but owing to the proximity of Lake Winnipeg which is late of breaking up, the weather is always variable until the middle of May. The slightest breeze from the north or northwest, blowing over the frozen surface of that inland sea, has an immediate effect on the temperature during the Spring months. On the other hand, the Fall is generally open, with mild, dry, and pleasant weather."

Red River of the North.

This interesting section of country being closely connected with the Upper Lakes, and attracting much attention at the present time, we subjoin the following extract from "MINNESOTA AND DACOTA," by C. C. Andrews:

"It is common to say that settlements have not been extended, beyond Crow Wing, Minnesota. This is only technically true. A few facts in regard to the people who live four or five hundred miles to the north will best illustrate the nature of the climate and its adaptedness to agriculture.

"There is a settlement at *Pembina*, near the 49th parallel of latitude, where the dividing line between British America and the United States crosses the Red River of the North. Pembina is said to have about 600 inhabitants. It is situated on the Pembina River. It is an Indian-French word meaning '*Cranberry*.' Men live there who were born there, and it is in fact an old settlement. It was founded by British subjects, who thought they had located on British soil. The greater part of its inhabitants are half-breeds, who earn a comfortable livelihood in fur-hunting and farming. It is 460 miles northwest of St. Paul, and 330 miles distant from Crow Wing. Notwithstanding the distance, there is considerable communication between the two places. West of Pembina, about thirty miles, is a settlement called *St. Joseph*, situated near a large mythological body of water called *Miniwakin*, or Devil's Lake.

"Now let me say something about this RED RIVER of the North, for it is beginning to be a great feature in this upper country. It runs north and empties into Lake Winnipeg, which connects with Hudson Bay by Nelson River. It is a muddy and sluggish stream, navigable to the mouth of the Sioux Wood River for vessels of three feet draught for four months in the year, so that the extent of its navigation within Minnesota alone (between Pembina and the mouth of Sioux Wood River) is 400 miles. Buffaloes still feed on its western banks. Its tributaries are numerous and copious, abounding with the choicest kind of game, and skirted with a various and beautiful foliage. It cannot be many years before this magnificent valley (together with the Saskatchewan) shall pour its products into our markets, and be the theatre of a busy and genial life.

"*Red River Settlement* is seventy miles north of Pembina, and lies on both sides of the river. Its population is estimated at 10,000 souls. It owes its origin and growth to the enterprise and success of the Hudson Bay Company. Many of the settlers came from Scotland, but the most were from Canada. They speak English and Canadian French. The English style of society is well kept up, whether we regard the Church with its bishop, the trader with his wine-cellar, the scholar with his library, the officer with his sinecure, or their paper currency. The great business of the settlement, of course, is the fur traffic.

"An immense amount of Buffalo skins is taken in summer and autumn, while in the winter smaller but more valuable furs are procured. The Indians also enlist in the hunts; and it is estimated that upward of $200,000 worth of furs are annually taken from our territory and sold to the Hudson Bay Company. It is high time indeed that a military post should be established somewhere on Red River by our government.

"The Hudson Bay Company is now a powerful monopoly. Not·so magnificent and potent as the East India Company, it is still a powerful combination, showering opulence on its members, and reflecting a peculiar feature in the strength and grandeur of the British empire—a power which, to use the eloquent language of Daniel Webster, 'has dotted over the whole surface of the globe with her possessions and military posts, whose morning drum-beat following the sun, and keeping company with the hours, circles the earth daily with one continuous and unbroken strain of martial music.' The company is growing richer every year, and its jurisdiction and its lands will soon find an availability never dreamed of by its founders, unless, as may possibly happen, *popular sovereignty steps in to grasp the fruits* of its long apprenticeship."

The Charter of the Hudson Bay Company expired, by its own limitation, in 1860, and the question of annexing this vast domain to Canada, or forming a separate province, is now deeply agitating the British public, both in Canada and in the mother country.

TABLE OF DISTANCES,

From Fort William, SITUATED AT THE MOUTH OF THE KAMISTAQUOIAH RIVER, **to Fort Alexander,** AT THE HEAD OF LAKE WINNIPEG.

		Miles.
FORT WILLIAM..		0
Parapliue Portage...		25
(8 Portages)		
Dog Portage	51	76
(5 Portages)		
Savan or Swamp Portage*.....................................	54	130
Thousand Islands Lake.......................................	57	187
(2 Portages)		
Sturgeon Lake...	71	258
(4 Portages)		
Lac La Croix....	25	283
(5 Portages)		
Rainy Lake..	40	323
Rainy Lake River..	38	361
Lake of the Woods...	83	444
Rat Portage...	68	512
FORT ALEXANDER..	125	637

From Fort Alexander to For t Garry

OR RED RIVER SETTLEMENT, BY WATER.

		Miles.
To Pointe de Grand Marais.....................................		24
" Red River Beacon...	25	49
" Lower Fort...	23	72
" FORT GARRY..	24	96

From FORT ALEXANDER to NORWAY HOUSE, passing through Lake Winnipeg, 300 miles.

From NORWAY HOUSE to YORK FACTORY, passing through Oxford Lake and Hayes River, 400 miles.

* Summit, elevated 840 feet above Lake Superior.

THE NEW NORTH-WEST.

EXCURSION FROM DULUTH TO THE RED RIVER OF THE NORTH AND THE UPPER MISSOURI RIVER,

Via THE NORTHERN PACIFIC RAILROAD.

ON leaving DULUTH for MOORHEAD, Minn., situated on the east bank of the Red River of the North, or BISMARCK, on the Upper Missouri River, you pass Fond du Lac, near the Dalles of the St. Louis River, and proceed to THOMSON, 23 miles. Then cross the St. Louis River to the JUNCTION, 1 mile, where commences the NORTHERN PACIFIC RAILROAD — the first twenty-four miles of railroad track being over the *Lake Superior and Mississippi Railroad*, 155 miles in length, extending to St. Paul.

Leaving the *Junction*, elevated 400 feet above Lake Superior, bound for the Red River, or farther westward, you pass over a fine road-bed, with an ascending grade for a number of miles. A small growth of pine and birch is encountered, with some burnt timber, until you reach

NORMAN, Carlton Co., 33 miles from Duluth. This is a small settlement, at present surrounded by some Indian wigwams. A small growth of burnt timber is next passed, then small lakes on both sides of the railroad.

LONG SWAMP is surrounded by a stunted growth of pines. Here piles are driven 50 or 60 feet deep, to obtain a foundation for the road-bed.

ISLAND LAKE, 46 miles. Here are several small lakes, surrounded by a small growth of trees. Soon you pass the divide of waters running south-east into Kettle River and the St. Croix, and other waters running northward into tributaries of the Mississippi River.

SICOTTE'S, 58 miles from Duluth. Near this station is a tamarack swamp, into which an engine was lately tumbled by running off the track. It soon sank out of sight, and was brought to the surface by great labor, showing the treacherous character of these sloughs. Near by we passed a solitary "Lo," who was standing on a pile of wood smoking a pipe, while apparently looking with wonder on the passing train, being a fit representative of the surrounding country. The wild game here is said to consist of deer, bears, mink, otter, and muskrat, which are mostly taken by Indians during the fall and winter months.

KIMBERLEY'S, or *Rice River*, is next passed, 76 miles from Duluth.

AIKEN, Aiken Co., 88 miles, is situated on Mud River, emptying into the Mississippi about a quarter of a mile north. Here is a depot building, a post-office, store, and a few dwellings, with several Indian wigwams. This is a lumber region, the pine logs being floated down the Mississippi River to St. Anthony Falls.

WITHINGTON, 98 miles, is situated on Reno Lake. Here maple and oak trees abound of a fair growth. The railroad thus far is apparently constructed in the most substantial manner, costing about $20,000 a mile from the Junction to the Mississippi River, 91 miles. Beautiful lakes are next passed, abounding with wild ducks and geese during the spring and fall months. A few scattered pines are passed, together with birch and poplar trees, as you approach the "Father of Waters," here a comparatively small stream.

Brainerd, 115 miles from Duluth, the capital of Crow Wing County, is favorably situated on the east bank of the Mississippi, 1200 feet above the Gulf of Mexico, on a beautiful elevation, 50 or 60 feet above the river. It contains a court-house and jail, 1 bank, 3 churches, 6 hotels or taverns, 12 stores, 1 printing-office, 2 steam saw-mills, 1 planing-mill, 1 iron foundry, 1 sash, door and blind factory, a brewery and brick-yard, a portable house manufactory, and an extensive machine shop and roundhouse, belonging to the Northern Pacific Railroad. Here is also a reception house for emigrants and a hospital, erected by the above Company. The railroad bridge at this place is a substantial structure 600 feet in length. The river is navigable for a small class of steamers for 200 miles above to the *Pokegema Falls,* and 30 miles below to the Little Falls, passing Crow Wing and Fort Ripley.

A Branch Railroad is being constructed to extend from Sauk Rapids to this place, running on the east side of the Mississippi River, forming a direct railroad line between St. Paul and Brainerd, making it an important railroad station, where passengers now stop daily to dine when going east or west on the trains.

Climate of Northern Minnesota.

In speaking of the peculiar climate of *Fort Ripley,* situated about 20 miles south of Brainerd, the surgeon of the post remarks: "This section of Minnesota is, no doubt, influenced by cold currents of air descending from Hudson Bay and the Arctic regions, while the western and more southern portions of the State are, no doubt, favored by a climatic influence proceeding from the Pacific coast, across the Rocky Mountains, in British America, *hence the favorable climate of the Red River country of the North.*

"Different kinds of pines, oaks, and birch constitute the prevailing forest growth of this region. The sugar maple abounds in some places. The chestnut, walnut, and beech are unknown, as is every species of fruit trees, wild or cultivated. The soil is generally a sandy alluvium. The land, at least when first cultivated, is more productive than might be supposed, being what farmers term 'warm,' and adapted to the short summers. The average depth of snow during winter is from two to three feet, which lies for about five months,— from November to April.

"The phenomena of spring, when once begun, often progresses with great rapidity; and from the climate of winter, the region sometimes seems to pass at once into that of mid-summer. Wild strawberries, which are found here in great abundance, ripen from the 20th to the last of June. During the summer months, and September and October, the weather is generally clear and delightful, this whole region being favored with a healthy atmosphere."

Snow Line — Rainfall.

The "Maximum Snow Line," the St. Paul (Minn.) *Pioneer* says: "During the past season (1872–73) the centre of precipitation has corresponded nearly with the north boundary of Iowa and Nebraska. Southern Minnesota has had more snow than the central part of the State, and that more than the north end, with the single exception of parts bordering on Lake Superior, or within the influence of Lake Winnipeg. The railways south have been often, and much worse, blockaded than those north of us. Indeed, the North Pacific Railroad has been open and in operation for weeks together, while lines from Chicago eastward were completely buried. The reason of all this is that the great air currents which carry the water of the sea to all this section, during the severity of the winter, in coming up from the south unload their burden before, or by the time, they reach this latitude; and as they move on northward nearly every particle of moisture is wrung out of them. Our cold winds are almost always dry. The extreme North has more snowless territory than the South has rainless."

Starting from the Gulf of Mexico, west of the Mississippi River, in Louisiana, 18 inches of rain falls during the *winter months*, 16 inches in Southern Arkansas, 12 inches in Northern Arkansas, 8 inches in Missouri, 6 inches in Iowa, 4 inches in Southern Minnesota, and 2 inches in Northern Minnesota and in the Valley of the Red River of the North; in the northern region the moisture falling as snow.

GULL RIVER, 7 miles, is next crossed, and Chippewa Prairie is entered and passed.

PILLAGER STATION is 12 miles from Brainerd. The Pillager Indians are a small band residing some distance north of this station. It was near here that *Hole in the Day*, a celebrated Indian Chief, was assassinated by some of his own people, in 1868.

Crow Wing River is next seen, on the south of the railroad track, flowing from the west, and a small prairie passed, covered with grass and flowers.

MOTLEY, 22 miles from Brainerd, is a station near where Crow Wing River is crossed. Here is a settlement surrounded by a small growth of pines and birch trees. Wet prairie, with dwarfed willows and some good grass, is next passed.

Small lakes, in part surrounded by wet prairie, is also passed, while maple and bass-wood is seen on the ridges.

HAYDEN, 28 miles, is another small settlement. Pine trees of middling size, with some burr oak, is passed. Here two deer were seen near the railroad track. Soon pass the divide between the waters that flow east into the Mississippi and west into the Red River of the North.

ALDRICH, 36 miles, is surrounded by good land, with oak and birch trees. A rolling prairie is next passed, interspersed with clumps of small trees and some good grass land, where flowers of different colors abound.

WADENA STATION, Wadena Co., 46 miles from Brainerd, is surrounded by a rolling prairie. Many permanent settlements and cultivated fields begin to appear.

LEAF RIVER STATION, 51 miles. Here is a natural meadow, producing good hay. A number of farm-houses are seen on the prairie. Pass a stretch of land covered with oak, bass-wood, birch, and maple. Here the summer days become warm, ranging from 70° to 90° Fahrenheit. Natural meadows and small lakes are seen, with some good upland, producing corn, wheat, and vegetables in abundance.

A fine prairie of about 15 miles in length is next reached, being in part settled by German families. Here they are preparing to plant forest trees. Soon cross Otter Tail River, which was full of saw logs, and a steam saw-mill in operation. A charming prospect is here afforded of the surrounding country, the prairie being covered with grass and flowers of different hues.

PERHAM, 70 miles from Brainerd. This is a small village, surrounded by rich prairie land, and lying almost as level as a barn floor. A number of farm-houses are seen in the distance. Pass an extensive wheat field, owned

17

and worked by a German woman and her two daughters. A rolling prairie with clumps of trees is next passed as you proceed westward. Here are corn and wheat fields inclosed by fences.

For about one hundred miles east of Moorhead, on the Red River, and for several hundred miles north and south, is said to be a productive prairie region, while west of the river the region is vastly increased, extending north-west into the Valley of the Saskatchawan.

HOBART, 81 miles, is a small village, surrounded by lovely lakes and a rolling prairie, which is very fertile, producing corn, oats, wheat, and vegetables of almost every variety. A field was passed with corn standing 8 or 10 feet high, while cattle are seen feeding in rich natural meadows, affording a most pleasing prospect. A growth of hard wood is passed, with some tamarack, before coming to *Detroit Lake*, an extensive and beautiful sheet of water, surrounded by finely-wooded banks. Pike, pickerel, black-bass, and other fish, besides wild game, are here taken in large quantities.

FRAZEE CITY is a new settlement, about one mile from Hobart.

Detroit, capital of Becker Co., Minn., 92 miles from Brainerd and 47 miles from Moorhead, is a flourishing town, settled mostly by a colony of New Englanders, numbering some 400 or 500 souls. Good farming lands surround this favored place, no doubt, soon destined to raise into importance as a centre for the purchase and shipment of the products of the adjacent country.

OAK LAKE, 96 miles, is another fine body of water, lying two miles north of the railroad. Here is a settlement surrounded by a rolling prairie, with several small lakes in view.

AUDUBON, 99 miles, is situated on the side of a beautiful lake, surrounded by prairie and woodland. This place is on the line of northern settlements at the present time; beyond is an Indian Reservation of considerable extent. Here are to be seen productive corn and wheat fields. Wild ducks and prairie chickens are numerous.

LAKE PARK, 105 miles, is the name of a settlement surrounded by a good section of country. A rolling prairie, with haycocks, cattle, and dwellings, are seen in the distance. Then comes a small stretch of prairie, surrounded by steep bluffs or hillocks of poor land as you approach Buffalo River, a tributary of Red River, which is here a small stream. Buffalo bones are often met with, but the wild animal has not been seen east of Red River for the past eight or ten years. Antelopes and elk are occasionally taken in the winter and spring months.

HAWLEY, 116 miles, is a settlement situated on Buffalo River. Here is a beautiful rolling prairie, with small trees skirting the river banks. Ducks, geese, and grouse are to be seen in large numbers in the autumn months.

MUSKODA, 120 miles, is surrounded by a rolling prairie, where are to be found prairie chickens in large numbers. Cross and re-cross the Buffalo River, here lined with handsome trees, hanging over the stream. Much of the land is under cultivation, stretching as far as can be seen southward. Fierce, driving cold winds are said to sweep over these prairies during the winter months, while fires rage in the autumn, running with the wind like a race-horse.

Glyndon, Clay Co., Minn., 244 miles west of Duluth, is handsomely situated on a rich prairie, 10 miles east of Red River. The *Branch Railroad* here crosses the Northern Pacific, running from St. Cloud to Pembina, situated on the northern limits of the United States. Here is a land-office, a hotel, printing-office, and several stores, together with a number of dwelling-houses. Population, 500 or 600, being mostly from the Eastern States.

A level, open prairie extends for many miles north and south, which is passed before reaching the Red River, and is fringed by a belt of timber as far as the eye can reach, the land gently falling as you approach the stream.

Moorhead, 137 miles west of Brainerd, and 252 miles from Duluth, situated on the east bank of Red River, at the head of navigation, is the capital of Clay Co., Minn., lying near the 47th parallel. Here are located a depot building, storehouses, 2 churches, 6 taverns, 12 stores, a printing-office, and a number of dwelling-houses. Population about 1000, of different nationalities, the whole partaking the appearance of a frontier settlement. Here is a substantial railroad bridge, about 1,200 feet in length, spanning the river bottom, the stream, at low water, being 100 feet wide, with a gentle current flowing northward. Steamers of a light draught run to Fort Garry, Manitoba, 220 miles. Flat-boats and rafts also run from the head of the river to Moorhead, Pembina, and Fort Garry.

Fargo, Dakota Territory, lies directly opposite Moorhead, on the west bank of Red River. Here is built a station-house and work-shops by the N. Pacific R. R. Co., it being intended as their headquarters for the Dakota Div. Hotels, stores, and dwellings are already erected, and others in the course of rapid construction, bidding fair to become a town of considerable importance. A new and large hotel has here been erected by the Railroad Company. It is north-west of Fargo that most of the poor lands of Dakota lie, being in the vicinity of Minnewakan, or Devil's Lake.

Climatic Influence.

It is difficult to convey to strangers the great difference in *climate* that exists, during the summer months, between the Red River of the North and Duluth, situated at the head of Lake Superior, on the same line of latitude, being distant 250 miles.

Starting from Moorhead, Minn., lying near the 47th parallel of latitude, and proceeding eastward, you usually encounter hot weather in June, July, and August, but soon after crossing the Mississippi River at Brainerd you encounter a perceptible change, the air becoming cooler and crisp-like — the same difference about as is felt between the hot days of August and the cool days of September or October in other parts of the country. This steady summer heat of the Red River country, and farther North, forms a Climatic Influence that causes the NEW NORTH-WEST to be blessed with a *Siberian Summer*, thereby redeeming millions of acres of land from the chilling influence that comes sweeping down from Hudson Bay, crowding down the thermometer or Isothermal line from the head of Lake Superior to the eastern Atlantic coast. Then, again, this favored section of country has less fall of moisture, rain or snow, during the winter months than falls in the Eastern States, the usual difference being as *two* is to *ten* inches of moisture or water. This same favorable climatic influence extends north-west to the Upper Missouri and the Valley of the Saskatchawan, in. British America.

Route from St. Paul to the Red River of the North.

The most direct route from ST. PAUL to the Red River is by the *St. Paul and Pacific Railroad,* via Minneapolis to Breckinridge, Minn, 217 miles. This route crosses the State of Minnesota, running in a north-west direction, passing Lake Minnetonka and the Big Woods before reaching the Red River, 35 miles north of Traverse Lake, the head of the stream.

TABLE OF DISTANCES FROM DULUTH TO BISMARCK, DAKOTA,

VIA NORTHERN PACIFIC RAILROAD.

Minnesota Division.

Miles.			Miles.
448	**DULUTH**.................		0
447	Rice's Point..................		1
444	Oneota	3	4
433	Fond du Lac................	11	15
425	Thomson	8	23
424	JUNCTION, *Lake Superior*		
	and Miss. Railroad......	1	24
390	Sicotte's ....,.........	34	58
372	Kimberly..................	18	76
360	Aiken......................	12	88
333	**Brainerd**	27	115

Mississippi River.

311	Motley	22	137
286	Wadena	25	162
273	N. York Mills..............	13	175
263	Perham	10	185
252	Hobart.....................	11	196
241	DETROIT...................	11	207
237	Oak Lake..................	4	211
234	Audubon	3	214
228	Lake Park.................	6	220
217	Hawley.....................	11	231
204	GLYNDON, crossing of		
	Pembina Branch R. R...	13	244
195	**MOORHEAD**..........	9	253

Red River of the North.

Dakota Division.

Miles.			Miles.
194	**FARGO, Dak.**	1	254
183	Maple River............	11	265
137	Wapeton	46	311
122	Eckelson	15	326
101	JAMESTOWN............	21	347
65	Crystal Springs.........	36	383
0	**BISMARCK**	65	448

Missouri River.

Northern Pacific Railroad Continued — New Route.

1,590	**Bismarck**............		448
1,588	Crossing Missouri Riv.	2	450
1,383	YELLOWSTONE RIV....205		655
	Crossing Dakota into Montana.		
1,043	YELLOWSTONE DIV....340		995
	Passing up Yellowstone Valley.		
845	ROCKY MT. DIV.*....198		1,193
563	CLARK'S FORK DIV... 282		1,475
	Western Slope, Rocky Mountains.		
355	PEND D'OREILLE DIV. 208		1,683
	To mouth Snake River.		
105	COLUMBIA RIV. DIV...250		1,933
	Mouth Snake River to Kalama.		
0	PACIFIC DIVISION †...105		2,038
	Kalama to Tacoma, Puget Sound.		

* Boseman's Pass, elevated about 6,000 feet above the Ocean.
† Finished; leaving 1,483 miles of Railroad to be completed.
NOTE.—This new route from Bismarck to Puget Sound increases the former estimates about 250 miles, thereby obtaining a better grade.

STAGES run daily from Breckinridge to Moorhead, 40 miles, connecting with KITTSON'S RED RIVER TRANSPORTATION LINE OF STEAMERS, forming a speedy and direct route to FORT GARRY.

DISTANCE FROM ST. PAUL

To Breckinridge, *railroad* 217 miles.
" Moorhead, *stage*....... 40 "
" Pembina, *steamboat*... 156 "
" FORT GARRY, " ... 67 "

Total.............. 480 "
Usual time, 4 days; fare, $23.

Red River of the North.

This remarkable stream, like the "Father of Waters," attracts the attention of the observing traveller with great astonishment. Like the Mississippi, it has its bottom land and islands, being fringed with a heavy growth of forest trees, while running northward through an immense prairie region, formerly the favorite abode of the buffalo, who have receded westward with the onward march of civilization.

Its principal head source is Lake Traverse, 45° 30′ north latitude, and its outlet into Lake Winnipeg,* through which its waters flow, is in 51°, thus extending through five and a half degrees

* Lake Winnipeg receives also the waters of the Saskatchewan, emptying its accumulated waters into Nelson River, and thence into Hudson Bay, in about 57° N. latitude.

of latitude, having a very winding course almost directly northward. For its whole length it is surrounded by fertile prairies, which are now coming into favorable notice by the American and Canadian public.

Manitoba is the field for Canadian emigrants, while to the South the hardy sons of New England and foreigners from Northern Europe are flocking in great numbers.

The completion of the *Northern Pacific Railroad* to the Red River, and farther westward, has given an impetus to emigration in this direction. Villages and settlements are springing up with great rapidity, which is destined to extend northward and westward as facilities for travel increase, ultimately stretching north-westward across the Rocky Mountains to the Pacific Ocean.

The good soil and favorable climate of Manitoba has long been known to the Canadian public; but the want of railroads has heretofore prevented its rapid settlement. Now, in addition to the *Northern Pacific*, a branch railroad is being constructed to run direct from St. Paul, via St. Cloud, to Pembina, situated near the 49th parallel, thus, in connection with the *St. Paul and Pacific Railroad*, affording facilities to reach any part of the Red River Valley, it being now reached by three great Lines of Travel. In addition to the above, the *Winona and St. Peter's Railroad* is pushing its line north-westward to the sources of the Red River.

Red River Transportation Line, 1874.

MOORHEAD TO MANITOBA, VIA

Freight and Passenger Steamers—International, Dakota, Sheyenne, Selkirk, and Alpha.

During navigation this line will be ready to carry freight and passengers from MOORHEAD, Minn., where the *Northern Pacific Railroad* crosses the Red River of the North, to FORT GAR-

RY, Manitoba, stopping at the intermediate landings.

Passengers from ST. PAUL have the choice of *two* routes to Moorhead, etc., viz., the *St. Paul and Pacific Railroad*

to BRECKINRIDGE, 216 miles, thence by stage to Moorhead, 40 miles farther; or by the *Lake Superior and Mississippi Railroad* to Northern Pacific Junction, thence by *Northern Pacific Railroad* to Moorhead, a total distance of 358 miles, passing through Brainerd, situated on the Mississippi River, 137 miles east of Moorhead.

DISTANCES FROM MOORHEAD TO FORT GARRY,

Via Red River of the North.

Landings.		Miles.
MOORHEAD........................		0
Probstfields		3
Georgetown.....................	12	15
Elm River..	13	28
Goose Rapids	14	42
Frog Point.......................	10	52
Grand Forks.......	22	74
PEMBINA.........................	80	154
Mouth Assiniboine.		
FORT GARRY	67	221

Distance from DULUTH to MOORHEAD, 252 miles. Passengers will save expense and trouble by purchasing *through tickets*, on sale at principal railroad offices in Canada and the States.

OBJECTS OF INTEREST.

The varied scenery on this route, by rail and steamer, is of the most interesting character, passing through a wild and sparsely populated region, between the St. Louis and Mississippi Rivers, then coming to a better section of level country, until the prairie region is reached, when a lovely view is presented to the observing traveller, the country being dotted with small lakes, forest trees, and extensive prairies before reaching the Red River of the North, then beautiful river scenery for the remainder of the distance.

INDUCEMENTS FOR PLEASURE TRAVELLERS.

During the season of 1874 the facilities for reaching FORT GARRY, Manitoba, and Lake Winnipeg will be complete, opening a new and healthy region, abounding with different kinds of game, while enjoying an invigorating atmosphere.

The route to the Upper Missouri River is also completed, distant about 200 miles west of Moorhead. This noble stream being reached by the "*Iron Horse*," the field for adventure is illimitable. The tourist can proceed to "*End Track*," running toward the Yellowstone River, or take a steamer and proceed to FORT BENTON, Montana, or still farther west to the *National Park* among the Rocky Mountains, near where are deposited *golden treasures*, besides wild river and mountain scenery — rapids, falls, and spouting springs of fabulous extent and variety.

From this time forward Lake Superior, the Upper Mississippi, the Red River of the North, and the vast region drained by the Missouri River, will be the great camping-ground of hunters and sportsmen seeking health and enjoyment during the summer and autumn months. Outfits will have to be prepared, and in many instances faithful guides obtained in order to secure safety and comfort. A canoe or an Indian pony would, in many instances, be a necessary appendage, not forgetting a *mosquito-net*.

Letter from James River.

"DEAR STAR: Sure to my promise, I will give you a brief description of this town and adjoining country, and will begin with the beauty of the scenery with which nature has seen fit to embellish this till lately unknown country. Standing on the bluffs that guard the town on all sides, one has spread out before him a panoramic view impossible to describe with justice. To the north may be seen the clear waters of the beautiful stream, as it wends its way in serpentine folds over the magnificent prairie, with its banks adorned with the graceful elms that tower their giant forms far over the waters, and are reflected from its placid surface in magnified loveliness. In spots of many feet the foliage is so luxuriant that not a glimpse of the river can be obtained till one stands close on its banks, while at others the waters bask in the uninterrupted rays of the noonday sun, and sparkle with seeming delight as the golden showers fall on its ripples. Then again the eye can but at intervals penetrate the verdure to have the waters meet the gaze, sparkling like silver stars and lending to the scene a quiet beauty one does not often meet with. And so the streamlet glides in alternate beauties till it is lost behind the hills, far, far to the south. *Jamestown,* 94 miles west of Moorhead, is situated on a level plain of many acres, entirely surrounded by hills and trees that will form great protection in the winter months, besides giving the town a look of beauty that surpasses anything I have seen on the line; and since, I am informed by those that ought to be posted, that the soil for miles about is of good quality, I do not hesitate to bespeak for this place a prosperous future. The immediate vicinity, and as far as the valley is concerned, I am certain that vegetables, and, in fact, all else, will flourish as well as at any point on the line."

Dakota, or *James River,* flows south into the Missouri near Yankton. It is a long and large river, being crossed by the railroad about one hundred miles west of Red River, and in another hundred miles the Missouri River is reached, near the mouth of Apple Creek.

Dakota and its Agricultural Capacities — The City of Bismarck.

Correspondence of the Detroit Post.

BISMARCK, DAKOTA TERRITORY, July 31, 1873.

Before the commencement of operations on the line of the Northern Pacific Railroad, notwithstanding portions of it are nearer the East than any other Territory, Dakota was really a *terra incognita.* No precious metal had been discovered, so, in the absence of railroads to connect it with civilization, its wealth lay all hidden and undeveloped; but now it is feeling the touch that is quickening the whole North-west, and claims the attention of all who are desirous of being fully acquainted with the resources of their country.

Here is an empire of 150,000 square miles, as large as New York, Pennsylvania, and the New England States combined, an unusual proportion of which is not only tillable, but highly productive. The north-eastern portion has been quickened into life and activity, and measurably brought to public notice by the Northern Pacific Railroad, which now spans 200 miles of its area. Much has been said about the wonderful fertility of the Red River Valley in Minnesota, and the general impression has obtained that that valley is confined to our sister State, but the truth is it also embraces a portion of this territory, 200 miles long by 40 miles wide, fully equal in agricultural wealth to that portion east of the Red River. With the exception of the north-western quarter of the Territory, and the Black Hills with their abundant pine, agriculture must be our main reliance, and for this, notwithstanding some few portions may depend upon irrigation, and others be less valuable because of the presence of alkali in the soil, few areas of equal extent are so well adapted to settlement. The remarkable net-work of living brooks, lakes, streams, and navigable rivers, with which the region is supplied, is, perhaps, its most striking feature, and furnishes the basis for a simple, natural, and economical system of irrigation for the fertile farming lands of the interior. Where systematic irrigation has been practised, it seems to be the uniform testimony that the greatly increased yield, the absolute certainty of regular crops, and exemption from risks of damage by bad weather in harvest time, more than compensates for the cost of irrigating ditches.

Minnesota has outlived the reputation of being an inhospitable hyperborean region, and its fertility and productiveness are now admitted by all. With an average temperature considerably milder than that of its sister, nothing can prevent Dakota taking an equally high agricultural rank, if it be established that there is a sufficient amount of moisture to insure success to the labors of the farmer.

Gen. Rosser's Official Report.

Gen. Rosser, in charge of the railroad survey of the Stanley Yellowstone Expedition, has submitted to the authorities of the Northern Pacific Railroad Company, in New York, his official report of the results thus far accomplished by the expedition. He finds the new and final route across Western Dakota, from Missouri to the Yellowstone River, entirely practicable and satisfactory, it being greatly superior to those of former days. The distance is 205 miles, 21 miles shorter than the survey of 1871. The gradients are moderate, the average of work per mile is considerably less, the number of important bridges is reduced two-thirds, the Little Missouri River, which former surveys crossed 11 times, is crossed once on the line so located, and the route runs immediately through only one mile of the bad clay lands just east of the Little Missouri. With few exceptions, the country is rolling prairie, sometimes rising into low hills, the grass being ex-

cellent and the soil good. Good water was found the entire distance. Coal outcrops at various points in veins several feet in thickness, and timber is more abundant than on former routes. The report states that the main body of General Stanley's expedition accompanied the Scientific Corps, and most of the press correspondents did not accompany the engineers, who were escorted by General Custer's detachment, but followed the old abandoned route south of Heart River, hence descriptions of the region traversed by the main command do not apply to the country traversed by the new route for the railroad. The directors of the company have accepted the new line recommended by Gen. Rosser from Bismarck, the present end of the track, to the Yellowstone crossing, and have called for proposals to grade and bridge this section of 205 miles. The expedition is now prosecuting the survey westward up the left bank of the Yellowstone to Pompey's Pillar, where it will join the survey made last year from the west, and thus complete the surveyed line across the continent.

Bismarck, Burleigh Co., Dakota, situated on east bank of the Missouri River, opposite Fort Abraham Lincoln, is the present terminus of the *Northern Pacific Railroad,* 450 miles west of Duluth. It lies about 800 miles above Sioux City, by water, and 1,100 miles below Fort Benton, Montana. The distance to St. Louis by the Missouri River is about 1,800 miles, making 2,900 miles navigation from head of navigation to its mouth, by a circuitous route. The town contains a church, a bank, several public houses, stores and warehouses, and about 1,000 inhabitants.

The distance from Bismarck to St. Paul, via Moorhead, is 460 miles; St. Paul to Chicago, 410 miles; Chicago to St. Louis, 280 miles, making the total railroad route from Bismarck to St. Louis

1,150 miles, or 1,770 miles to the city of New York, *all rail.* Thus opening, by means of steamers, an immense region of country during the season of navigation, which usually lasts from the middle of April to the middle of November.

Steamers Running on the Upper Missouri River.

There are now two Lines of Steamers running from Bismarck, Dakota, to Carroll and Fort Benton, Montana, carrying passengers and freight, besides other steamers from Omaha, Sioux City, and ports on the Lower Missouri.

The steamers leaving Bismarck, the present terminus of the *Northern Pacific Railroad,* run to *Carroll,* 280 miles below Fort Benton, connecting with a wagon road extending to *Helena,* Mont., 250 miles. This new route is now becoming very popular, reducing the time and expense about one-half over the old route via the Union Pacific Railroad, and by stage road from Corinne, Utah, to Montana.

Table of Distances

FROM BISMARCK, DAKOTA, TO HELENA, MONTANA, BY STEAMBOAT AND STAGE.

		Miles.
BISMARCK		0
Fort Clark		60
Fort Stevenson	54	114
Fort Berthold	15	129
Little Missouri River	40	169
White Earth River	80	249
Big Muddy River	70	319
Fort Buford	75	394
Mouth Yellowstone	6	400
Fort Union	5	405
Little Muddy River	25	430
Dawson's Bend	38	468
Big Muddy River	13	481
Fort Peck	25	506
Poplar River	55	561
Disaster Bend	20	581
Mitchell's Bend	16	597
Milk River	63	660
Dry Fork	20	680
Harvey's Point	50	730
Round Butte	50	780
Muscle Shell River	60	840
CARROLL, *steamboat landing,* 280 miles below Fort Benton	40	880

FORT BENTON, Mont.	1,160

The Missouri and Yellowstone Rivers.

The MISSOURI RIVER, says the late Prof. J. W. FOSTER, is the longest affluent of the Mississippi — though the volume of water discharged is not so great as that of the Ohio — and by reason of its length ought to be regarded as the main stream. It has its sources in longitude 112° and latitude 47°, where they nearly interlock with those of the Columbia, — the only river which rises in the Rocky Mountains and breaks through the lowest Ranges, in the vast extent from British Columbia to Mexico." Standing on the summit at any point, Captain Mullan remarks, you can see the waters that flow into two oceans ; and no where on the continent do we find such a perfect network of water-courses.

The YELLOWSTONE, rising in the Rocky Mountains, in Montana and Wyoming Territories, is the principal affluent of the Missouri, whose volume, as estimated by Gen. Warren, is as large as that which is considered the main stream ; and Jefferson, Madison and Gallatin Forks, are by no means inconsiderable streams. In connection with Yellowstone Lake and the Geysers, here is a field for exploration far more important, to Americans, than the sources of the Nile or the mountains of the Moon in Africa.

ACCESS TO THE YELLOWSTONE PARK.

COPIED FROM PROF. F. V. HAYDEN'S REPORT, 1873.

" As the PARK will soon become an object of general interest, and be the resort of thousands of visitors, the question of proper and convenient access is of great importance at present. A journey by wagon from the Central Pacific or Northern Pacific Railroads would prove long and tedious; we must, therefore, look for a railroad to carry tourists within a much shorter distance from the Park, allowing, if any, but one or two days journey by wagon. A project for such a railroad has already been conceived, but encountered pecuniary difficulties sufficient to impede its immediate construction. This line would become the main route connecting the Central Pacific or Union Pacific Railroad with the Northern Pacific Railroad, and also furnish the best means of reaching the land of wonders," the great *National Park*, situated in the north-west portion of the Territory of Wyoming.

The best route at the present time is by stage from CORINNE, Utah, to Yellowstone Lake.

The following are the DISTANCES and ELEVATIONS on this Route:

	Miles.	Miles from Corinne.	Elevation in feet.
Corinne to Henry's Fork Valley	0	145	5,130
Falls River	13	158	5,670
Entrance to Pass	20	178	6,950
Beulah Lakes	6	184	7,525
Union Fork	6	190	7,800
Lewis Lake	11	201	7,828
Hot Springs, Yellowstone Lake	8	209	7,788
Yellowstone Falls	30	239	7,700
Lower Geyser Basin, by another route	...	275	7,260
Upper Geyser Basin, " " "	...	283	7,390

For further information, see the Sixth Annual Report of the United States Geological Survey of the Territories, by Prof. HAYDEN, United States Geologist.

PENNSYLVANIA
RAILROAD.

THE GREAT TRUNK LINE

AND

United States Mail Route

BETWEEN THE

ATLANTIC SEABOARD AND THE MISSISSIPPI VALLEY; ALSO, THE GREAT LAKES AND CANADA,

BY MEANS OF ITS CONNECTIONS.

This is the most splendidly equipped and best constructed

RAILWAY LINE IN THE WORLD.

It is Double Track and Stone Ballasted, and is laid with Solid Sleepers and heavy Steel Rails between New York and Pittsburgh. The Bridges are of Iron and Stone, and all material used in construction is subjected to the closest inspection and highest tests. The Westinghouse Air Brake is attached to all Passenger Trains, and the system of Safety Signals is perfect.

Pullman Drawing-room, Sleeping, and Parlor Cars are run on all Express Trains from New York and Philadelphia to Chicago, Cincinnati, Louisville, Indianapolis, and St. Louis without change.

The scenery on this route, for grandeur, beauty and variety, is unsurpassed in America. All who contemplate a *Trip Across the Continent,* should travel over this famous Line.

THROUGH TICKETS

For sale at the LOWEST RATES, at all the principal Ticket Offices of the Company.

A. J. CASSATT, General Manager.
D. M. BOYD, Jr., Gen'l Pass. Agent.

PHILADELPHIA, May, 1874.

268

FOR LAKE SUPERIOR.

WARD'S
CENTRAL AND PACIFIC
LAKE COMPANY.
SEASON—1874.

This Line, composed of ten first-class STEAMERS, will run from *Buffalo to Duluth*, and intermediate points, and make five departures each week from each end of the route. Connects at Buffalo with the New York Central Railroad, for all points East, and at Duluth with the Northern Pacific and Lake Superior and Mississippi Railroads, for all points in Minnesota, Manitoba, and the North-west.

One of the Passenger Steamers will leave *Buffalo* every Sunday, Tuesday, and Thursday—*Cleveland*, every Monday, Wednesday, and Friday—*Detroit*, every Tuesday, Thursday, and Saturday, and touch at PORT HURON, SARNIA, and Grand Trunk Railroad, on the mornings following the departures from DETROIT.

For information apply to

CHARLES E. SLACK, Foot of Mississippi St., Buffalo, N. Y.
L. L. DAVIS & CO., Foot of St. Clair St., Cleveland, Ohio.
BUCKLEY & CO., Foot of First St., Detroit, Mich.
JOHN GORDON, Duluth, Minn.

Or to

EBER WARD, Manager,
OFFICE, Foot of First St., Detroit, Mich.

LAKE SUPERIOR ROYAL MAIL LINE,

IN CONNECTION WITH THE

Northern Railway of Canada.

This LINE embraces *Three Magnificent, First-Class, Upper Cabin,* Side-Wheel Steamers, viz.:—

CHICORA, CUMBERLAND, and FRANCES SMITH.

Leaving COLLINGWOOD every Tuesday and Friday for *Owen Sound, Byng Inlet, Killarney, Little Current, Spanish River, Bruce Mines,* **Sault Ste. Marie,** *Point Aux Pins, Batchewanung, Michipecoten, St. Ignace, Pic, Neepigon, Silver Islet,* **Fort William,** MARQUETTE, and **Duluth.**

Connecting at SAULT STE. MARIE, with the American Daily Line Steamers for all ports on South Shore of Lake Superior. Connecting at DULUTH with the *Lake Superior & Mississippi Railway,* for *St. Paul and Minneapolis,* and with the *Northern Pacific Railway* for MOORHEAD, on *Red River;* thence by *Kittson's Line of Steamers* for **Fort Garry** and *Province of Manitoba.*—Connection also at FORT WILLIAM with the *Canadian Government Overland Route to* FORT GARRY, via Lake of the Woods.

EXCURSION RATES.

During the Summer Months, greatly reduced rates from *Toronto to Silver Islet, Fort William, Neepigon, &c.,* will be made.

This Line and Route possess great attractions. For the Tourist and Pleasure-Seeker the trip is one of the finest that can be made; the scenery is most picturesque in character, and is not excelled by any in America; whilst the pure, bracing atmosphere offers great attraction to the Southern Tourist.

The Round Trip of *Lake Superior,* by this Route, can be made by transfer from the *American Line,* at *Sault Ste. Marie, Marquette,* or *Duluth* to the *Canadian Line,* for the *North Shore;* and thence, via COLLINGWOOD and the *Northern Railway,* to *Toronto,* and *East and South* by *Lake Ontario and River St. Lawrence* for *Niagara, Montreal, Quebec, White Mountains, Boston, New York, &c.,*—affording one of the most picturesque and attractive routes on the **American Continent.**

For further information, address

ADAM ROLPH,

General Agent Northern Railroad, Toronto, Can.

270

𝔇uluth 𝔏ake 𝔗ransportation ℭo.
LAKE SUPERIOR

1874  **1874**

South Shore Line Passenger Steamers
THE SPLENDID PASSENGER STEAMERS

MANISTEE and METROPOLIS,
Capt. J. McKAY, **Capt. B. ATKINS,**

WILL FORM A SEMI-WEEKLY LINE BETWEEN

Duluth, Bayfield, Ashland, Ontonagon, Isle Royale, Eagle River, Eagle Harbor, Houghton, Hancock, L'Anse, and Marquette.

THIS IS THE ONLY LINE

That makes all LANDINGS each way at points where COPPER and IRON MINES are located, thereby affording Pleasure-seekers and Business Men facilities with much less Lake travel than by other Lines.

TICKETS

Can be purchased in St. Paul, Minneapolis, and Stillwater, of Agents of the Northern Pacific and L. S. & M. Railroad Companies.

☞ For further information, passage or freight, apply to or address

W. R. STONE & CO., Agents, Duluth,
F. B. SPEAR & CO., Marquette,
JAMES A. CLOSE, Hancock,
J. B. STURGIS, Houghton,
BALEY & SHAPLEY, Eagle Harbor,

WILLARD & MERCER, Ontonagon,
THOS. S. KING, Minneapolis, No. 5, under Nicollet Hotel,
CHAS. THOMPSON, West Wisconsin Office, cor. Third and Jackson Sts., St. Paul, or to

JOHN GORDON, Manager,
DULUTH, MINN.

271

272

MICHIGAN EXCHANGE,

JEFFERSON AVENUE, ... DETROIT, MICHIGAN.

EDWARD LYON, } Proprietor.

This is a large and well-kept HOTEL, situated near the Steamboat Landings.

ROSSIN HOUSE,

TORONTO, ONTARIO.

This Favorite and First-Class HOTEL, situated on the

CORNER OF KING AND YORK STREETS,

Is a large and roomy edifice, affording every comfort desired by the Travelling Public.

An Omnibus runs to and from the Railroad Depots and Steamboat Landings.

GEORGE P. SHEARS,

Proprietor.

MISSION HOUSE,

MACKINAC, MICH.,

E. A. FRANKS, Proprietor.

———•◄●►•———

This old and favorite Hotel is most delightfully situated on the romantic Island of Mackinac, within a short distance of the water's edge, and contiguous to the **Arched Rock, Sugar Loaf,** and other Natural Curiosities in which this famed Island abounds ; being alike celebrated for its pure air, romantic scenery, and fishing grounds.

Mackinac, *July,* 1874.

CHIPPEWA HOUSE,

SAUT STE MARIE,

MICHIGAN.

This favorite Hotel is pleasantly situated, near the Steamboat Landings, at the mouth of the **Ship Canal,** and in the immediate vicinity of Fort Brady.

No section of country exceeds the Saut and its vicinity for

Fishing, Hunting, or Aquatic Sports.

The table of the Hotel is daily supplied with delightful White Fish, and other varieties of the season, no pains being spared to make this house a comfortable home for the pleasure-traveler or man of business.

H. P. SMITH, Proprietor

First National Bank,

NEGAUNEE, MARQUETTE CO., MICHIGAN.

H. E. HAYDON, President.　　　　JOHN B. MAAS, Vice-Pres.
F. E. SNOW, - - Cashier.

CAPITAL, ONE HUNDRED THOUSAND DOLLARS.

This Bank is now ready for business, and solicits the patronage of the public. Its affairs will be conducted in a manner strictly in accord with safe banking.

A SPECIALTY WILL BE MADE OF

FOREIGN EXCHANGES AND PASSAGE TICKETS.

Exchanges available at all points East and West.

☞ Collections entrusted to our care will receive prompt and careful attention.

. HAYDON'S

NEGAUNEE BANK,

H. E. HAYDON, President.　　　　NEIL CAMPBELL, Cashier.

NEGAUNEE, (L. S.,) MICHIGAN.

Foreign and Domestic Exchange

FOR SALE AT LOWEST CURRENT RATES.

PASSAGE TICKETS

To and from the Old Country by all the First Class Lines.

Parties wishing to send money abroad, or who desire to bring out their friends, will always save money by purchasing DRAFTS or TICKETS at this Bank.

A GENERAL BANKING BUSINESS TRANSACTED.

J. B. SMITH & CO.,
Merchants, Forwarders, and Agents,
L'ANSE, LAKE SUPERIOR, MICHIGAN.

GOODS AT WHOLESALE AND RETAIL.

Stock consisting of Dry Goods, Groceries, Provisions, Boots and Shoes, Clothing, Hats and Caps, Furniture, Crockery, Sash and Doors, Lime, Grain, Feed, Hay, and a large and complete Stock of Mining Supplies of every description; Oak Lumber, Stove and Steamboat Coal, Fire Brick, Common Brick, and all heavy articles freighted by vessel, sold in

LARGE OR SMALL QUANTITIES.

SPECIAL AGENTS

For Hazard's Rifle and Blasting Powder, Superior Safety Fuse, St. Clair River Turned Mining and Axe Helves, Steinway Pianos at N. Y. prices, etc.

J. B. SMITH & CO.,
L'ANSE, HOUGHTON CO.

C. E. HOLLAND. J. N. SCOTT.

HOLLAND & SCOTT,
DEALERS IN
SHELF AND HEAVY HARDWARE,
SHIP CHANDLERY GOODS,
AND MANUFACTURERS OF
TIN, COPPER AND SHEET-IRON WARE,
L'ANSE, L. S., MICHIGAN.
BRANCH STORE, MICHIGAMME, MICH.
Agents for Fairbank's Standard Scales, Broad's Patent Cant-Hooks, Pike Poles, Hall's Safes, etc.

L. J. DECOTEAU & CO.,
DEALER IN
GENERAL GROCERIES, PROVISIONS
AND ALL KINDS OF
CANNED GOODS AND DRIED FRUITS,
GLASSWARE AND CROCKERY.
VESSELS SUPPLIED.
L'ANSE, MICHIGAN.

279